KU-282-724

Quaternary Environments

M. A. J. WILLIAMS
D. L. DUNKERLEY
P. DE DECKKER
A. P. KERSHAW
T. J. STOKES

Edward Arnold
A division of Hodder & Stoughton
LONDON NEW YORK MELBOURNE AUCKLAND

Cover photograph

The desert north of Aquaba, Wadi Rum, Jordan. During the early Holocene this region was inhabited by Neolithic herders and farmers. The now-abandoned Nabatean city of Petra in nearby Wadi Musa controlled the lucrative caravan trade in this area some 2000 years ago, when the climate was somewhat less arid.

Photograph: M. A. J. Williams, May 1975.

HERTFORDSHIRE
LIBRARY
ARTS & INFORMATION

H50 117 689 X

Class   551 - 79

Supplier | Price £16.95 | Date 29.3.94

© 1993 M. A. J. Williams, D. L. Dunkerley, P. De Deckker, A. P. Kershaw and T. J. Stokes

First published in Great Britain 1993

Distributed in the USA by Routledge, Chapman and Hall, Inc. 29 West 35th Street, New York, NY 10001

British Library Cataloguing in Publication Data

Williams, M. A. J.
  Quaternary Environments
  I. Title
  551

ISBN 0–7131–6590–1

All rights reserved. No part of this publication may be reproduced or transmitted in any form or by any means, electronically or mechanically, including photocopying, recording or any information storage or retrieval system, without either prior permission in writing from the publisher or a licence permitting restricted copying. In the United Kingdom such licences are issued by the Copyright Licensing Agency, 90 Tottenham Court Road, London W1P 9HE

Typeset in Great Britain by Wearset, Boldon, Tyne and Wear
Printed and bound in Great Britain for Edward Arnold, a division of Hodder and Stoughton Limited, Mill Road, Dunton Green, Sevenoaks, Kent TN13 2YA by Butler & Tanner Ltd, Frome, Somerset

Contents

List of Tables

List of Figures

The Authors

MARTIN WILLIAMS is Professor of Geography and Environmental Science at Monash University and a graduate of Cambridge University and the Australian National University. He has carried out extensive fieldwork in Australia, Africa, India and China and is the author of over 100 research papers on landscape evolution, climatic change and prehistoric environments in Australia, Africa and India. He co-edited *Evolution of Australasian Landforms* (Canberra: ANU Press, 1978); *The Sahara and the Nile* (Rotterdam: Balkema, 1980); *A Land Between Two Niles: Quaternary Geology and Biology of the Central Sudan* (Rotterdam: Balkema, 1982); *Monsoonal Australia* (Rotterdam: Balkema, 1991) and *The Cainozoic in Australia* (Sydney: Geological Society of Austrlia, 1991).

DAVID DUNKERLEY is a Senior Lecturer in Geography and Environmental Science at Monash University. He has monitored chemical and physical denudational processes throughout Australia as well as in the equatorial lowlands of Papua New Guinea and Borneo, and has particular interests in global tectonic processes and karst geochemistry.

PATRICK DE DECKKER is a Senior Lecturer in Geology at the Australian National University, Canberra. He has studied lakes in Australia, Antarctica, South America, Europe, Africa and China and has helped to pioneer the quantitative evaluation of water temperature and salinity using ostracods in marine and non-marine sediments. He co-edited *Limnology in Australia* (Melbourne: CSIRO; and Dordrecht: Junk, 1986), *Ostracoda in the Earth Sciences* (Amsterdam: Elsevier, 1988) and *The Cainozoic in Australia* (Geological Society of Australia). He is author of over 80 scientific papers covering numerous aspects of Quaternary and modern environments and biota.

PETER KERSHAW is Director of the Centre for Palynology and Palaeoecology and Reader in Geography and Environmental Science at Monash University. His research on the Quaternary and Tertiary vegetation history of Australia has resulted in three books and over 70 papers. He is President of the Palynological and Palaeobotanical Association of Australia, and is presently involved in joint palynological research in China and the USSR.

TONIA STOKES is a graduate of Sydney University and a Research Assistant in the Department of Geography and Environmental Science, Monash University. Her research interests include Quaternary chronology, vegetation history and landscape evolution.

Preface

As University teachers and active researchers we have long been aware of the need for an up-to-date text dealing with the global and regional environmental changes associated with the Quaternary glaciations. This book is primarily aimed at second- and third-year undergraduate classes, but it will also be useful for graduate students seeking to enlarge their understanding of global change. We have tried out on our own students many of the ideas scattered through this book. Our bibliography does not purport to be exhaustive, and is mainly confined to sources in English, but does point the reader to some of the more comprehensive recent studies. Ultimately, the best way to learn is to go out and discover for yourself through fieldwork. It is also a lot more fun. As always, an ounce of practice is worth more than a pound of theory . . . but a little theory can and does help!

To attempt a global overview of Quaternary environments is a daunting task. Twenty years have elapsed since Richard Foster Flint wrote his unsurpassed 'Glacial and Quaternary Geology' (Wiley 1972) and Karl Butzer his magisterial 'Environment and Archaeology' (Methuen 1971). We understand all too well why so few have followed the difficult trail they blazed, but difficulties seldom disappear unless confronted.

Between us, we have carried out field research on every continent, including Antarctica. In this book we have tried to adopt a world view. There remain some inevitable gaps in our coverage but our preference has been for a selective rather than an encyclopaedic approach.

Many people have helped to enhance our appreciation of Quaternary environmental fluctuations, on land and sea, around the globe. Before and during the conception of this volume, we have enjoyed the benefit of stimulating discussions with friends and colleagues around the world. They include Don Adamson, Stan Ambrose, Mike Barbetti, Raymonde Bonnefille, Jim Bowler, Karl Butzer, John Chappell, Desmond Clark, Jack Davies, Tom Dunne, Gerry Eck, Hugues Faure, Leon Follmer, Jean-Charles Fontes, Bob Galloway, Françoise Gasse, Alan Gillespie, John Gowlett, Dick Grove, Bernard Hallet, Jack Harris, Don Johnson, Hilt Johnson, Pete Lamb, Estella Leopold, Liu Tungsheng, Dan Livingstone, Virendra Misra, Dick Peltier, Nicole Petit-Maire, Steve Porter, S. N. Rajaguru, Pierre Rognon, Roman Schild, Asher Schick, Geoff Spaulding, Alayne Street-Perrott, Minze Stuiver, Maurice Taïeb, Mike Talbot, Claudio Vita-Finzi, Donald Walker, Wang Pinxian, Andrew Warren, Link Washburn, Bob Wasson, Fred Wendorf, Tim White, Herb Wright, Karl-Heinz Wyrwoll, and Aaron Yaïr. We thank them all.

Special thanks go to Gary Swinton, who drafted every figure; to

Jan Liddicut, who handled numerous drafts with aplomb; and to Tim Barta, Sharon Davis, Alan Fried, Kate Harle, Kim Newbury and Helen Quilligan for their cheerful help with figure compilation and reference checking.

Martin Williams
David Dunkerley
Patrick De Deckker
Peter Kershaw
Tonia Stokes

Melbourne, March 1992

1 Quaternary Environments
An Introduction

Il n'y a pas de fait pur; mais toute expérience, si objective semble-t-elle, s'enveloppe inévitablement d'un système d'hypothèses dès que le savant cherche à la formuler.

Teilhard de Chardin (1881–1955),
Preface to *Le Phénomène Humain*, 1947.

The Quaternary period spans roughly the last 2 million years of geological time (Figure 1.1) and is of critical importance in earth history. It was during this period of remarkably frequent and rapid changes in world climate (Figure 1.2) that bipedal, toolmaking, fire-using hominids emerged from Africa and gradually moved out to occupy Eurasia, Australia and the Americas, as well as distant oceanic islands around the globe. The Quaternary is thus not simply the coda to 4.5 billion years of earth history but is also the time during which we became fully human.

One lesson we are slowly learning after the long saga of continuous human interaction with our environment is that we ourselves are an integral part of that same environment, and that we are the custodians rather than the owners of the lands we now inhabit. We return to this theme in the final chapter of this book when we consider the past, present and possible future impact of our species upon the air we breathe, the water we drink, and the land and the sea which sustain the plant and animal life upon which we depend for our survival.

The aims of this book are to examine some of the global environmental fluctuations of the last 2 million years, to analyse some of the more important evidence used in reconstructing Quaternary environments, and to consider some of the ways in which living organisms (including humans) have responded to past environmental changes. We also believe that a knowledge of the past, besides being intrinsically interesting, is also our only real guide to what may befall us in the future.

Prelude to the Quaternary

An accurate long-term perspective on global climatic change has now become possible owing to recent advances in our under-

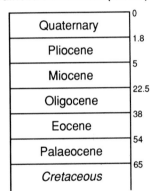

Cainozoic time scale (Ma BP)

	Ma BP
	0
Quaternary	
	1.8
Pliocene	
	5
Miocene	
	22.5
Oligocene	
	38
Eocene	
	54
Palaeocene	
	65
Cretaceous	

Fig. 1.1 Cainozoic time scale (Ma) (modified from Cowie and Bassett 1989)

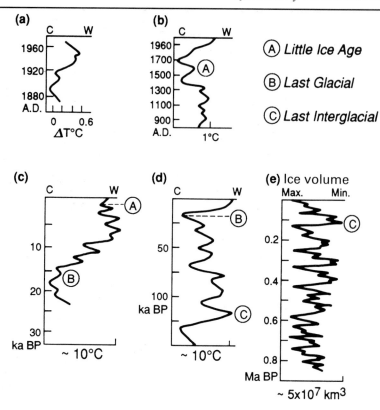

Fig. 1.2 Climatic variability at different time scales during the last 0.9 Ma of the Quaternary (adapted from Australian Academy of Science 1976)

standing of world tectonic history. The combined evidence from deep-sea drilling, seismic surveys and palaeomagnetic studies has allowed reconstruction of sea-floor spreading history, and of continental apparent polar wandering curves. The data from land and sea are impressive and persuasive. The timing of late Cainozoic ice build-up in the two hemispheres is now known, as are some of the associated changes in oceanic and atmospheric circulation, which are in turn related to the origin and expansion of the deserts and the contraction of the tropical rainforests. A proper understanding of Quaternary climatic changes therefore requires some appreciation of the legacy of the Tertiary.

The Tertiary and Quaternary periods together comprise the Cainozoic era and embrace the past 65 million years (Figure 1.1). The present geographical distribution of land, sea and ice (Figure 1.3) and of the corresponding morphoclimatic regions shown on Figure 1.4 are the end-product of Mesozoic and Cainozoic lithospheric plate movements and sea-floor spreading. A number of major regional episodes, including Himalayan uplift, Antarctic ice accumulation, closure of the Panama isthmus, build-up of the North American ice sheets, intertropical cooling and desiccation, and expansion of savanna at the expense of tropical rainforest, were all closely linked with the global tectonic events of the Tertiary and are the subject of Chapter 2.

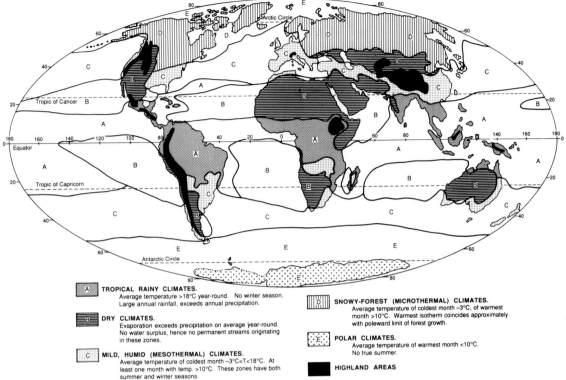

Fig. 1.3 Present-day global climates (after Tricart and Cailleux 1972; Bartholomew *et al.* 1980; and Strahler and Strahler 1987)

A TROPICAL RAINY CLIMATES.
Average temperature >18°C year-round. No winter season. Large annual rainfall, exceeds annual precipitation.

B DRY CLIMATES.
Evaporation exceeds precipitation on average year-round. No water surplus, hence no permanent streams originating in these zones.

C MILD, HUMID (MESOTHERMAL) CLIMATES.
Average temperature of coldest month −3°C<T<18°C. At least one month with temp. >10°C. These zones have both summer and winter seasons.

D SNOWY-FOREST (MICROTHERMAL) CLIMATES.
Average temperature of coldest month −3°C, of warmest month >10°C. Warmest isotherm coincides approximately with poleward limit of forest growth.

E POLAR CLIMATES.
Average temperature of warmest month <10°C. No true summer.

HIGHLAND AREAS

Quaternary glaciations

Ice began to accumulate on Antarctica well over 20 million years ago. Ice build-up came much later in the Northern Hemisphere, and it was not until 2.4 million years ago that major ice sheets began to grow rapidly in North America. For reasons which remain obscure (but which appear to be closely related to cyclical changes in the earth's orbital path around the sun and in the tilt of the earth's axis), the great ice sheets of the Northern Hemisphere in particular developed a characteristic cycle of slow build-up to full glacial conditions, followed by rapid ice melting and deglaciation. These topics are the focus of Chapter 3.

Quaternary sea-level changes

The larger of the Quaternary ice caps were up to 4 km thick. As the ice caps slowly built-up to attain their maximum thickness, the underlying bedrock was progressively depressed beneath the weight of accumulating ice. When the ice melted, the crust slowly rose again to its pre-glacial level. These 'isostatic readjustments' to the waxing and waning of the great Quaternary ice sheets caused changes in the relative levels of land and sea. During

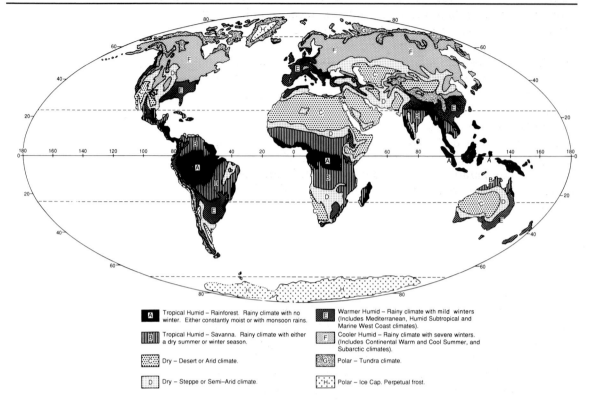

Tropical Humid – Rainforest. Rainy climate with no
winter. Either constantly moist or with monsoon rains.

Tropical Humid – Savanna. Rainy climate with either
a dry summer or winter season.

Dry – Desert or Arid climate.

Dry – Steppe or Semi–Arid climate.

Warmer Humid – Rainy climate with mild winters
(Includes Mediterranean, Humid Subtropical and
Marine West Coast climates).

Cooler Humid – Rainy climate with severe winters.
(Includes Continental Warm and Cool Summer, and
Subarctic climates).

Polar – Tundra climate.

Polar – Ice Cap. Perpetual frost.

Fig. 1.4 Present-day morphoclimatic regions (adapted from Tricart and Cailleux 1972; and Strahler and Strahler 1987)

glacial maxima, roughly 5.5 per cent of the world's water was locked up in the form of ice. (The corresponding value today is 1.7 per cent, or roughly three times less.) As the ice sheets grew, so the level of the world's oceans fell by up to 150 m, depending upon total ice volume. With deglaciation and rapid melting of the ice caps, sea level rose once more to about present levels. These 'glacio-eustatic' sea-level fluctuations are analysed in Chapter 4, together with the influence of 'isostasy' and other tectonic movements upon global and local sea levels.

Evidence from the oceans

Reconstruction of past sea-level fluctuations can throw useful light on the rate of accumulation and the rate of melting of global ice, but well-dated Quaternary sea-level histories only extend back to about 250 000 years ago, so that the first nine-tenths of the record must be sought elsewhere, most notably from deep-sea cores.

Inferences about changes or fluctuations in ocean circulation patterns used to depend very largely upon sedimentological and microfossil studies. Analysis of the oxygen isotopic composition

of the calcareous tests of suitable benthic and planktonic foraminifera now provides an additional and powerful means of assessing changes in ocean water temperature and salinity at depth and
near the surface (see Chapter 5). After allowing for local effects, it
is also possible to use this technique to estimate changes in global
ice volume. Deduced changes in regional surface salinity can also
indicate changes in runoff from major rivers, changes in evaporation, and changes in the amount of seasonal rainfall.

The record from deep-sea cores has the double advantage of
good global coverage and of spanning much of the Cainozoic.
There are comparatively few such long, continuous terrestrial
records, and those that do now exist are usually confined to
particular types of lake basin.

Rivers, lakes and groundwater

Although the oceanic record can provide unrivalled information
about the pattern and tempo of global climatic fluctuations in the
Quaternary, it is often more useful to know about the direct
changes to the landscape caused by local and regional hydrological fluctuations. Such changes are evident in the Quaternary
depositional legacy of rivers large and small, as well as in the
ever-changing response of lakes to local fluctuations in evaporation, precipitation and groundwater levels. Unfortunately for our
purposes, the alluvial history of most rivers can only be pieced
together from fragmentary and often poorly dated suites of sediments. However, as Chapter 6 points out, rivers and lakes together
can yield highly informative accounts of how certain regions
responded to the environmental vicissitudes of the
Quaternary.

Evidence from the deserts

A growing body of evidence from deep-sea cores, lake deposits
and ice cores shows that times of lowest world temperature
during the Quaternary ('glacial maxima') were times of greatest
aridity on land, with massive export of desert dust offshore, and
even to central Antarctica. Deserts are excellent geological and
geomorphological museums, for the very aridity to which they
owe their existence has minimized the destructive impact of
fluvial erosion and has helped to conserve an array of river, lake
and wind-blown deposits. These deposits sometimes contain
remarkably well preserved and occasionally, as in certain semi-
arid rift valleys in Africa, or the loess plateau in China, a nearly
continuous fossil vertebrate and invertebrate record spanning

most of the late Pliocene and Quaternary. Chapter 7 enlarges on these topics.

Evidence from terrestrial flora and fauna

The emergence of the plants and animals upon which humans have long depended for food and shelter took place against the environmental changes of the late Tertiary and was finally accomplished during the Quaternary. Changes in the non-marine plant and animal record provide an invaluable adjunct to the purely physical evidence furnished by landforms and sediments, and can be used to construct former temperature and rainfall fluctuations with great precision and accuracy. Some organisms are inherently sensitive to local changes in habitat, and may respond rapidly to external disturbance. Perhaps the most versatile and certainly one of the best tested methods used in Quaternary environmental reconstruction is the technique of 'pollen analysis', which is considered in some detail in Chapter 8, along with other more circumscribed techniques.

Human origins, innovations and migrations

As the great Cainozoic ice caps waxed and waned, and deserts expanded and contracted, a small-brained vegetarian hominid left its footprints clearly visible in a carbonatite ash which was laid down during a volcanic eruption near Laetoli in Tanzania nearly 4 million years ago. This creature, *Australopithecus afarensis*, was fully bipedal, and may well be the ancestor from which later hominids, including the genus *Homo*, were to derive. Chapter 9 describes the slow progression from user of tools to toolmaker, discusses the development and refinement of stone-knapping techniques, and concludes with a short analysis of the origins of plant and animal domestication. The food-producing economy of the Neolithic saw the virtual demise of most hunter-gatherer societies around the world, and the inception of modern urban civilization.

Atmospheric circulation during the Quaternary

The cultural development of our human forebears took place against a background of ever-changing global climate. In the intertropical zone, for instance, cold, dry and windy glacial maxima alternated with warm, wet interglacials. Regions delineated as arid on Figures 1.3 and 1.4 were sometimes studded with deep freshwater lakes; areas now under rainforest were sometimes covered in savanna, or partly mantled with wind-blown sand.

Chapter 10 is an attempt to explore some of the changes in global atmospheric circulation patterns during the Quaternary, particularly the terminal Pleistocene towards 18 000 years ago, and the early Holocene towards 9000 years ago. We do this for two very good reasons. Firstly, the last 20 000 years contain the best dated, best preserved and most abundant palaeoclimatic evidence with which to test global atmospheric circulation models. Secondly, the two time spikes considered coincide, respectively, with the last glacial maximum (18 000 years ago) and the so-called early Holocene 'climatic optimum' of 9000 years ago, which we prefer to regard as simply the postglacial antithesis of the full glacial climate. Between them, they encompass a substantial component of climatic range of the most recent glacial–postglacial cycle.

Environmental changes: past, present, future

Throughout the Quaternary there has been a prolonged series of interactions between hominids (ancestral humans) and their environment. Stone toolmaking dates back to about 2.5 million years ago, and fire was being used in Africa about a million years later. The question of how far prehistoric hunters contributed to the demise of certain species of animals is a vexed one, as is the related question of the role of burning in bringing about plant extinctions. With the advent of Neolithic food production, and accelerated clearing of the natural vegetation, the degree of human impact upon the biosphere and hydrosphere began to increase dramatically. By alternating plant cover, we may increase runoff, and thereby accelerate soil erosion. There is a delicate balance between the different components of the hydrosphere (Figure 1.5) and the atmosphere (Figure 1.6). Since the Industrial Revolution, in particular, we have begun to interfere with that balance by unwittingly altering some of the feedback loops which are an integral part of the global climate system (Figure 1.6). Chapter 11 discusses these issues in greater detail.

Quaternary chronology

There has long been controversy over the exact duration of the Quaternary. Some workers espouse a long chronology starting as early as 3.5 Ma. Others prefer a shorter chronology, beginning at 2, 1.8 or 1.6 Ma. We tentatively opt for 1.8 Ma (Figure 1.1), which also coincides reasonably well with the Olduvai palaeomagnetic event, an interval with normal magnetic polarity bracketed by K/Ar dates of 1.87 and 1.67 Ma (see Appendix). An equally good case may be made for placing the Pliocene-Pleistocene boundary

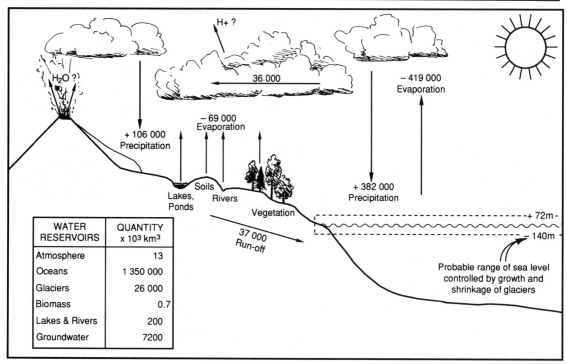

Fig. 1.5 The hydrosphere (after Bloom 1978; and Strahler and Strahler 1987)

at 2.5 Ma, when there was a rapid build-up of ice in the Northern Hemisphere. The choice of Quaternary boundary is very much a matter of personal taste, and has often generated more heat than

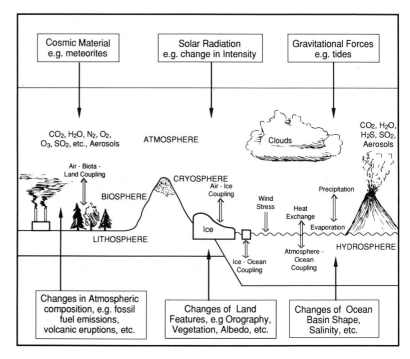

Fig. 1.6 Feedbacks in the global climate system (after Bloom 1978; Bach 1984; and Bradley 1985)

light. We likewise favour a simple fourfold subdivision of the Quaternary (Figure 1.7) into Lower Pleistocene (1.8–0.75 Ma), Middle Pleistocene (750–125 ka), Upper Pleistocene (125–10 ka) and Holocene (10–0 ka), while noting that none of these somewhat arbitrary divisions or ages is particularly sacrosanct.

Reconstructing Quaternary environments

A knowledge of past events and processes can offer useful insights into both present and future environmental changes, but a few preliminary words of caution are necessary here. Earth history is a tale of constantly varying interactions over time between lithosphere, atmosphere, hydrosphere (including cryosphere) and biosphere. Present world landscapes (Figure 1.4) reflect the influence of past as well as present-day processes. Theoretical constructs about the relation between present-day weathering processes and climate (or latitude), depicted below in Figure 1.8, are only useful if we are fully aware of their limitations.

Table 1.1 shows some of the types of evidence commonly used to reconstruct Quaternary environments and climates. Each is useful for a specific purpose, and for a particular area or time. Difficulties arise immediately when we use Procrustean tactics to force the data to yield palaeoenvironmental information at par-

Quaternary time scale (ka BP)

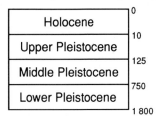

Fig. 1.7 Quaternary time scale (modified from Shackleton and Opdyke 1977; and from Cowie and Bassett 1989)

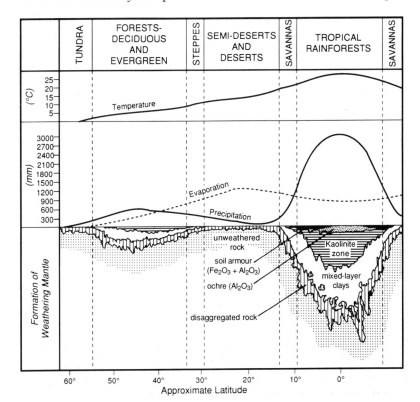

Fig. 1.8 Present-day weathering processes in relation to climate and latitude (after Strakhov 1967; and Bloom 1978)

Table 1.1: Sources of data used to reconstruct Quaternary environments (after National Academy of Sciences 1975; Bradley 1985; and Williams 1985)

Proxy data source	Variable measured
Geology and geomorphology – continental	
Relict soils	Soil types
Closed-basin lakes	Lake level
Lake sediments	Varve thickness
Aeolian deposits – loess, desert dust, dunes, sand plains	
Lacustrine deposits and erosional features	
Evaporites, tufas	Age
Speleothems	Stable isotope composition
Geology and geomorphology – marine	
Ocean sediments	Ash and sand accumulation rates
	Fossil plankton composition
	Isotopic composition of planktonic and benthic fossils
	Mineralogical composition and surface texture
	Geochemistry
Continental dust	
Biogenic dust: pollen, diatoms, phytoliths	
Marine shorelines	Coastal features
	Reef growth
Fluviatile inputs	
Glaciology	
Mountain glaciers, ice sheets	Terminal positions
Glacial deposits and features of glacial erosion	
Periglacial features	
Glacio-eustatic features	Shorelines
Layered ice cores	Oxygen isotope concentration
	Physical properties (e.g., ice fabric)
	Trace element and micro-particle concentrations
Biology and biogeography – continental	
Tree rings	Ring-width anomaly, density Isotopic composition
Fossil pollen and spores	Type, relative abundance and/or absolute concentration
Plant macrofossils	Age, distribution
Plant microfossils	
Vertebrate fossils	
Invertebrate fossils: mollusca, ostracods	
Diatoms	
Insects	Type, assemblage abundance
Modern population distributions	Refuges
	Relict populations of plants and animals

Proxy data source	Variable measured
Biology and biogeography – marine	
Diatoms	Faunal and floral abundance
Foraminifera	Morphological variations
Coral Reefs	
Archaeology	
Written records	
Plant remains	
Animal remains, including hominids	
Rock art	
Hearths, dwellings, workshops	
Artefacts: bone, stone, wood, shell, leather	

ticular scales in space or time for which those data are totally inappropriate. It is essential always to take due note of the time scales at which the different processes involved in environmental change normally operate (Figure 1.9). A related issue is the precision available in dating the proxy data or samples used in reconstructing past events (Figure 1.10). In many cases we are still

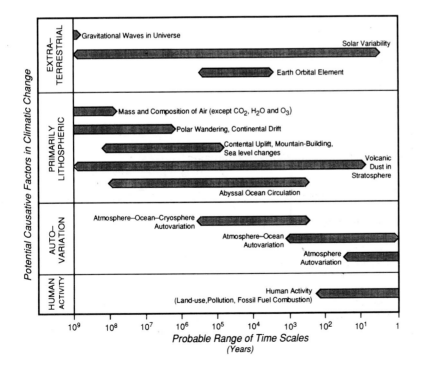

Fig. 1.9 Processes involved in environmental change and their time scales (after National Academy of Sciences 1975; Bloom 1978; Goudie 1983; and Bradley 1985)

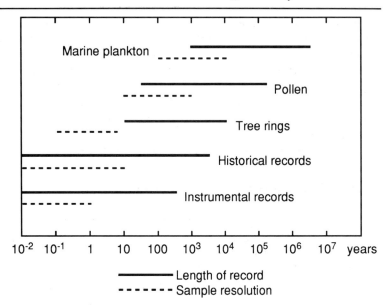

Fig. 1.10 Sample resolution and length of potential palaeoclimatic record from various independent lines of evidence (after Birks 1981; modified from T. Webb, unpublished diagram)

limited by the imprecision of existing dating methods, which is often a function of the half-life length of the particular radioactive isotopes involved (see Appendix). Given all of the above caveats, it would seem that the task of Quaternary environmental reconstruction is still more of an art than a science. Such a conclusion is in no way dismissive of some of the excellent progress made in quantifying past fluctuations in temperature and salinity, on land and in the sea, using stable isotopes and trace element composition of the calcareous shells of ostracods and forams. However, we still have a very long way to go to gain the spatial and temporal resolution necessary to test existing models of global atmospheric circulation in the Quaternary (Chapter 10).

Quaternary environmental analogues

It is always tempting to use past climatic events as analogues for possible future climatic changes. For example, some workers have suggested that future global warming linked to the greenhouse effect may have an early Holocene climatic analogue. We consider Holocene boundary conditions including sea level, the extent of the cryosphere, terrestrial albedo, and sea-surface temperatures may have been very different from those used to model future change. Of greater value in understanding possible future change is the geological and biological evidence of past hydrological events. Rather than seeking past climatic analogues to a warmer earth, it may be more useful for us to focus on how various elements of the biosphere and hydrosphere have responded to former climatic fluctuations. What were the directions and rates of response? What were the thresholds? Was the response synchronous or time-transgressive? How did one set of changes (e.g., deforestation) repercuss upon the rest of the landscape? If we adopt this approach then it is possible to argue that an appreciation of past Quaternary events can provide us with insights about possible future events which are unattainable by any other means (De Deckker, Kershaw and Williams 1988; Wasson 1990).

Practical relevance of Quaternary research

The Quaternary legacy is ubiquitous. Many of our soils formed during the Quaternary, as did many of the depositional features created by moving ice, wind and rivers. Human activities in the last few centuries have served to accelerate many natural processes, including soil erosion by wind and water. Some modern rates of soil erosion are several orders of magnitude greater than the long-term geological rates for that region. One reason for this discrepancy may simply be the ease with which unconsolidated Quaternary sediments can be mobilized by present-day runoff, but another may be destruction of the vegetation cover which increases the vulnerability of the soil surface to the erosive impact of rainsplash and runoff.

A knowledge of the rates and magnitudes of past and present change is essential to our understanding of the world we live in. Planners and policy makers are becoming increasingly attuned to the relevance of Quaternary studies to agricultural and resource management. For instance, Quaternary research can contribute its unique historical perspective to a sensible policy of long-term management of soil erosion, desertification, salinization, coastal erosion, floods and droughts, and biological conservation. Recent experience shows all too well that to ignore the past is to court future land-use problems. Present rates of plant and animal population changes mean very little unless set in historical context, in this case Quaternary palaeoecology. The long-term development and preservation of our soil, plant and groundwater resources thus requires a balanced understanding of recent Quaternary environmental changes as well as a thorough knowledge of present-day geomorphic, ecological and hydrological processes.

Further reading

Bowen, D. Q. 1978: *Quaternary Geology. A Stratigraphic Framework for Multi-disciplinary Work.* Oxford: Pergamon Press, 221. (A good clear account with an emphasis on stratigraphy and correlation.)

Bradley, R. S. 1985: *Quaternary Palaeoclimatology. Methods of Paleo-climatic Reconstruction.* Boston: Allen and Unwin, 472. (An excellent review of the scope and limitations of many of the methods used to reconstruct former climates.)

Butzer, K. W. 1974: *Environment and Archaeology. An Ecological Approach to Prehistory* (Third edition). Chicago: Aldine. (A masterly and advanced overview of the methods used to reconstruct Quaternary prehistoric environments.)

Flint, R. F. 1971: *Glacial and Quaternary Geology.* New York: Wiley, 892. (The classic text on glacial geology by one of the great masters. Still well worth consulting.)

Goudie, A. 1983: *Environmental Change* (Second edition). Oxford: Clarendon Press, 258. (A clear, concise and eminently readable undergraduate text on Quaternary environments.)

Lowe, J. J. and Walker, M. J. C. 1984: *Reconstructing Quaternary Environments.* London: Longman, 389. (A useful account, with examples drawn mainly from Europe and North America.)

Vita-Finzi, C. 1973: *Recent Earth History.* London: MacMillan, 138. (A thoughtful, concise and often witty analysis of relative and absolute dating methods used in late Quaternary research.)

West, R. G. and Sparks, B. W. 1977: *Pleistocene Geology and Biology, with special reference to the British Isles* (Second edition). London: Longman, 440. (A comprehensive text by two highly experienced practitioners; useful well beyond the British Isles.)

2 | Prelude to the Quaternary

Who can tell us whence and how arose this universe?
Anon (c. 1000 BC),
Rigveda-Samhita.

Introduction

In the Quaternary period, the slow and uneven progression toward cooler conditions which had characterized the earth for tens of millions of years (and about which we shall have more to say shortly) gave way to extraordinary climatic instability. Temperatures swung repeatedly from values like those of the present day to levels many degrees colder, with the growth of enormous ice sheets on land. In the cold phases, sea level fell dramatically, while on land, mountain tree lines shifted to much lower elevations, grasslands replaced forests, ground ice and associated periglacial processes became more common, and dust blew extensively on cold winds. Simultaneously, human cultural development began with the early use of stone tools and of fire, culminating in the Holocene cultivation of crops and the herding of domestic animals. These developments fostered the growth of urban civilization and the development of writing.

The environments of this remarkable period of 1.8 Ma developed from those of the Tertiary, and are best understood with a knowledge of the global environment as it was prior to the onset of the marked Quaternary instability. We therefore begin our exploration of the Quaternary with a brief investigation of longer-term environmental trends and of the global processes responsible for them.

The lithospheric plates and their motion

The environment at the surface of the earth is controlled by a suite of external and internal influences. A dominant external influence is the amount of solar radiation received through the atmosphere, which varies in the long term as the sun ages, and more rapidly because of changes in the geometry of the earth's orbit around it; events associated with the impact of comets provide another.

One of the principal internal influences on the environment is the geometry and movement of the lithospheric plates. These plates, fragments of the 'lithosphere' (the crust together with the

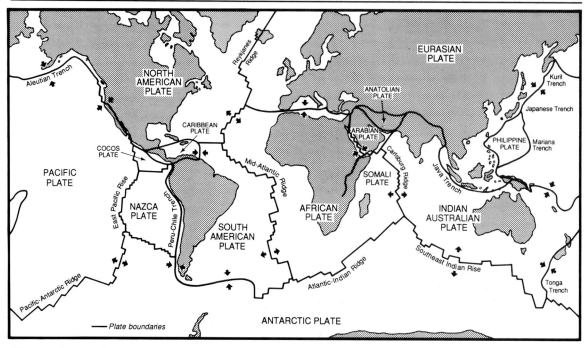

Fig. 2.1 Present-day configuration of the lithospheric plates. Note the relative absence of plate boundaries crossing continental crust, and the remarkable symmetry of the Atlantic Ocean about its axial mid-oceanic ridge system (after Press and Siever 1986; and Strahler and Strahler 1987)

uppermost part of the mantle), move over the earth's surface because of gradients set up by thermal isostatic processes. Carrying the continents with them, the moving plates determine the form of the ocean basins, and the distribution of land and water.

The present configuration of the lithospheric plates is shown in Figure 2.1. This pattern has developed by the disintegration of the supercontinent of Pangea (and the superocean Tethys, or Panth-alassa), which existed for some hundreds of millions of years before commencing to break up in the Jurassic, around 180 Ma BP. Early rifting opened the proto-Atlantic Ocean, and the development of the other modern oceans followed during the Mesozoic era, concluding with the rapid development of the Southern Ocean after about 50 Ma BP, almost entirely within the Cainozoic era. During this process, the mantle warmth built up under Pangea was allowed to dissipate, so that the smaller continental fragments subsided as they moved; simultaneously, the old Tethyan sea floor was replaced with younger, more thermally buoyant oceanic crust, and the average elevation of the sea floor rose. An inescapable consequence of these changes acting in concert was major marine transgression on a global scale.

Associated with the continental fragmentation were other changes, however. The system of oceanic circulation in the Tethys was gradually replaced by the contemporary pattern of smaller, constrained gyres carrying sensible heat across the modern oceans. Since the transport of heat by ocean currents is a major component of the thermal balancing of the global climate, significant regional climatic changes were produced; some of these will be described below. As might be expected, various seaways opened and closed during the redistribution of the continents, major events including the closure of the Panama isthmus, and the opening of the Drake Passage between South America and Antarctica, and the major Southern Ocean Passage between Australia and Antarctica (see Figure 2.2). The relatively rapid readjustment of circulation systems which must have followed these events is presumably one of the factors contributing to the uneven, stepped history of climatic deterioration in the Cainozoic.

The break-up of Pangea was associated with elevated levels of igneous activity along the rift systems and the new continental margins. This increase was responsible for a parallel rise in the CO_2 concentration in the atmosphere both by direct emission, and because the global transgression already referred to reduced the area of exposed land on which chemical weathering could take up CO_2 from acidified rain. Positive feedback of many kinds existed in this environmental change (as in so many environmental processes), acting to reinforce the initial tendency. For example, the greenhouse warming produced by the CO_2 would be likely to warm the surface of the oceans; this would reduce the solubility of the gas and further enhance the warming, as would the increased atmospheric water vapour concentration resulting from evaporation from the warm sea surface. Likewise, the warming would cause thermal expansion of sea water, further supporting global transgression and the reduction in area of exposed land. In turn, the growing global proportion of sea (and the diminishing land fraction), would, because of the lower albedo of the ocean surface, further reinforce a warming trend. Through mechanisms such as this, significant climatic change may be triggered by seemingly relatively minor events.

Tectonic control of the oceans and continents

In their present-day configuration, the lithospheric plates vary in their geography; some, like the Pacific plate, are largely covered by ocean. Some (such as the Eurasian) are largely occupied by land; others, like the Indo-Australian plate, are of intermediate character.

Certain features stand out in the fundamental structure of this assemblage of plates: the ocean basins contain in addition to the seawater, the bulk of 'divergent' plate boundaries (the mid-oceanic ridge system), and their underlying crust is of basaltic composition.

The continents, those parts of the surface of the earth not water

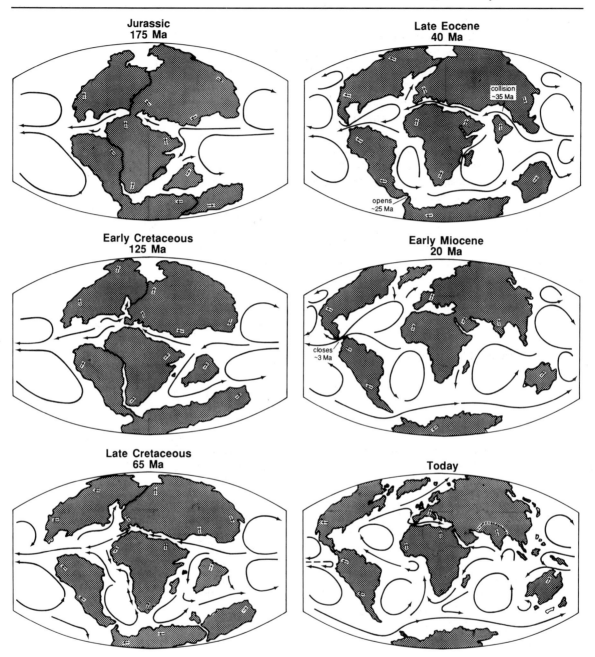

Fig. 2.2 Schematic
reconstruction of the pattern
of major ocean currents during
the most recent cycle of
supercontinent disintegration.
The timing of certain major
events in the development of
the oceanic circulation is
indicated in the diagram (after
Van Andel 1985; and Strahler
and Strahler 1987)

covered, are in addition different in composition: they contain
assemblages of light rocks dominated by aluminosilicate minerals
of lower density than those found in the basalts of the oceanic
crust; further, very few plate boundaries cross continental crust.

An explanation for these features of the planet is readily
available. The underlying mantle supports, by its own deformation in adjustment to loads applied from above, the mass of the
crust. Rocks forming the crust of the earth sit at two distinct
elevations, as a function of their composition: the continents, with
a mean elevation of 840 m, and the ocean floors, lying at an

average depth of 3800 m (Gross 1982). This bimodal distribution of elevations would exist whether the oceans contained water or not (although the ocean floors, without the weight of water, would rise somewhat). Water condensing at the earth's surface through time has inevitably accumulated above the lower-lying areas of oceanic crust. Thus, areas of basaltic crust are oceans and areas of continental (granitic) composition are land areas. There are several additional reasons why the continental crust stands high. Water and sediment are continually shed into the oceans as a result of river runoff and erosional processes affecting the land. The considerable weight of terrestrial and marine sediments accumulating in the oceans causes mantle subsidence and corresponding uplift of the land. Further, as a general rule, the larger the continent, the higher its average elevation. This phenomenon is not completely understood; it may reflect accumulation of heat and consequent isostatic rise of larger continents, or it may relate to the thickness of the crust, smaller fragments of continental crust perhaps being thinner in general, and hence lying at lower elevations.

Effects of plate evolution on the atmosphere and oceans

The episode of Mesozoic and Cainozoic continental dispersal described earlier must have been associated with many changes in the mechanisms which control the global climate.

Marine regression and transgression clearly alter the proportions of land and water at the earth's surface, and hence the overall planetary albedo. Times of high sea level would, all else being equal, tend to be associated with enhanced global warming.

The movement of the continents has the potential to alter global heat balance in other ways. Land located at high latitudes can support snow and ice cover if the regional temperature is sufficiently low, reinforcing cooling by the albedo mechanism. Land of sufficient elevation, located at any latitude, and produced perhaps by uplift following continental collision, can act in the same way. It is possible that the movement of land into higher northern latitudes during the episode of Mesozoic and Cainozoic drift acted in just this way, and so contributed to the Cainozoic cooling. A second albedo-based mechanism is possible in view of the fact that continental movement between 100 Ma BP and the present considerably increased the area of land in the northern sub-tropics. Aridity here, produced by subsiding air, results in high terrestrial albedo, and this effect too may have supported Cainozoic cooling (Kennett 1982).

The distribution of heat within the earth-atmosphere-ocean system is also of fundamental importance to global and regional

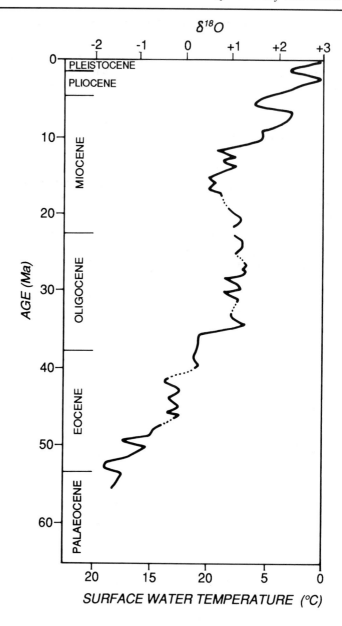

Fig. 2.3 Recontruction of the surface water temperatures of high-latitude Southern Hemisphere oceans during the Cainozoic era. Early Cainozoic warmth gradually gives way to the cold conditions of the Quaternary and the present day (from Shackleton and Kennett 1975)

climates. Changes in the configuration of land or oceans which affect the winds or the oceanic currents thus may also be involved in climatic change. The uplift of major mountain belts, particularly if they are meridionally aligned, can alter the long-wave structure of the atmospheric circulation, perhaps directing polar air into a more southerly course, and influencing the production and survival of snow and ice.

Global cooling and growth of the Antarctic ice cap

At the start of the Cainozoic, Antarctica was ice free. Oxygen isotope palaeotemperature analysis shows that the nearby oceans were quite warm, perhaps 18°C (see Figure 2.3). Forests grew even at very high latitudes.

Continental movement in key locations may have been pivotal in the subsequent cooling. As Australia moved north after about 50 Ma BP, opening a significant Southern Ocean seaway, the westerly wind circulation was able to establish the major Antarctic Circumpolar Current (ACC). This current completely encircles the Antarctic continent, and acts as a barrier preventing warm currents from lower latitudes reaching the Antarctic coast. The development of this current, therefore, must have substantially reduced the transport of 'sensible' heat to the high southern latitudes, and contributed to growing cold there. The tropics would presumably have been essentially unaffected by these events, so that the equator-pole temperature gradient must have begun to increase, thus invigorating the atmospheric circulation.

In the cooling trend, a series of quite rapid temperature drops can be discerned (see Figure 2.3). Each may relate to steps in the development of the modern configuration of land and ocean, which we will briefly review.

During the Palaeocene and Eocene, it would have been possible for major equatorial ocean currents to encircle the globe, with passages open between the Americas, and between India and Europe, and Australia and Indo-China. The climate must have been generally warm during these periods. As the Eocene progressed, the Southern Ocean began to widen, but the ACC was not able to develop to its full depth because shallow barriers still existed here and in the Drake Passage. However, oceanic cooling proceeded in these latitudes because of the restriction posed by the developing ACC to the poleward transport of sensible heat.

One of the major cooling episodes referred to earlier took place at about 38 Ma BP, at the end of the Eocene. Deep-water temperatures dropped by 4–5°C, and this is taken to indicate the development of freezing conditions at sea level around Antarctica, with extensive sea ice. Temperatures over land in the high latitudes of each hemisphere may have dropped by 10°C. Development of the ACC in the Oligocene, enhanced by the opening of the Drake Passage at about 30 Ma BP, produced further cooling and the progressive development of ice on Antarctica. The very rapid cooling at the end of the Eocene probably reflects the influence of positive feedback effects relying on the albedo mechanism as land ice grew in Antarctica, with sea ice possibly also extensive.

By the early Oligocene, the approach of India to the Eurasian land mass must effectively have closed off the equatorial currents of the Paleocene and Eocene. The widening of the Southern Ocean and Drake Passage links during much the same time permitted full development of the ACC, including bottom water circulation.

By the middle Miocene, extensive Antarctic ice was present, and significant cooling is shown in the palaeotemperature record (Figure 2.3). Ice-rafted debris is found in deep-ocean sediments after this time, reflecting the arrival of glacial ice at sea level. It is not clear why it took until the middle Miocene for the ice cap to develop to its present size. A clear consequence of this development, however, was that the system of ocean circulation intensified at this time because of the steepening equator-pole temperature gradient and the resulting strengthened winds. The intensified water circulation was in turn associated with in-creased upwelling and biological activity in the surface layers.

Further significant global cooling followed in the later Miocene. Cooling and marine regression occur together at this time, the latter involving a fall of 40–50 m (Kennett 1982). This regression produced extraordinary effects in the Mediterranean Sea: en-hanced perhaps by secondary tectonic effects, it isolated the Mediterranean from the rest of the world oceans except for a small inlet. Evaporation proceeded to lower the level in the closed basin, concentrating salts and eventually setting down enormous thicknesses of evaporite minerals. Because of the destruction of marine biota which this caused, the event is known as the 'Messinian salinity crisis' (the Messinian is the name of a local stratigraphic stage); it occurred in the period 6.2–5 Ma BP. Other oceans were similarly affected, including the Persian Gulf and the Red Sea (Benson 1984).

The volume of salts laid down represents the evaporation of about 40 times the water volume held by the Mediterranean (Kennett 1982). Water spilling into the basin and being evaporated there concentrated significant amounts of the global salt store into the Mediterranean, so that the salinity in the open sea fell by about 6 per cent. This would have had effects on the global climate because the less saline sea water would freeze at a higher temperature, permitting more sea ice to develop in the high latitudes, and reinforcing cooling by yet another positive feedback mechanism.

Onset of Ice Age conditions in the Quaternary

After a slightly warmer interlude in the early Pliocene, conditions cooled still further. Finally, in the late Pliocene (around 3 Ma BP) climatic conditions in the Northern Hemisphere also cooled sufficiently for significant ice growth. Ice-rafted debris is found in sediments of the North Atlantic after this time (Kennett 1982). This marks the beginning of the phase of repeated ice sheet growth and decay which characterizes the Quaternary. The reason for the lateness of ice growth in the Northern Hemisphere relative to the Southern remains unknown. The final closure of the seaway between the Americas may have been involved, possibly through a strengthening of the Gulf Stream, delivering more moisture to feed snow accumulation at high latitudes. It has also been suggested that mountain building may have triggered the ice growth, through the provision of sufficient land at high elevations to permit the survival of summer snow, and the development of albedo feedback reinforcement of the cooling.

Moderate-sized ice sheets had grown in the Northern Hemisphere by 2.4 Ma BP, near the beginning of the Quaternary period (Ruddiman and Raymo 1988). Their effect can be seen in the marine sedimentary record, which shows at this time a dramatic but irregular fall in the proportion of carbonate materials caused by incoming ice-rafted continental siliceous debris.

In the Quaternary, evidence of various kinds shows that there were numerous rhythmic alternations from cold to warmer conditions. These alternations are shown in the oxygen isotope analyses from marine microorganisms (discussed in Chapter 5), in the record of global sea level (discussed in Chapter 4), and in various kinds of terrestrial evidence such as the periodic accumulation of wind-blown dust. During the early Quaternary, in the Matuyama 'magnetic chron', the warm to cool alternation appears to have had a periodicity of 41 ka (Ruddiman and Raymo 1988) which is the same as the period over which the obliquity of the earth's axis to the plane of the ecliptic cycles from 21.8–24.4° (Bradley 1985).

After 0.9 Ma BP, the amplitude of environmental fluctuations recorded in marine sediments increased, reflecting the growing size of ice sheets during cold phases. In addition, the fluctuations became less frequent, adopting a cycle time of about 100 ka. Interestingly, this once again correlates with the cycle time of an orbital parameter, but in this case it is the eccentricity of the earth's orbit around the sun, which varies from nearly circular to somewhat elliptical. This frequency became especially dominant in the last 400 ka or so, but there were still 41 ka cycles (and 23 ka cycles, close to that of the cycling or 'precession' of the equinoxes in the earth's orbit) superimposed on the slower variation.

The repeated coincidence between the frequency of environmental change preserved in the Quaternary fossil record and known parameters of the earth's orbit supports one of the major hypotheses seeking to explain the extreme instability of Quaternary climate: the 'astronomical theory' of climatic change. We shall consider this theory in more detail in later chapters. First, a number of important questions remains:

(i) If orbital parameters trigger climate instability, perhaps through slight changes in the amount of solar heating received by the earth, why did they not do so in earlier periods? In other words, why was the Quaternary environment so distinctively unstable?

(ii) We have already seen that during the Cainozoic, climatic change was perhaps more closely related to the progressive development of the tectonic plates, and to the changing configuration of the oceans and of ocean currents. These changes may have been partly responsible for the Cainozoic trend to cooler climates that culminated in the development of the Antarctic ice cap, and subsequently, of ice sheets in the Northern Hemisphere. What role, though, was played by external processes, such as the orbital variations, in the Cainozoic cooling?

(iii) Are there other categories of environmental change which need to be considered in a search for fundamental causes? Need we, for example, consider the parallel evolutionary developments occurring in the biological realm, including particularly the plants, which are the largest users of carbon and thus potential controllers of the concentrations of major global greenhouse gases?

Possible causes of the Quaternary instability

The cyclic alternation of warm and cold phases in the Quaternary, occurring over regular 23, 41 and 100 ka periods, hints at a regular underlying mechanism which it should be possible to discover. The mechanism involved is clearly a relatively complex one, however, involving probably a number of feedback processes. The present day is a Quaternary warm phase, or 'interglacial', but there is still substantial land ice. In the Quaternary glacial phases, the volume of land ice grew dramatically, its growth fostered by increasing global albedo, but never to completely cover the land. The climatic instability of the Quaternary was thus 'damped' or limited, such that global temperature only oscillated between values like those of the present and 5–9° cooler. There must therefore be some negative feedback mechanisms which act to damp the instability. Some possible mechanisms include the earth's own radiative cooling as a limit on temperature rise: terrestrial long-wave radiation increases in amount as the fourth power of the temperature, so that for each additional degree of global warming, much more heat must be trapped in the earth-atmosphere-ocean system. This may prevent warming in the short term more than say 5°C above present values. For a negative feedback process to limit cooling, there is the possibility that the cold oceans of a glacial phase release less moisture to the air, so that the developing land ice sheets are essentially starved of moisture and do not grow beyond the maximum size seen in the Quaternary record.

A further feature of the Quaternary instability that hints at a mechanism is the asymmetry of the changes. The progression into a glacial phase, at least in the upper Quaternary, was relatively slow, but the warming phase or 'termination' which ended each was far more rapid. Palaeotemperature curves through the glacial-interglacial cycles of the upper Quaternary thus show a sawtooth pattern. This hints at a positive feedback mechanism which comes into play during deglaciation but which does not act during cooling stages.

Tectonic changes as a precursor to Quaternary instability

One possible explanation for the Quaternary instability is that it was made possible by tectonic uplift generating land at high elevations. Several key locations have been identified which may be critical as sites for snow accumulation: Tibet, the Sierra Nevada and the Colorado Plateau in North America, and the Himalayan-Alpine belt. One possible role for uplift in these areas is to supply land on which, in response to a decline in solar heating, snow may fall earlier and/or survive later in the year, and so generate an albedo feedback mechanism to promote cooling. Areas of high mountains or plateaux may also act by deflecting major airstreams involved in the global atmospheric circulation. Model experiments indicate that the Tibetan plateau and the ranges of western North America do indeed act to lower atmospheric temperatures by bringing cold polar air southward and reducing summer ice ablation (Ruddiman and Raymo 1988; Molnar and England 1990).

It may be that there has been sufficient uplift in the last few million years in these areas to provide accumulation sites and sufficient modification to atmospheric circulation that the albedo feedback process has been triggered when orbital geometry reduced solar heating. The exact chronology of uplift remains to be resolved. It is also possible that the bulk of the uplift preceded the Quaternary, so that it may have contributed to the Cainozoic cooling but does not help to explain the onset of Quaternary climatic instability (Molnar and England 1990).

Possible roles for plants in climate change

Vegetation cover globally stores vast amounts of carbon, extracted from the atmosphere. Ecosystems dominated by angiosperm-deciduous plants regrow leaves and flowers annually, taking fresh nutrients from the soil, and hence are associated with higher weathering rates than are conifer-evergreen plant communities.

The nature of the soil atmosphere may be important in this difference.

By the early Tertiary, the evolutionary diversification of angiosperm-deciduous ecosystems had resulted in accelerated chemical weathering over large areas, and hence may have produced a trend to falling atmospheric CO_2 levels and thus to global cooling (Volk 1989). The exact contribution that this mechanism made to the Tertiary cooling already described remains uncertain.

Over Quaternary time scales, a somewhat different role for plants may be envisaged. Glacial phases were associated with lower temperatures, altered land climates, and about 15 per cent greater land area because of sea-level fall. Thus, we may speculate that the zonation of vegetation across the globe must have undergone major shifts also. Large areas that were ice covered would no longer have carried vegetation cover at all, but vast areas on the newly exposed continental shelves would. The net effect of these changes on the amount of carbon stored in plant tissue depends upon the area occupied by particular plant communities (forest, grassland, etc.). If large areas of forest occupied the exposed shelves in the still relatively mild low latitudes and equatorial zone, it is possible that significant additional CO_2 would have been taken from the atmosphere to build the new biomass, enhancing cooling. Estimates of the changed distribution of plant communities that would have existed at the height of the last glacial period, at 18 ka BP, do not clearly reveal the magnitude of the effect, but suggest the possibility that it may be a significant one (Prentice and Fung 1990).

Yet another role for plants may be mentioned. During marine transgression, the vegetation communities of the continental shelves would be inundated. Burial of the vegetation would both store carbon and possibly lead to anoxic conditions in the water column, reducing biological productivity and its associated carbon storage. Changes of this kind might be associated with every glacial termination. The role of anoxia and its possible effects on biological productivity in the oceans remains to be explored.

Volcanic activity and climate

An association between cooling and volcanic activity has long been considered possible. Eruptions releasing large ash clouds into the atmosphere for example might act to reduce the amount of solar radiation reaching the surface, and hence trigger cooling by albedo feedback effects from just a few years of enhanced snow survival. There is indeed an excellent correlation between glacial phases and preceding episodes of volcanic activity (Bray 1977). However, it has also been argued that the volcanic activity might be the result of the glaciations, not their cause. This could occur if the enormous ice sheets of the glacial phases produced sufficient stress and deformation in the crust or mantle that volcanic activity was triggered. Hence, as is so often the case, a good correlation does not necessarily prove that a causal link exists!

Meltwater and its effects

A final contributor to the complex pattern of environmental instability in the Quaternary that should be mentioned here is glacial meltwater. Land ice is, of course, essentially salt-free; evaporation from the oceans as ice ages developed in the Quaternary removed fresh water and left the oceans more saline than during interglacial times. At the termination of each glacial phase, enormous volumes of fresh water were poured back into the oceans as the land ice retreated. There were dramatic effects on land, because the meltwater was often temporarily dammed by ice or moraine, and then suddenly released by breaching of the barrier to generate truly catastrophic floods whose impact on the landscape is still evident. Enormous amounts of erosional work took place in just a few days during these catastrophic events, and their repetition, quite possibly at the end of each ice age, represents one of the major forces which has moulded the landscapes of areas lying around the Quaternary ice margins.

However, we need here to consider the effects that this fresh water would have had upon reaching the oceans. Being salt-free and less dense than sea water, it is possible that the incoming water produced stratification in parts of the ocean, with fresh water lying above salt, and restricting oxygenation and biological productivity of the lower layers. More importantly, the meltwater might have dramatically lowered the salinity of sea water, and hence weakened one of the major forces which causes poleward surface water currents to sink and strengthen the circulation of bottom water. Normally, continued evaporation from these currents results in a poleward increase in salinity and hence in water density, which results in subsidence and a return flow of bottom water. Dilution by glacial meltwater arriving at high-latitude coastlines may well have weakened or indeed cancelled this effect altogether, altering the strength of the oceanic circulation and hence its ability to transport sensible heat across the earth's surface.

Conclusions

The broad trend toward global cooling which preceded the Quaternary is well established. However, there seems little hope at present of unravelling the causes of this climatic deterioration from its effects. Undoubtedly, crustal processes and plate tectonic redistributions of land and water are involved; so too may modification of the global carbon cycle by evolutionary trends in plant communities. The role of atmospheric CO_2 is particularly complex and difficult to identify. Potentially, critical control on atmospheric temperature may be exerted by the biological productivity of the oceans, and the resulting flux of carbon into sedimentary storage. However, the productivity of the oceans is influenced by the strength and configuration of ocean currents which are influenced in turn by the condition of the atmosphere, so that it is once again difficult to escape from the web of inter-dependencies.

The excellent correspondence between the periodicity of climatic change, especially in the late Quaternary, and orbital parameters, leaves little doubt that variations in solar heating exerted the fundamental control on climatic instability. The precise mechanisms by which the small variations in solar heating resulting from orbital characteristics are translated into global glacial and interglacial cycling remain unresolved. So too do the important mechanisms which limit the magnitude of the climatic swings and determine the rates of warming and cooling.

In the remainder of this book, we will examine in more detail what is known of the environments of the Quaternary period. Only through continued research on the many fronts examined will it be possible to proceed further toward a true understanding of the earth's environment and its past (and future) variability. We begin in the next chapter by examining the nature of the Quaternary ice sheets, whose rhythmic growth and recession formed perhaps the most pervasive influence on Quaternary environments worldwide.

3 | Quaternary Glaciations

The ice was here, the ice was there
The ice was all around:
It cracked and growled, and roared and howled,
Like noises in a swound!

Samuel Taylor Coleridge (1772–1834),
The Ancient Mariner.

Introduction

The Quaternary involved many environmental changes, but none
more dramatic on land (nor more important in their impact on the
global climate) than the development of the enormous ice sheets
which grew and receded many times, episodically blanketing
many land areas in the Northern Hemisphere and to a minor
extent in the Southern. In this chapter we introduce some of the
evidence for these glaciations, and explore hypotheses seeking to
explain their occurrence.

The Cryosphere

Those parts of the earth subject to the effects of ice, including
glaciers, ice sheets, ground ice and sea ice, compose the 'cryosphere'. The Quaternary period involved considerable expansion
of the cryosphere, such that probably 40 per cent of the land area
was included, together with vast areas of the oceans. Some of the
land and oceanic areas affected were not actually ice covered, but
were subject to freezing effects of various kinds, such as permanent or seasonal 'permafrost' in which the soil and underlying
rock (or the materials of the sea floor) are frozen, often to great
depths. Cryospheric effects of this kind are grouped as 'periglacial' phenomena, a term which identifies the landscape features
and processes associated with the cold but non-glacial parts of the
earth.

The present-day cryosphere

Because of the behaviour of radiative heat transfer, planets of our
solar system relatively near to the sun are hot and those further
out are cooler. The coldest planets would require enormous

amounts of heat in order to support liquid water; the hot planets would require to be cooled massively. The earth, however, has an average temperature which is high enough for liquid water to exist, but only marginally, with a mean surface temperature of perhaps 15°C. The surface temperature is not uniform, of course, because of uneven solar heating and the heat transfer effected by the moving atmosphere and ocean currents. Therefore, in places the earth is even now too cold for liquid water, so that permanent ice cover exists (largely in Antarctica and Greenland), while the low latitudes of course are warm. Because the earth's mean temperature is only marginal for the support of liquid water, relatively little cooling is required for the amount of ice to grow substantially: the difference in mean temperatures between the present day and the coldest of glacial times only amounts to 5 or 10°C.

Ice and snow presently cover about 10 per cent of land area, which amounts to about 15×10^6 km²; in addition, sea ice covers about 7.3 per cent of the oceans. The distribution of contemporary glacial and periglacial environments is illustrated in Figure 3.1.

Fig. 3.1 Present-day glacial and periglacial zones of the world (after Davies 1969; Brown 1970; Péwé 1983; and Harris 1985)

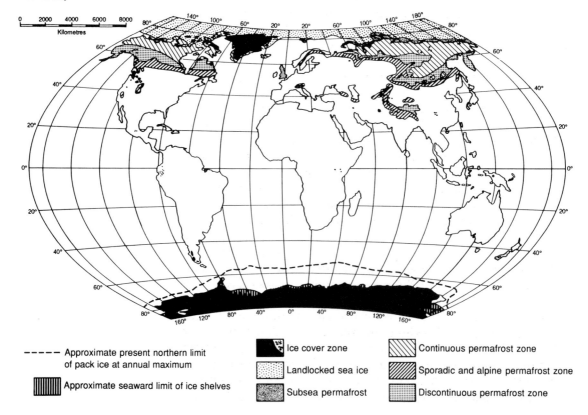

- - - - - Approximate present northern limit of pack ice at annual maximum

▓▓▓▓ Approximate seaward limit of ice shelves

■ Ice cover zone

▒ Landlocked sea ice

▒ Subsea permafrost

▨ Continuous permafrost zone

▨ Sporadic and alpine permafrost zone

▨ Discontinuous permafrost zone

The Pleistocene cryosphere

Enormous expansion of the cryosphere is one of the features
which characterizes ice ages like that of the Quaternary. During
such ice ages, the vast continental ice sheets come and go
repeatedly. Without knowing the future, it is not strictly possible
to label the present day (certainly a time of restricted ice extent) as
an interglacial, but many lines of evidence suggest that this is
indeed the case. It is also possible that the Quaternary ice age,
having lasted nearly 2 million years, is now over. Adding uncertainty to this issue is the further modern process of human
intervention with its effects on the composition of the atmosphere
(notably via release of greenhouse gases) and the albedo of large
parts of the earth's surface which are altered by deforestation,
agriculture, pastoralism and desertification. Advance knowledge
of the combined effects of these activities must be sought in
climate modelling, the only real alternative being a wait-and-see
approach which might prove to be foolhardy. Climate modelling,
in turn, must be guided by reconstructions of the ways in which
the earth's environment has responded to previous climatic
change, most importantly the most recent and relatively well-
dated glacial period.

During the coldest part of the last glacial period, when temperatures were lowered globally, ice covered nearly one-third of the
land area of the earth. Since Antarctica and Greenland (which
account for about 97 per cent of the area occupied by land ice
now) are effectively completely covered by ice, the additional area
covered by glacial ice was largely in parts of the globe now
ice-free. These include most of Canada, much of northern USA,
and large areas in Scandinavia and northern Europe. Figure 3.2
portrays the inferred extent of the affected area. The ice sheets
extended from high latitudes to about 36°N (the most southerly
point reached by the ice lobes flowing through the basin of present
Lake Michigan into Missouri), with alpine glaciers growing even
into the tropics where high mountains existed. The additional
ice-covered area in the last glacial was almost all in the Northern
Hemisphere, with less than 3 per cent in the Southern. It is
therefore reasonable to describe the growth of ice in the Quaternary as essentially a Northern Hemisphere phenomenon,
although of course lowered temperatures and the glacio-
eustatically lowered sea level were global in their occurrence. The
small area of ice in the Southern Hemisphere was largely made up
of the Patagonian ice cap in South America, with minor ice caps
and valley glaciers elsewhere, including Australia, Papua New
Guinea and Irian Jaya, New Zealand and East Africa, together with
substantial additional ice on the many sub-Antarctic islands. In
the Northern Hemisphere, nearly 60 per cent of the additional
ice-covered area was in Canada and the USA, and the bulk of the

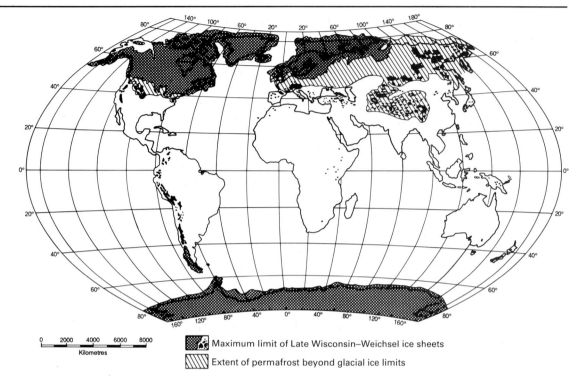

Maximum limit of Late Wisconsin–Weichsel ice sheets

Extent of permafrost beyond glacial ice limits

0   2000   4000   6000   8000
Kilometres

Fig. 3.2 Maximum extent of ice sheets and permafrost at the last glacial maximum (after Löffler 1972; Washburn 1979; Hollin and Schilling 1981; Denton *et al.* 1981; Anderson 1981; Hughes *et al.* 1981; Nilsson 1983; and Péwé 1983)

remainder in northern and western Europe and the Alps.

The thickness of the now vanished ice sheets was remarkable: in places, maximum depths of up to 4 km are inferred, but with more typical depths of 2–3 km. The landscape modification accomplished as the ice ground across the landsurface was considerable: Bell and Laine (1985) have estimated that on average, bedrock was stripped to a depth of 120 m over the area of the Laurentide ice sheet in Canada and North America. Much of the eroded material now lies as marine sediment offshore, but there are also vast areas blanketed by glacial debris on land. Significant changes were also produced by the massive meltwater streams that were fed as the ice retreated during climatic recovery. The ice sheets vanished with remarkable speed, being completely gone from large areas in only 8000 years after the process of deglaciation had begun. In parallel, recovery of vegetation proceeded as the environment recovered to the conditions of the present day. The transformation of the landscape of the glaciated areas from what it was at 20 ka BP to its present form (often to forest cover, croplands, and pasture) is so dramatic that it is not surprising that early researchers found it difficult to accept the evidence around them and the reality of the massive Quaternary ice sheets.

The development of ideas about the Quaternary glaciations

The realization that a large fraction of the land area had relatively recently formed part of the extended Quaternary palaeocryosphere came slowly and fitfully. This 'Glacial Theory' as it used to be known, evolved in the last decades of the eighteenth century and early in the nineteenth century. It is important to remember that at this time, neither the Greenland nor the Antarctic ice sheets were known at all well, the first crossing of the Greenland ice cap being made in 1888 and the first interior exploration of the Antarctic ice cap not until the first decade of the twentieth century. Consequently, the only glacial phenomena familiar to researchers were the smaller valley glaciers of places such as the European Alps and Scandinavia. Nonetheless, geologists and other naturalists correctly identified 'glacial erratics', large blocks of rock found hundreds of kilometres from their nearest bedrock outcrop. The idea that ice had been the agent of transportation slowly supplanted the formerly accepted view that the blocks had been carried in moving water, perhaps the biblical floods, or had been dropped from floating icebergs. Deposits of 'till', the unsorted debris carried and deposited by glaciers, were also common in many areas: in the early nineteenth century these too were explained as having fallen from floating icebergs, and are still often referred to in consequence as 'drift'. Eventually, though, the form of the glacial deposits, the striations left on the erratic blocks and on outcropping bedrock, led to the widespread acceptance of the idea that great ice sheets had formerly occupied the landscape.

Chronology

As noted in Chapter 2, substantial ice sheets had developed in the Northern Hemisphere by about 2.4 Ma BP. The marine isotope record suggests that until about 0.9 Ma BP, the dominant rhythm of growth and recession in these ice sheets was the 41 ka period reflecting the cyclic change in obliquity of the earth with respect to the ecliptic plane.

Glacial periods in the late Quaternary, however, occurred with a periodicity of around 100 ka, and displayed a slow and uneven cooling followed by a rapid deglaciation, as noted in Chapter 2. During the slow cooling, lasting perhaps 70–90 ka, there were numbers of slight returns towards warmth (called 'interstadials') as well as unusually cold periods, 'stadials'. Eventually, maximum cooling resulted in the occurrence of a corresponding maximum ice volume, reached at the glacial peak (called 'full-glacial conditions'); deglaciation then led rapidly back into interglacial conditions with warmth like the present. The period of ice

growth which ended about 10 ka BP and marked the onset of
Holocene interglacial conditions is thus merely the latest glacial
stage of the many which occurred during the Quaternary ice age.

There is a vast and confusing array of local names which is
applied to glacial periods, stadials, interstadials, and interglacials;
often these are locality names chosen from sites providing good
exposures of the field evidence. The naming of the last glacial is
typical: 'Wisconsin' in Canada and the USA, 'Weichsel' in western Europe, 'Devensian' in Britain, 'Würm' in the Alps, and so on.
Because of their dominance in terms of ice area and volume, the
North American and European names are here linked, as in
common practice, to identify this glacial period as the Wisconsin–
Weichsel.

This glacial stage is subdivided into an early substage, a middle
substage, and a late substage. The timing of these divisions is not
universally agreed upon but commonly for purposes of description they are taken as follows:

Late Wisconsin–Weichsel substage: 24–10 ka BP (including the glacial maximum at 21 ka BP)
Middle substage 74–24 ka BP
Early substage 117–74 ka BP

For some time prior to 117 ka BP, conditions were warm: this is
the last interglacial, termed the Eemian interglacial in Europe, and
the Sangamon in North America. Full glacial conditions were
reached at 21–17 ka BP, after cooling spanning the early and
middle substages: this late Wisconsin–Weichsel substage is best
known because the evidence is freshest, and lies within the span
of radiocarbon (^{14}C) dating (see Appendix). As we shall see, on the
basis of the fragmentary terrestrial record alone, relatively little is
known with certainty about events in the early substages, nor
about previous glacial stages in the Quaternary. The evidence
preserved in deep-sea cores is therefore an invaluable adjunct in
reconstructing Quaternary environmental changes, since it is
usually more complete than the land record (see Chapter 5).

Evidence of glaciation

Scientific study of the contemporary cryosphere and of the
landscape legacy of the Quaternary palaeocryosphere has
advanced enormously in the last 150 years. Precise isotope dating
techniques, the palynological reconstruction of former vegetation
communities (discussed in Chapter 8), and analysis of data from
realms other than the crysophere but which shed light on ice age

events (e.g., the oceans and their sediments, discussed in Chapter 5), have all assisted in this process.

In the modern landscape, including adjacent continental shelves which were crossed by ice during the glacial low sea levels, many ice-related features have been described and their origins accounted for. These are too numerous to recount here, but we can note that the poorly sorted rock debris forming glacial till and the ice-shaped moraines formed from it, often still standing as barriers marking the limit of ice advance, are vital features enabling the areal extent of ice to be worked out. Use is also made of 'glacial outwash' deposits, the material carried from the ice margins by meltwater streams, and often set down to great depths further down valley or in other ice-marginal areas, and 'loess', the fine rock dust deflated from outwash surfaces and carried by the wind to blanket the landscape beyond the ice margin (see also Chapter 7). Many effects in soil and rock, associated with frost shattering, are also known: the movement of rock and soil by growing ground ice, and the production of the resulting 'patterned ground' displaying sorted stone polygons and the like, are very common in areas around the margins of the former ice sheets. Seasonal thawing of the upper layers of the frozen soil permitted the sodden mass to move downslope under the influence of gravity to produce a wide range of distinctive lobed 'solifluction' forms. Over vast areas, landscape features such as rounded and polished rock outcrops, bearing crescentic ice chatter marks and linear striations, are abundant. Datable materials are found within glacial deposits, and in the fossil-bearing sediments that accumulated in lakes associated with ice sculpture of the landscape and with meltwater. Indeed, in some lakes a chronology can be built up by counting the annual layers ('varves') which result from seasonal changes in the nature of the sediment being set down. Pluvial lakes of Quaternary basins show, in some areas, lake-floor deposits which connect laterally with the deposits resulting from ice advance from nearby uplands, and the glacial debris can be dated by its association with the lake sediments.

In the case of the major continental ice sheets, quite detailed reconstructions can be made on the basis of the morphology of the glaciated surfaces. These large ice sheets were sufficiently thick that, like the present Greenland and Antarctic ice sheets, they adopted a smoothly domed form which blanketed the underlying topography, with thickest ice in some central location (see Figure 3.3). Large ice sheets have complex temperature characteristics, in which the ice would melt by friction where movement is fast, and in which the temperature at which the ice would melt is lowered where the ice is thicker and exerts greater hydrostatic pressure at its base. Near the centre of an ice sheet, the ice is nearly stagnant and is frozen to the underlying rock, so that only minimal erosion occurs. As the flowing ice accelerates in its outward movement

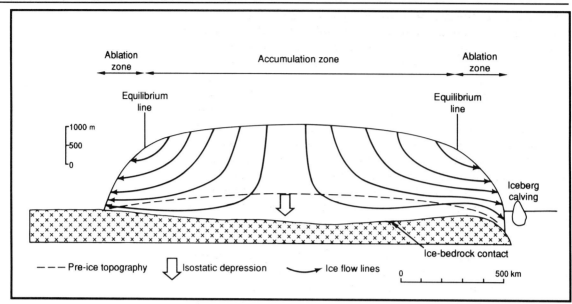

Fig. 3.3 Flow lines and zones of accumulation and ablation in a large ice sheet (modified from Flint 1957; Sugden and John 1976; West 1977; and Reeh 1989)

further away from the centre, more heat is generated and basal meltwater results, permitting the lubricated ice to slide, pluck bedrock, and grind and erode the surface. Further out still, towards the ice margin, ice movement slows as the ice thins in the zone of ablation; freezing once again dominates because of the reduced frictional heating. Between the frozen and thawed zones are zones where the outward-moving ice either becomes progressively warmer (a melting zone) or cooler (a freezing zone). The kinds of deposits left allow these zones to be mapped for the Wisconsin–Weichsel ice sheets, and hence permit reconstruction of the locations of their feeder ice domes, and their patterns of thickness and rate of movement.

As an example, we may consider the Laurentide ice sheet of Canada and North America. The major central ice dome for this sheet was evidently over Hudson Bay, which was consequently isostatically depressed under the weight of ice, and even though this area now displays rapid postglacial rebound, it remains a marine embayment (see Chapter 4). In a zone around the Hudson Bay centre, basal melting would have uncoupled the ice from the bedrock, allowing considerable glacial erosion. This erosion stripped the younger Palaeozoic sedimentary cover and exposed the ancient Canadian Shield rocks in an arc around the presumed ice dome. Bedrock hummocks (around which pressure melting point was not reached, as the overlying ice was thinner) would have generated frozen patches where erosion was restricted; hollows in the subglacial topography (where the ice was thicker and basal melting resulted) would have favoured erosion and deposition. Thus, the deglaciated landscape has the distinctive appearance of numerous lakes (representing thawed patches under the ice sheet) and fields of mounded till deposits ('drumlins') located in the lee of the bedrock high points. In zones of basal melting near the ice-sheet margins, hydrostatic pressure is low because of the thinner ice, and meltwater can gather into subglacial streams. After deglaciation, their sediment load may be left as linear 'eskers' (snaking ridges of alluvium) which are distinctive of this zone. Material delivered to the outer freezing zone near the ice margin will be set down during the retreat of the ice sheet to form a 'till blanket', partly modified by meltwater processes taking place during the retreat. In the case of the Laurentide sheet, this area is the zone of water-filled depressions around the margins of the Canadian Shield, such as Lakes Huron, Michigan, Athabasca, Great Slave, and Great Bear, and Coronation Gulf and the Gulf of Boothia.

Finally, along the warmer equatorward margin, a melting zone existed which set the southern limit reached by the ice. Here, lubrication of the ice base by water permitted acceleration of the ice into 'ice streams', which poured out to form lower-lying 'ice lobes' around the ice-sheet margin, with lobate moraines left after ice retreat. The Des Moines lobe downstream of Lake Winnipeg and the James lobe below Lake Manitoba were major lobes fed from the Laurentide ice sheet in this way.

Quaternary cryosphere reconstruction

What then do we know about the geographical extent of the glaciers, ice sheets, and periglacial phenomena of the Quaternary? Likewise, what is known of the chronology of the growth and decay of the vast ice sheets? The answer is a complex one, with great variation from region to region. As we shall see, there are time-lags in ice-sheet growth and shrinkage following a climatic change which are not constant but depend upon the actual bulk of the particular ice sheet, the extent to which it is confined by ridges or mountain ranges along its margins, and the nature of the climate change which has occurred. These parameters are likely to be different for each of the principal Quaternary ice sheets and for each major climatic swing, so that we should expect the record of ice advance and retreat to be far from simple and certainly not in strict agreement as to timing at different sites.

Considering the area affected in a glacial stage on all continents together provides an indication of the size of the palaeocryosphere and the volume of ice involved in glaciated regions. The total ice-covered area at a typical glacial maximum amounts to about 40×10^6 km^2, compared with the present 15×10^6 km^2. The volume of water stored as ice at a glacial maximum amounts to about 90×10^6 km^3, compared to the present 30×10^6 km^3. Thus, the ice volume is tripled, while the area covered increases by a little more than 2.5 times. In addition, there is a very substantial area affected by periglacial conditions in both glacial and interglacial times: at present this amounts to about 20×10^6 km^2 (in addition to the ice-covered areas). There is still uncertainty about the reconstruction of conditions at the last glacial maximum, because the field data are not absolutely unambiguous and it is still possible that numbers of individual ice sheets which have been inferred were part of one much larger system. Resolving this requires considerable understanding of the dynamics of ice sheets which extended much closer to the equator and were larger than anything presently in existence, and is the subject of continuing research. Some uncertainty also remains about the extent of ice on the now drowned continental shelves: these areas, for obvious reasons, are harder to explore and their deposits are less well mapped and dated.

We will now briefly review the extent of Quaternary ice, especially in the late Wisconsin–Weichsel glacial, in several key areas of the world.

Alpine glaciations

It is difficult to reconstruct glacial history in valley glacier areas such as the Alps because major ice growth can erase the sedimentary record left by earlier ice, especially in a rugged environment such as the Alps where high erosion rates exist. Loss of evidence is particularly likely in the case of the minor advances associated with stadial phases and with short interstadials. Minor fragments of the glacial moraines, or interstadial lake sediments from these times may exist, but be exceedingly hard to find or to date.

Despite its relatively small extent (less than 0.2 per cent of the additional ice-covered area at glacial maximum), the area of Quaternary glaciation in the European Alps is the starting point for our discussion of ice extent for three reasons:

(i)   The Alps, an area rich in erratics, striated bedrock, and well-preserved moraines, provided the field area which led to the early advancement of the 'glacial theory' already mentioned.

(ii) Research in the Alps led to the early suggestion, now known to be incorrect, that four major glacial episodes had taken place.

(iii) Refinement of knowledge of the field evidence in this area makes it clear that there are problems with attempting to reconstruct ice-age chronologies on the basis of evidence from valley glaciers, a conclusion pertinent to analysis of glacial records from other parts of the world also.

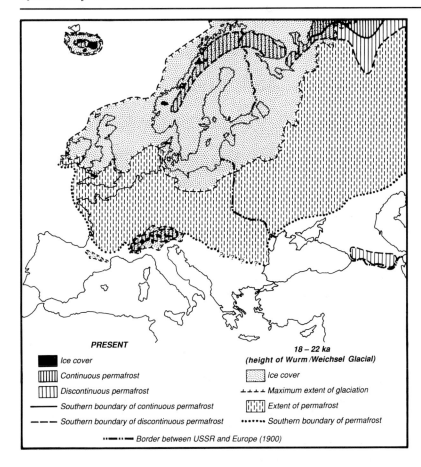

PRESENT

- ■ Ice cover
- IIIII Continuous permafrost
- IIIII Discontinuous permafrost
- —— Southern boundary of continuous permafrost
- – – – Southern boundary of discontinuous permafrost

—··—··— Border between USSR and Europe (1900)

18 – 22 ka
(height of Würm/Weichsel Glacial)

- Ice cover
- ⊥⊥⊥⊥ Maximum extent of glaciation
- Extent of permafrost
- •••••• Southern boundary of permafrost

Fig. 3.4 Glacial and periglacial zones of Europe and Scandinavia at present and at the last glacial maximum (from Maarleveld 1976; Anderson 1981; Baulin and Danilova 1984; and Harris 1985).

The small area of Alpine glaciation can be seen in Figure 3.4, which also shows the extent of glacial and periglacial phenomena affecting Britain and Europe during the last glacial maximum.

In the Alps, some earlier Quaternary glaciations were evidently more extensive than the most recent, their terminal moraines lying further downvalley than those of the late Würm. It is important to remember therefore that reference to the glacial maximum of the last glacial stage does not imply that this represents the maximum Quaternary extent of ice.

Early this century, Penck and Brückner (1909) identified a sequence of four sets of putative glacial sediments in the Alpine foreland or piedmont zone. These were represented by the older and younger 'Deckenschotter' (cover gravels) and the 'Hochterrassenschotter' and 'Niederterrassenschotter' (high and low terrace gravels). They were taken to represent four sequential cold periods, which were named after four small tributaries of the Danube west of Munich, the Günz, Mindel, Riss, and Würm. The older cover gravels are high-level remains found on plateau surfaces; the materials are strongly weathered glaciofluvial sediments, and have been linked to deeply weathered tills ascribed to the Günz and Mindel glacial stages. The terrace gravels are set within broad valleys cut lower in the landscape. The high terrace gravels have been connected with Riss moraines, and the low terrace with Würm moraines. The materials of the terrace gravels are substantially fresher than those of the 'Deckenschotter'. The fact that the terrace gravels are set into broad valleys suggested to the early workers that a substantial period of incision had followed the cold phase which resulted in the formation of the younger cover gravels, since these lie above the valley sides on the upland surfaces. The period of incision was taken to represent a long interglacial.

Mapping in the Alps has shown that in the late Würm, the snowline was about 1200 m lower than at present. During the older of the glacial events allegedly responsible for the 'Deckenschotter', the snowline was lower by an additional 100–200 m (Nilsson 1983). During the glaciations, ice occurred largely in the form of separate valley glaciers, not an integrated ice sheet. Some glaciers descended into the plains to the north and south, to form large ice lobes. The location of these is very clearly preserved by lobate terminal moraines, such as those on the Po plain of northern Italy; the scoured rock basins lying inside the moraines now hold major lakes, such as Lago Maggiore and Lago di Como, north of Milan in Italy, and Lakes Lucerne, Zürich and Geneva on the north flank of the Alps. These basins would have been sculpted to a degree by each glacial, with their final form being a result of the most recent action of the Würm ice. Downvalley from the ice lobes, streams were fed with sediment-laden meltwater. These overloaded streams proceeded to aggrade, and to build up floodplains. During interglacial times, as the ice retreated up-valley, and vegetation re-established itself, sediment delivery was reduced, and the under-supplied streams proceeded to incise, creating the terraces. The upper surface of each terrace therefore dates from the start of an interglacial stage. Uplift of the Alps has ensured that the chronologic sequence of terraces from old to young is also a height sequence, old remnants at the greatest elevations, with fresher materials nearer present river level. The outwash trains, as the alluviated valleys are called, supplied extensive dusts which blanket the landscape as loess. Soils developed slowly in these loess deposits, and now provide one of the kinds of field evidence used to recognize and define interglacial conditions.

The effects of four major glacial phases do indeed dominate landscapes around the Alps. However, this is not to say that we still accept, as Penck and Brückner did, that this means that there were only four glacials. Indeed, revision of the system of four glaciations has proceeded slowly ever since it was proposed. Detailed mapping has suggested to subsequent workers that each of the glacial stages was indeed a compound event of two to four separate ice advances delimited by interstadials. The interstadials and interglacials in turn were recognized not to be simple single events, on the basis of multiple loess layers, separated by palaeosols, in the windblown deposits which cap the moraines. Sediments taken to date from the last interglacial (i.e., post-Riss and pre-Würm) are consistent with warmer conditions than present, with *Quercus* (oak) forest subsequently replaced by *Picea* (spruce) and cold-tolerant species as the next glacial developed (Husen 1989). The fossil fauna of the interglacials is also revealing, with elephant, rhinoceros and beaver. In cooler early Würm interstadials which followed, the fauna includes reindeer, giant deer, and woolly mammoth. The disappearance of these animals is considered further in Chapter 8.

In addition to subdividing the glacial and interglacial phases, modern work in the Alps has suggested the existence of older glacials than the Günz (they have been named the Donau and the Biber). The chronology of these older events is not well established. Normal magnetic polarity is displayed by the older 'Deckenschotter' sediments of the Günz glacial, as well as by Günz tills, on which basis it has been argued that the younger four main glacials at least occurred within the Brunhes magnetic chron, i.e., within about the last 730 ka (Kohl 1986).

Recession of the Würm glaciers in the Alps was underway by 15 ka BP, and it has been suggested that the retreat was very rapid, with glaciers contracting to half their full-glacial length in only 1–2 ka (Husen 1989). The retreat is dated on the basis of evidence such as ^{14}C dated tree stumps, in growth position, standing within the Würm moraines. In the French and Italian piedmonts, deglaciation was evidently interrupted by two major readvances. By 10 ka BP the main valleys were ice-free and indeed ice became less extensive than today by 8.4 ka BP (Billard and Orombelli 1986). There were minor Holocene readvances, mentioned later in this chapter.

North America and Greenland

The largest ice accumulations of the late Wisconsin–Weichsel glaciation occurred as a series of domes and ice sheets which, at their maximum areal extent, covered essentially all of Greenland, all of Canada and offshore islands, and many of the northern states of the USA (see Figure 3.5). This amounts to about 16×10^6 km^2, about twice the area affected in Europe (6.7×10^6 km^2) and slightly more than the present Antarctic ice sheet (nearly 14×10^6 km^2). In the east, major ice lobes extended equatorward through the low-lying parts of the Mississippi basin, reaching to just past St Louis at 36°N. Over most of the southern margins of this ice, the late Wisconsin–Weichsel

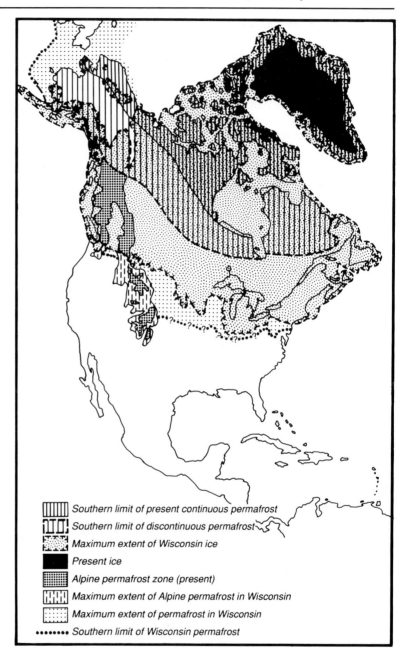

Fig. 3.5 Glacial and periglacial zones of North America and Greenland at present and at the last glacial maximum (from Davies 1969; Prest 1969; Brown 1970; Hamilton and Thorson 1983; Péwé 1983a, 1983b; and Harris 1985)

Southern limit of present continuous permafrost
Southern limit of discontinuous permafrost
Maximum extent of Wisconsin ice
Present ice
Alpine permafrost zone (present)
Maximum extent of Alpine permafrost in Wisconsin
Maximum extent of permafrost in Wisconsin
Southern limit of Wisconsin permafrost

advanced to about the same point as earlier Quaternary glacial stages. Sea ice was also more extensive during the Wisconsin–Weichsel, as shown in Figure 3.6. As we shall see later, the greater extent of sea ice has an important bearing on the availability of the moisture required to sustain the terrestrial ice sheets.

The land ice cover is envisaged by many to have consisted of multiple ice domes and accumulation centres. In the west, a

Cordilleran ice sheet was supported by the Rocky Mountains, and flowed north and south to produce major lobes spreading onto lower elevations of Alaska and Washington State, as well as generating offshore ice shelves along the Pacific coast and glaciers draining to the east to link with the Laurentide ice sheet.

In the Queen Elizabeth Islands, an Innuitian ice sheet seems probable. This, however, is one of the geographic regions where the true extent of late Wisconsin ice remains uncertain. In terms of annual precipitation much of the area rates as very dry desert today. Thus, it seems unlikely that ice could accumulate rapidly now. The greater cold of the last glacial, surprisingly, might not have helped: it would have permanently frozen many of the surrounding oceanic moisture sources, and it is therefore reasonable to imagine that the region would consequently have been even drier. As a result, the chronology of glaciation may be unusual in areas like this, with most ice growth occurring early on in the glacial, perhaps even with retreat occurring in the coldest phases, and possibly renewed growth as conditions warmed somewhat and precipitation increased.

The largest of the ice sheets was the Laurentide, which was probably composed of multiple ice domes. These may have included a Keewatin ice sheet covering much of western Canada (and adjoining the Cordilleran ice sheet), a Labrador ice sheet in the south-east, and a Foxe ice sheet lying over Baffin Island and the Foxe Basin (between Baffin Island and Hudson Bay). A single dominant ice dome located over Hudson Bay is viewed by some as better describing the state of the ice sheet. The Laurentide ice may have reached maximum thicknesses of nearly 4 km; others favour lesser values of 2–3 km, with quite thin margins composed of ice lobes perhaps 500 m thick. Many such ice lobes have been inferred from the pattern of moraines along the southern margin of the ice, including the major Des Moines lobe of Iowa already referred to, and lobes produced by ice moving through the low-lying basins of the Great Lakes. Some of the ice movement toward these marginal lobes must have involved flow up the topographic gradient from the most isostatically depressed area around Hudson Bay. Along the southern margin of the ice, a major zone of periglacial conditions existed. Casts of 'ice wedges', ice-filled frost cracks resulting from repeated freeze–thaw alternation of water moving in from the surface soil, and filled by inwashed sediments as the ice melted, are abundant indicators of this zone; their distribution along the former ice margin is as distinctive as that of ice-margin moraines. The presence of ice wedges is taken to mean the deeper soil, below a surface 'active layer' which may have thawed each summer, was permafrost.

In Greenland, the present extensive ice sheet was enlarged somewhat at the glacial maximum. Uncertainty remains about the

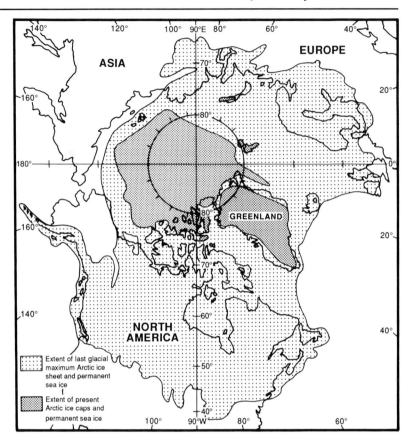

Fig. 3.6 Distribution of Arctic sea ice and adjacent land ice at present and at the last glacial maximum (adapted from Bartholomew *et al.* 1980; Denton and Hughes 1981; and COHMAP 1988)

extent of ice movement onto the continental shelf. In eastern Greenland, there is evidence for restricted ice growth, possibly because of limited moisture availability, as in the case of the Queen Elizabeth Islands.

The chronology of ice growth and retreat over this vast area is not straightforward. The areal extent of ice was greatest in the early Wisconsin–Weichsel, and the middle substage was a time of extensive deglaciation in some areas. Ice readvanced in the late substage, generally reaching its maximum for that advance in the period 21–17 ka BP. The southern margin of the Laurentide ice sheet certainly reached its maximum extent in this period, and readvanced close to this again at about 15–14 ka BP (Denton and Hughes 1981). The Des Moines lobe, which flowed into Iowa, did not reach its maximum extent until about 14 ka BP, a date similar to that of 15 ka BP for the time of maximum southward extent of the Cordilleran ice. Relatively little is known of the detailed chronology of ice retreat, except that it was proceeding by 14 ka BP, and was fastest along the southern ice margins. Ice-margin recession here reached speeds of a few hundred metres per year (i.e., a few hundred kilometres per ka). By 11 ka BP the ice had contracted to the margins of the Canadian Shield, and the Canadian Plains and Cordillera were completely deglaciated by 10 ka BP (Fulton 1989); by 8 ka BP the sea invaded Hudson Bay and the Laurentide ice sheet ceased to exist as one continuous ice mass. Ice lasted in the Foxe Basin until about 7 ka BP, and remnants of Labrador ice until about 6.5 ka BP (Fulton 1989). This record of ice retreat really only tells us about the diminution in the area covered by ice, and does not shed light on whether the ice sheet became thinner also as it retreated. Simultaneous thinning would result in a much greater rate of ice-volume loss than would simple retreat of the ice margin. This distinction is an important one to which we will return later in this chapter.

Europe

In northern Europe, the late Wisconsin–Weichsel ice did not cover quite so extensive an area as did some of the earlier Quaternary glaciations, older moraines lying several hundred kilometres further south in some areas. The largest ice sheet was the Scandinavian, the terrestrial extent of which is well known but which also extended onto the Norwegian continental shelf. Smaller ice sheets occurred in other areas, including the Siberian Plateau, Svalbard, Franz Joseph Land, Novaya Zemlya, and Severnaya Zemlya (see Figure 3.7).

The late Wisconsin–Weichsel ice did not reach Holland, permitting the accumulation there of one of the best sedimentary records of glacial times. There is a good record in Denmark, which lay near the southern ice limit during the Wisconsin–Weichsel. The whole of Denmark was, however, ice covered in earlier glacial stages: four episodes of ice growth are recognized. In the last glacial, ice did not reach Denmark until after 25 ka BP, reaching its maximum extent in the period 20–18 ka BP (Lundqvist 1986).

In Norway and Sweden, the record of earlier glaciations is very fragmentary, as both areas were completely glaciated in the late Wisconsin–Weichsel. The Swedish and Norwegian west coasts show evidence of several phases of ice retreat during the last glacial, whereas the central areas show only one early Wisconsin–Weichsel retreat. Regional variation in the glacial record of this kind must be expected, as a consequence of Quaternary gradients in temperature, elevation, and moisture availability. The southern ice margin was in retreat by 15 ka BP, and a significant cooling of the North Atlantic is inferred to have resulted at that time from the influx of ice and meltwater (Lehman *et al.* 1991).

In the British Isles, the area affected by late Wisconsin–Weichsel ice was less extensive than that covered by some earlier glacial episodes. In the Devensian, as the last glacial stage is known in Britain, ice spread from various upland centres, most importantly the Scottish highlands. At glacial maximum, ice

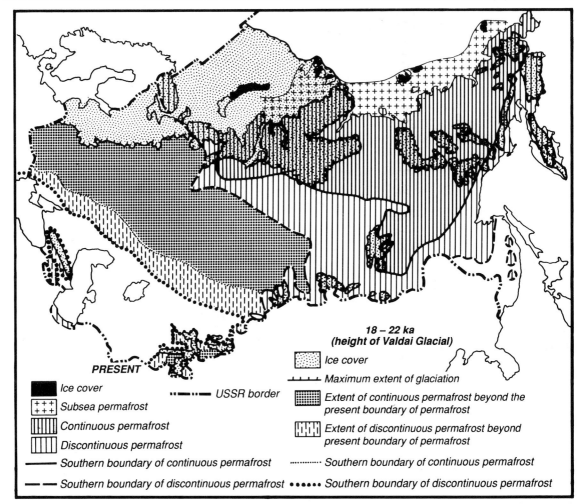

18 – 22 ka
(height of Valdai Glacial)

- Ice cover
- Maximum extent of glaciation
- Extent of continuous permafrost beyond the present boundary of permafrost
- Extent of discontinuous permafrost beyond present boundary of permafrost

PRESENT

- Ice cover
- Subsea permafrost
- Continuous permafrost
- Discontinuous permafrost

- USSR border

- Southern boundary of continuous permafrost
- Southern boundary of discontinuous permafrost
- Southern boundary of continuous permafrost
- Southern boundary of discontinuous permafrost

Fig. 3.7 Glacial and periglacial zones of the USSR at present and at the last glacial maximum (adapted from Davies 1969; Washburn 1973; Andersen 1981; Péwé 1983; Baulin and Danilova 1984; and Harris 1985)

covered most of Scotland, the northern parts of Ireland and Wales, as well as a large part of England, notably the Pennine uplands (see Figure 3.8). The ice extended south to about the latitude of the Thames valley, but the northern ice margin was sinuous, running northward in a broad loop to skirt the lower-lying Vale of York, and thus leaving much of eastern England ice-free but affected by permafrost. An ice stream moved southward through the Irish Sea, while in the east movement of the ice was deflected into northern and southern branches by the adjacent Scandinavian ice sheet. The exact location of the ice front linking the ice in the British Isles with that of the Scandinavian ice sheet remains uncertain.

There is no firm evidence of glaciation in the early or middle substages of the Devensian, although it is likely that ice existed in upland areas in the west and north. Major glaciation began in the Dimlington stadial, early in the late Devensian. Ice growth is

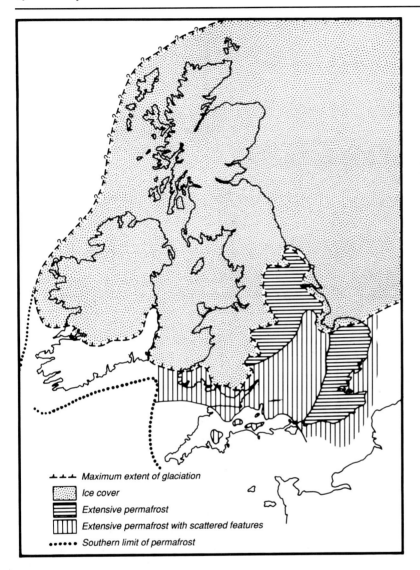

Fig. 3.8 Glacial and periglacial zones of the British Isles at the last glacial maximum (from Williams 1969; Sparks and West 1972; Maarleveld 1976; and Andersen 1981)

Legend:
- ▲▲▲ Maximum extent of glaciation
- Ice cover
- Extensive permafrost
- Extensive permafrost with scattered features
- ●●●●● Southern limit of permafrost

inferred to have begun after 26 ka BP, reaching its maximum extent in the period 18–17 ka BP, and retreating from most areas by about 14.5 ka BP (Bowen *et al.* 1986). After the Windermere interstadial, during which ice completely disappeared, a Loch Lomond stadial has been identified. This renewed ice growth was largely restricted to the Scottish uplands, where many valley and corrie glaciers advanced. This stadial is correlated with a widespread cooling, termed the 'Younger Dryas' event, which has been identified in both hemispheres. The name of this stadial comes from a group of Arctic and alpine plants belonging to the genus *Dryas*.

Southern Hemisphere

We will only briefly consider the evidence from the Southern Hemisphere, which is more fragmentary and less well dated. To the extent that it is known, the chronology of the last glacial stage here is very similar to that of the Northern Hemisphere already described, and there is no doubt that the Southern Hemisphere ice growth reflected control by the same instability of global climate.

The largest area of ice growth that is well known was in South America, where there were locally enlarged glaciers on high peaks of the Andes even in the tropical north. However, the largest fraction of the ice developed in the far south of the Cordillera, in Chile and Argentina. In the more northerly parts of this zone, the ice mostly took the form of valley glaciers which poured down towards the Pacific continental shelf, where they calved into the sea. Further south, over Tierra del Fuego, ice accumulated over about 480 000 km^2 to form the Patagonian ice cap; this reached a thickness of about 1.2 km.

On the Australian mainland and on the island of New Guinea, snowlines were lowered and small cirque and valley glaciers developed, together with more extensive periglacial phenomena. The snowline in tropical New Guinea was lowered by about 1000 m, an amount comparable to that recorded for Northern Hemisphere sites. In Tasmania, a small ice cap of about 8000 km^2 reached its maximum extent at about 19–20 ka BP, and had receded by 10 ka BP.

The largest ice-covered area in the south-west Pacific region occurred in the Southern Alps in the South Island of New Zealand. Here the most recent glacial stage is called the Otiran; it is equated to the Wisconsin–Weichsel of the Northern Hemisphere. At its maximum, which occurred just before 18 ka BP, ice covered about 40 000 km^2. The ice was relatively thin, however, possibly no more than 100–200 m on average.

The extent of ice growth in Antarctica during Wisconsin–Weichsel time remains uncertain. Denton and Hughes (1981) argue for a significant growth of the West Antarctic ice sheet outward to the edge of the continental shelf, together with lesser extension in the volume of the main East Antarctic ice. Their estimate of the maximum probable growth in ice volume amounts to more than a 50 per cent increase over the volume of the present ice sheet, but the field evidence is not yet resolved and this may prove to be a substantial overestimate. The timing of ice retreat around Antarctica also remains unresolved; Peltier (1988) has argued that a late onset of deglaciation here (perhaps beginning 7 ka later than in the Northern Hemisphere) is required in order to explain the global pattern of Holocene sea-level rise. The time of most rapid melting, however, is probably coincident in both areas, at around 11 ka BP. No good explanation for the lag is available, although it is known that the conditions of the present day do result in different circumstances of sea ice melting in the Arctic, where melt ponds develop and lower the albedo, than in the Antarctic, where a reflective snow cover persists on the sea ice until the air temperature increases above freezing in summer (Andreas and Ackley 1982). Different behaviour might therefore have been displayed during Quaternary episodes of deglaciation.

As was the case in the Arctic Ocean and North Atlantic, sea ice around Antarctica during the last glacial maximum was considerably more extensive than at present. Significant seasonal expansion and contraction would have occurred during glacial times, as at present, with contraction of the sea ice during the southern summer and renewed expansion in winter (see Figures 3.9 and 3.10).

Dating cryosphere growth and retreat

One of the major benefits of the documentation of the extent of the Quaternary cold climates through the use of the kinds of field evidence just referred to is the ability to use this information to make inferences about past temperatures, snowlines, and the like. The use of evidence from the cryosphere to shed light on past

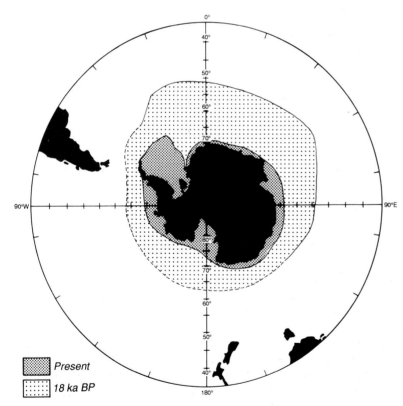

Present

18 ka BP

Fig. 3.9 Estimated extent of Antarctic summer sea ice at present and at the last glacial maximum (from Hays 1978; and Bowler 1978)

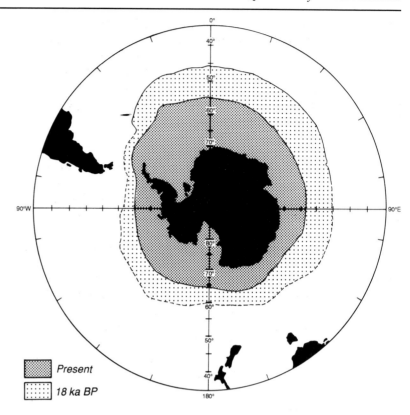

Fig. 3.10 Estimated extent of Antarctic winter sea ice at present and at the last glacial maximum (from Hays 1978; and US Navy Hydrographic Office 1961)

Present
18 ka BP

climates is most useful if the timing of events can be established. We need to know, as precisely as possible, when the climatic cooling that triggered a particular advance of ice sheets itself began, and when climatic warming reversed the trend. A major problem here is that the enormous Quaternary ice sheets, by virtue of their bulk, presumably could not respond rapidly to changes in global or regional climate. The movement of great masses of ice is a response to the mechanical weakness of crystalline ice and the resulting gravity flow processes. Even if there was a sudden warming, such that no further snow fell on a major ice sheet like that of Greenland, its outlet glaciers would continue to flow for considerable periods as the kilometres-thick accumulated ice slowly dissipated by outward flow. Thus, we cannot take a date identifying the onset of glacier retreat also to identify the onset of climatic warming: this may well have been considerably earlier. Similarly, the time elapsing during the accumulation of sufficient snow and ice to establish flow, and the further time taken for the advance of the ice margin to locations hundreds or thousands of kilometres distant, mean that we cannot take the date of ice advance recorded at a site to represent the date of climatic deterioration. Small valley glaciers can respond much more quickly, and in principle could provide useful data on the timing of climate swings. The problem with such small valley glaciers, however, is that in repeatedly advancing and retreating along their valleys, former deposits are destroyed so that the record of earlier fluctuations becomes incomplete or absent altogether. This is the case with the glacial record of the Alps, already described. Additional problems arise in attempting to date cryosphere instability on the basis of terrestrial evidence. Because of the ^{14}C half-life (see Appendix), the timing of the growth and recession of the continental ice sheets is only adequately known from terrestrial evidence for the late Wisconsin–Weichsel, since only materials from this period lie within the range of ^{14}C dating. However, it must be remembered that not only is the chronology blurred by the time-lags in ice response mentioned above: in addition the radiometric dates really only bracket the time of ice advance or retreat, yielding maximum or minimum ages. For example, the age of interstadial plant remains overlain by till only sets a maximum age for the ice advance; materials from meltwater lakes only set a minimum age for ice retreat. Trees in growth position upvalley of a moraine simply indicate that ice retreat had occurred at some earlier time; how much earlier remains unclear. In a similar way, dated raised shorelines, used to infer the onset of isostatic rebound following deglaciation, and hence the timing of the deglaciation itself, really only indicate a minimum date for the ice retreat, which might have begun gradually at an earlier time.

Therefore for the most reliable chronology of climatic fluctuation, we must in general rely on environments with shorter lag times. Living organisms, which may in a single lifetime reflect the characteristics of their environment (say, the salinity of the water in which they grew), are vital here, especially the microorganisms of the oceans, whose remains have accumulated in marine oozes throughout the Quaternary. The microfossil remains of these organisms now provide one of the most sensitive chronometers of Quaternary environmental change (see Chapter 5), and a palaeoclimatic record of much finer time resolution is obtained from them than from the terrestrial record of the Quaternary cryosphere. The use of oceanic evidence, however, raises the additional issue of whether the two environments, terrestrial and marine, experienced the same environmental fluctuations. Certainly, the controls on conditions in the oceans (including factors such as salinity, speed and depth of currents, and degree of oxygenation of the water) are not the same as those (largely temperature, wind, and precipitation) which dominate conditions on land, and we must interpret the separate records with care.

It is nonetheless the case that terrestrial evidence of ice sheets or ground ice is definitive of cold climatic conditions. Knowledge of the areal extent and general chronology of these features remains vital in our developing picture of the timing of the

Quaternary glacial and interglacial periods. The extent of glacial conditions at the Wisconsin–Weichsel peak in particular is employed in climate modelling on the assumption that this represented a few thousand years of steady-state conditions, in which the ice was neither advancing nor retreating. In such steady-state conditions, it is reasonable to take the extent of the ice to be a reflection of approximately stable environmental conditions reflecting the full-glacial state.

Ice-core records

In recent times, new techniques have made it possible to extract quite sensitive records from the large ice sheets still in existence in Greenland and Antarctica despite their relatively sluggish response to climatic change. This is done by examining the layers of ice, corresponding to each year's precipitation, that are preserved in the cold accumulation zones of the ice sheets. Major ice cores have been recovered by drilling programmes from various sites in Greenland, the Canadian Arctic, and Antarctica. Cores of more than 2 km have been collected, with the ice at the base ranging in age up to 160 ka, but typically only to 100–130 ka (the age of the last interglacial). In addition to ice, the annual snow accumulation layers contain a wealth of other materials including dust, sea salts, pollen, volcanic debris, cosmic particles, and isotopes from weapons testing.

The layering representing the snow from each year is progressively compacted as it is buried, and undergoes the transition through firn to dense ice. As this happens, air is trapped as small bubbles within the ice, and this can be extracted to provide samples of air roughly as old as the ice itself. Analyses of the air can provide indications of the composition of the atmosphere at the time the firn was finally compacted to dense ice.

Layers in the ice sheet can be dated by counting downward from the surface, the annual layers being recognized by seasonal variations in dust content, or by patterns of acidity or isotope content. This can be done with great precision over at least 10 ka; deeper, older ice is more difficult to date precisely because compaction leaves the layers increasingly thin and ice movement deforms them.

One of the principal records extracted from the long ice cores is that of the isotopic composition of the oxygen in the water molecules. Of the two more abundant isotopes of oxygen, ^{16}O is enormously more abundant than ^{18}O. It also forms slightly lighter water molecules, which therefore evaporate more easily than do those of the heavier $H_2^{18}O$. As moisture-bearing air is cooled, water with the heavier oxygen isotope condenses preferentially, leaving the remaining water relatively richer in ^{16}O. It is well known that the abundance of these two isotopes in snow within an ice sheet depends largely on the mean annual temperature of the location. Sites further inland (and therefore at higher elevations on the ice dome, and therefore colder) receive snow enriched in the lighter water which has survived the journey inland. The amount by which water or snow deviates from the isotope proportions found in the modern oceans, taken as the standard, is expressed as the $\delta^{18}O$ value (the δ stands for 'difference'). The differences are small, and so are expressed in parts per thousand (per mille [‰]) rather than parts per hundred (per cent). Water depleted in ^{18}O receives negative scores; the more negative the value, the lower is the proportion of ^{18}O. Ice sheets are isotopically light (and so display negative values of $\delta^{18}O$, down to $-40‰$ or so); the ice age oceans are left isotopically heavy (and so display $\delta^{18}O$ values of up to say $+5‰$). An isotope ratio deviation of 0.6–0.8‰ is produced by a temperature change of 1°C at the site of condensation of water vapour. Analysis of this ratio on samples of ice from within the cores reveals the temperature of the snow accumulation site at the time the snow fell. There are some effects which could act to interfere with this analysis:

(i)   If the samples represent a growing ice cap, whose height will therefore be increasing, the one ground point will experience increasingly cold conditions, and the temperature record derived from it will show cooling. This, however, is not external global climate cooling, but merely reflects the normal decline of temperature with elevation above the earth's surface.

(ii) If the warmth of the moisture source changes, or the ice cap is fed from winds blowing from a different direction, then the isotopic composition of the snow may change. Such a change could be produced as a glacial stage progressed by the development of sea ice, for example, or by shifts in the general circulation of the atmosphere.

Deuterium, the heavier isotope of hydrogen, is employed in a manner similar to that described for the oxygen isotopes. Water molecules containing deuterium are heavier, and display a lowering of vapour pressure similar to that resulting from the presence of heavier oxygen isotopes: thus, water containing deuterium is preferentially lost as moist air proceeds inland. Concentration differences with respect to modern sea water are in this case written as δD, and show a good relationship to mean air temperature at the point where the water containing deuterium falls as snow. Typical δD values in ice from the last glacial are around $-480‰$; the value rises to between -420 and $-440‰$ for interglacial ice.

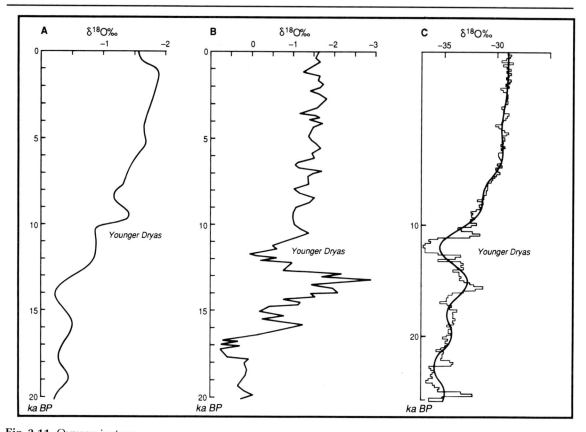

Fig. 3.11 Oxygen isotope records for the last 20 ka indicating the Younger Dryas stadial.
(A) Composite equatorial Atlantic Ocean record, based on an assumed constant sedimentation rate (from Berger *et al.* 1985).
(B) Record from marine sediments in the Gulf of Mexico dated by [14]C (from Leventer *et al.* 1982).
(C) From the Greenland Dye-3 ice core dated by counting annual accumulation layers (from Dansgaard *et al.* 1985; and Dansgaard *et al.* 1989)

Greenland ice-core records

Despite the possible complications outlined above, ice cores evidently provide extremely valuable records of temperature reflecting external global climate change. The core from Camp Century at 77.2°N in northern Greenland, for example, shows the transition from the last glacial stage to interglacial conditions at a depth of 1150 m; here the $\delta^{18}O$ values shift from the very light values of $-40‰$ which characterized the glacial to $-29‰$ or so. By counting the layers, the date of the warming at the end of the Younger Dryas stadial, which may be taken as the transition from glacial to interglacial stages, can be dated to 10.750 ka BP $\pm$ 150 a (see Figure 3.11). The final warming marking this upper boundary of the glacial stage appears to have been extremely rapid: in the ice cores, the transition is reflected in only 2 m of ice, which corresponds to only about 100 years! Of course, the form of the ice sheets themselves could not possibly respond to such rapid climatic change: lag times of thousands of years are likely to be involved.

Employing the conversion to temperature mentioned earlier, the isotope record indicates that the glacial maximum was about 11°C colder at this site than the early part of the Holocene. In the

Greenland Dye-3 core, the glacial maximum (identified by the $\delta^{18}O$ minimum in the ice) can be placed by layer counting at 17.2 ka BP (Paterson and Hammer 1987). There appears to have been a linear warming trend from glacial maximum until about 14 ka BP, during which time about 35 per cent of the full glacial-Holocene warming occurred. More rapid warming is indicated beginning at 13 ka BP, but this gave way to the Younger Dryas stadial at 11 ka BP; in this stadial, the $\delta^{18}O$ values fall back to levels comparable to those of the glacial maximum. The rapid warming referred to above then followed at 10.75 ka BP, marking the end of the late Wisconsin–Weichsel and the transition to the Holocene. Warming continued for 1–1.5 ka after the boundary, reaching levels typical of the Holocene by about 9 ka BP.

Former precipitation is also estimated from ice-core records. This technique is based upon measurements of the isotope ^{10}Be, which is produced in the atmosphere by cosmic radiation, and then washed out attached to aerosol particles. If the intensity of the cosmic radiation remains constant, so too will the abundance of ^{10}Be. The abundance of the isotope in ice will then be inversely proportional to the precipitation. Years of abundant snowfall, which produce a thick snow layer, will be associated with low average concentrations of the isotope in that snow; years which only produce a thin snow layer will result in higher ^{10}Be concentrations in the smaller volume of ice that would result.

The pattern of precipitation revealed in this way for the last glacial stage over Greenland shows a value only about 30 per cent of the present; thus, as would be expected, the extremely cold glacial maximum was also dry. This result is matched by a similar thinning of the annual accumulation layers during the glacial stage. Precipitation reached the typical Holocene value by about 13 ka BP. The record for earlier parts of the Quaternary shows relatively high precipitation in the intervals 125–115 ka BP, 80–60 ka BP, and 40–30 ka BP. These time periods match those deduced from the oceanic isotope record to be times of continental ice growth (see Chapter 5).

Dust concentration in ice cores generally shows levels up to 70 times higher during the last glacial stage than during the postglacial. The additional dust is thought to represent two conditions: deflation from dry continental interiors (such as the Australian deserts), from glacial outwash deposits and from exposed continental shelves. Increased wind strength may have assisted in the deflation of materials from these sites.

One of the most interesting records obtained from ice cores is provided by analysis of the gas samples preserved in air bubbles. The air is not of exactly the same age as the ice containing it, because both new snow and firn are porous, and allow exchange of gases with the atmosphere. It is only once the firn is transformed to ice, and the pores are no longer connected to one another, that the air samples are finally locked in. The actual age of the air is then younger than the ice, by an amount that depends on the ice-accumulation rate. In Greenland, the air is 100–400 years younger; in Antarctica, the difference in the Vostok core is 3–4 ka (Barnola *et al.* 1987; Paterson and Hammer 1987). Nor is the gas at any level in an ice core all of the same age, since the pores are pinched shut at different times. The analyses which are obtained thus represent averages spanning perhaps a few centuries. Carbon dioxide concentrations at the glacial maximum turn out to be about 190–200 ppmv (parts per million by volume); in the Holocene these values rise to 260–280 ppmv. Thus, greenhouse warming because of the higher CO_2 levels must have been a significant process fostering deglaciation. The ice cores also show rapid fluctuations in CO_2 content during the Wisconsin–Weichsel, some of up to 60 ppmv in 100 years. Some of these apparent jumps may reflect the melting of snow while at the surface, and resulting gas concentration, but others are considered to be real. The mechanisms which control these variations in atmospheric CO_2 abundance, and which accounted for the major postglacial increase, are not known. It is likely that they relate to processes in the oceans, which contain vastly more CO_2 than the atmosphere.

Antarctic ice-core records

The Vostok core from 78.5°S in east Antarctica shows changes in CO_2 concentration similar to those described from Greenland (refer to Figure 3.12). Analysis of ^{10}Be also shows that precipitation here during glacial conditions was only about 50 per cent of that received during the last interglacial or the present (Yiou *et al.* 1985).

Ice-core analyses have also provided data on former atmospheric methane (CH_4) concentrations. Methane in the atmosphere also acts as a greenhouse gas, and it has been suggested that it too may have an important role in the periodic development of glaciation and deglaciation, which will be mentioned shortly. Atmospheric concentrations of methane are much lower than those of CO_2, amounting to about 1700 ppbv (parts per billion by volume) at present. The Vostok core shows that much lower values, near 350 ppbv, occur under glacial conditions, with interglacial values reaching 650 ppbv (Chappellaz *et al.* 1990). The record from the last glacial termination shows a sharp drop in concentrations at 11 ka BP, corresponding to the Younger Dryas stadial recognized in the terrestrial record of the Northern Hemisphere. Increases this century, presumably anthropogenic, have generated higher concentrations than occurred at any time in the

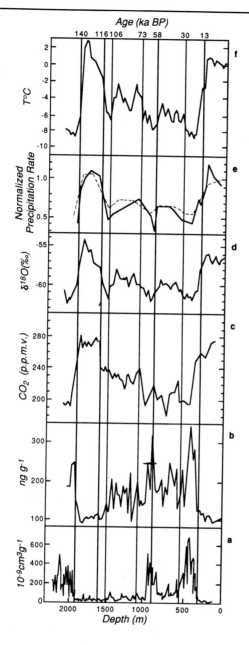

Fig. 3.12 Records from the Antarctic Vostok ice core.
(a) Volumetric abundance of particulates (from Petit *et al.* 1990).
(b) Non-seasalt sulphate content (from Legrand *et al.* 1988).
(c) CO_2 content (from Barnola *et al.* 1987).
(d) $\delta^{18}O$ (from Lorius *et al.* 1985).
(e) Estimated precipitation rate (from Yiou *et al.* 1985).
(f) Inferred palaeotemperatures (from Jouzel *et al.* 1987)

last 160 ka, and also the highest rate of increase in concentration recorded.

The deuterium temperature record deduced from the Vostok ice is based on sampling at 1 m intervals throughout the 2083 m long core. It shows that at this site, the last glacial maximum was about 9°C colder than the average for the Holocene (Jouzel *et al.* 1987). The transition to Holocene warmth is revealed as a two-step process, beginning at about 15 ka BP and interrupted by about 1 ka of cold conditions at 12–11 ka BP (probably reflecting the same events as the Younger Dryas stadial of Europe). The Holocene is shown to have been warmest early on (around 9 ka BP) and to have cooled subsequently. Temperatures in the last interglacial were about 2°C warmer than present for a period of about 5 ka. Cooling toward last glacial maximum was interrupted by major interstadials at 106–73 ka BP, when temperatures peaked at about 6°C above those of the glacial maximum, and at 58–30 ka BP, when warmest temperatures were only about 4°C warmer than the glacial maximum.

Many complications arise in the interpretation of the ice-core records, especially those of temperature derived from the oxygen isotope analyses. One of the major problems to be evaluated is the degree to which the inferred postglacial warming described above might be due to effects other than temperature rise. The obvious effect to be considered is that during deglaciation, the ice cap might become thinner; the snow would then be falling at lower (and warmer) elevations, so that the temperature record subsequently extracted would partly be a record of elevation change, and not simply temperature. The unravelling of this effect requires that the amount of lowering be known separately, but this is rarely the case. A similar effect arises because, unless the sampled site has always been located exactly on an ice divide, the ice will have been moving down the slope of the ice dome since its accumulation. The older layers at the base of the core will then had time to move further than the younger materials nearer the top. This means that the ice sampled down a vertical core will not all have accumulated at that point: the basal material will be from snow which fell higher up the ice cap. Thus, the indicated temperatures at the base are likely to be colder than present, whether or not there was an external climate change. Correcting for this effect again requires that the rate and direction of ice movement be known for the whole period represented in the core, so that a correction can be worked out and applied to the isotope record. There is often insufficient information to allow this to be done with complete confidence.

Holocene glacier records

In addition to shedding light on the glacial and interglacial stages of the Quaternary, study of valley glaciers has the potential to reveal the chronology of climatic fluctuations over shorter time periods. The use of the pattern of advance and retreat of valley glaciers has mostly been applied within the Holocene with the aid of ^{14}C dating of moraines and through the use of historical records, sketches, and photographs (see Le Roy Ladurie 1972; and

Grove 1988). Glacial episodes of the Holocene are generally termed 'neoglaciations'.

The reconstruction of Holocene environments from glacial evidence is based upon the fact that the downvalley extent of glacial ice in a particular glacial valley reflects to some degree the 'mass balance' of the glacier, which controls the physical form that is adopted to balance snow accumulation in the snowfields above and ablation losses at the melting snout of the glacier. A climatic change resulting in greater snowfall or lessened ablation can result in thickening of the glacier, and its advancement downvalley. Changes in the opposite direction result in thinning or recession upvalley. Stable environmental conditions result in essentially constant glacier form, once sufficient time for adjustment has elapsed.

The evidence of glacier fluctuations is mostly in the form of moraines, and analysis of these is beset by the problems already described for the Würm glacial in the Alps. Additionally, there is again the problem of lags in glacier response, such that downvalley advance may not begin until some decades after the climatic change has triggered greater snow accumulation; indeed, during rapid climatic changes, insufficient time may elapse for the equilibrium glacier form to be established at all.

The pattern of Holocene glacier advance and retreat therefore shows an unsurprising variability. In part this reflects the differing lag times of large or small glaciers, those on steep versus gentle slopes, and variation in ice temperature, which enable some glaciers to be advancing while others are in recession. The numerous records of Holocene glacier fluctuation indicate that ice growth was particularly marked in the second half of the Holocene, earlier millennia presumably being somewhat warmer.

A major period of ice advance is known in the 'Little Ice Age' of the sixteenth to mid-nineteenth centuries (i.e., 0.1–0.4 ka BP). Much support for the climatic cooling interpreted from glacial evidence is provided by indirect historical records dating from this time. Records of grain and wine harvests show substantial reductions during this period, and stock losses in some areas were considerable; increasing winter sea ice made navigation difficult or impossible in places. Interestingly, the Little Ice Age is reflected in records extracted from the Quelccaya ice cap in the mountains of tropical Peru, at latitude 13°S, indicating that this period of cooling must have been essentially global in its effect. (Some of the records extracted from the Quelccaya ice are presented in Figure 3.13.) The global temperature drop in this cold period however is inferred to have been less than 2°C (e.g., see Grove 1988). Denton and Karlén (1973) identified an earlier period of Holocene glacial advance dating from 3.3 to 2.4 ka BP, and a later, less significant one, at 1.25–1.05 ka BP, relating these to variation in solar output. They suggested that indeed there may

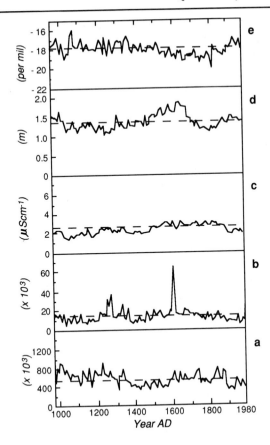

Fig. 3.13 Records from the Peruvian Quelccaya ice core (from Thompson *et al.* 1986). Note the depression in the ^{18}O values (curve [e]) in the last few centuries, corresponding to the Little Ice Age.
(a) Particulate abundance per ml of sample.
(b) Abundance of large particles (over 1.59 μm in diameter) per ml of sample.
(c) Electrical conductivity.
(d) Accumulation rate (m per year).
(e) $\delta^{18}O$

be a regular periodicity of about 2.5 ka in the occurrence of ice growth. A major period of glacier recession is also known, dating from the second half of the nineteenth century and running up to the present day. The extent to which this may reflect anthropogenically produced climatic warming, following the Industrial Revolution, remains to be resolved.

Causes of ice-sheet growth and decay

Having examined something of what is known about the areal extent of the cryosphere during the last glacial stage, and the timing of the ice growth and recession, we can now turn to the complex issues involved in identifying the causes of the glacial and deglacial episodes.

Ice-sheet growth and glacial inception

The exact growth mechanisms for the major ice sheets remain unclear. Insolation changes during summer at high northern latitudes appear to be the 'forcing function' for climate change, as described in the Milankovitch or 'astronomical theory' (mentioned in Chapter 2). It is envisaged that lower solar heating resulting from the orbital characteristics (primarily obliquity and precession) periodically permits summer snow to survive without ablation, with albedo feedback from the growing snow cover reinforcing atmospheric cooling. We would thus expect slow accumulation of snow and eventually glacier growth from initial accumulation areas on uplands in the high northern latitudes, with a gradual extension of the ice-covered area. It is also possible that ice growth accelerated quickly once the regional snowlines began to fall in response to climatic cooling, because of the albedo feedback mechanism, so that snow persisted through summer and accumulated annually and simultaneously over large areas. This has been termed the 'instant glacierization' hypothesis. Even in this case, ice accumulation would have begun first in upland regions such as those on Baffin Island, Labrador, and the Rocky Mountains, in the Alps, and the mountains of Scandinavia.

In addition to an astronomically caused drop in high-latitude summer insolation and suitable land area with uplands in the high latitudes, we can identify two further necessary precursors to the establishment of a glacial stage. These include:

(i) an adequate supply of moisture, which implies a sufficiently warm ocean located somewhere upwind; and

(ii) minimal loss of accumulated snow and ice. In particular, an inland area of uplands, with no glacier connection to the sea which could result in iceberg calving and reduced accumulation, would favour ice-cap development.

According to the oceanic oxygen isotope record, which is taken to be dominantly a reflection of global ice volume (see Chapter 5), there must have been rapid periods of ice volume growth (revealed by rapid increase in the foram $\delta^{18}O$ values) centred on 115 ka, 90 ka, 75 ka, and 25 ka BP. The two most important of these periods were at 115 ka BP and 75 ka BP; the phase of ice growth in each case lasted for about 10 ka. According to estimates by Ruddiman *et al.* (1980) these periods each contributed nearly half of the ice-volume growth of the last glacial stage. The net rate of ice-volume increase in these short growth phases is thus very rapid, amounting to 5 per cent per ka.

According to the temperatures indicated by foram assemblages, the first period of ice growth at 115 ka BP occurred before the sea surface at 40–45°N in the Atlantic had begun to cool significantly. Thus, we can conclude that ice growth precedes oceanic cooling, in this case by about 4–4.5 ka. The mechanism envisaged to explain this is that the ice growth was in an area not connected to the sea, so that despite major ice accumulation, no iceberg calving which might have chilled the oceans took place. This appears to confirm the idea of inland ice accumulation as envisaged in the list of preconditions mentioned above. Interestingly, 115 and 70 ka BP were the times of lowest summer insolation at 70°N, according to astronomical calculations, which gives us further confidence that the ice growth was driven by the Milankovitch mechanism. A likely candidate for the location of this ice is in the eastern sector of the developing Laurentide ice sheet of south-east Canada, adjacent to the warm ocean.

Ruddiman *et al.* (1980) reported a major increase in the abundance of ice-rafted sands in sediments of the North Atlantic after 75 ka BP, and concluded that iceberg calving began at about that time. Following this, the North Atlantic cooled rapidly. If the scenario outlined here is typical, we may conclude that ice growth is out-of-phase with oceanic cooling, preceding it by 4–5 ka; this is consistent with the suggestion made above that a sufficiently warm moisture source must be available to sustain ice growth. This immediately raises the question of how the ice sheets fared in terms of moisture nourishment after the oceans had cooled. We shall return to consider this shortly.

Ice-sheet retreat and deglaciation

In order to be able to document the decline in ice volume, it is once again necessary to refer to the marine isotope record, because the moraines and other features of the terrestrial record only indicate the decline in ice area; volumes can only be estimated by making assumptions about the thickness of the waning ice sheets. The rapidity of deglaciation is, however, clear in all the records: it was completed in only about 8 ka, which contrast with the 90 ka or so of the Wisconsin–Weichsel which elapsed before full-glacial conditions were reached.

The isotope record of deglacial times is unfortunately a complex encoding of many related processes, and the history it contains is still not resolved with any finality. Let us consider some of the processes which may confuse analysis of the deglacial record in marine isotopes.

The elevation of a growing ice sheet changes continually, as the ice thickens to its final 2–4 km. Thus, the snow which feeds a developed ice sheet is deposited from air which has had to rise higher than would have been the case earlier in the period of ice growth. As a result, the isotopic make-up of the ice in an ice cap will not be uniform through its depth, as noted earlier in this chapter, but is likely to be more negative in younger ice. This in turn implies that, depending on the age of the ice that is melting at any time during the deglaciation, meltwater of different isotopic character will be returned to the oceans, and the marine record will partly reflect the compositional layering of the ice sheets, not just their remaining volume. The actual effect will depend, for example, on whether most meltwater is being returned from ablation along the equatorward ice margin which lies at low elevation, or is mainly by iceberg calving from fast ice streams draining the poleward flanks of the ice sheet. The only way that these issues can really be approached is by glaciological modelling. Additional complication stems from the fact that the isotopic make-up of land ice varies with latitude, in the same way as temperatures vary, so that ice formed in lower latitudes is less isotopically negative. Thus, the effect of meltwater return on the isotope record of the oceans also depends upon the latitudinal distribution of the ice retreat. In all likelihood this would have involved a systematic tendency for most retreat in low latitudes first, followed by recession at more poleward sites.

Both the terrestrial and marine records indicate that ice retreat was episodic. Major moraine deposits located along the margins of the retreating Laurentide ice sheet, for example, are taken to indicate halts or reductions in recession rate following major ice loss. The marine isotope record from the North Atlantic suggests two or possibly three separate major episodes of ice-volume loss, one at 14–12 ka BP, another at 10–9 ka BP and possibly a third at 8–6 ka BP (Mix 1987; Jansen and Veum 1990). A very large fraction of the Northern Hemisphere ice volume, perhaps one-third, was lost during the first two episodes of retreat. Only relatively small areal contractions of the ice were coincidental with the first episode, according to the moraine evidence: the Northern Hemisphere ice sheets still occupied 75–80 per cent of their full-glacial area at 13 ka BP (Ruddiman and McIntyre 1981) so that much ice thinning must have occurred. Interestingly, the same timing of ice-volume loss is inferred from the sea-level record preserved in corals at Barbados. Here Fairbanks (1989) has identified two intervals of rapid sea-level rise. The first involves a very rapid rise of 24 m in less than 1 ka, centred on 12 ka BP; the second, lasting slightly longer and involving a 28 m rise, is centred on 9 ka BP. The period of slower sea-level rise separating these two intervals presumably reflects the Younger Dryas stadial. The dates for the period of rapid sea-level rise are very similar to those inferred from the marine oxygen isotope record to have been times of rapid ice-volume loss.

Mechanisms of deglaciation

It is necessary now to consider the mechanisms behind this deglaciation. Summer insolation over the high northern latitudes was a little higher than that of today at 17 ka BP but did not reach its maximum until 11 ka BP. Thus, while the deglaciation is centred on 11 ka BP, it began considerably earlier, being well underway by 14 ka BP. Now, the additional solar heating even at its maximum at 11 ka BP is only slight (just a few per cent), so that the fact that deglaciation began well before this slight push requires explanation. The most attractive idea is that, once a slight deglaciation had begun in response to the early minor increase in solar heating, one or several positive feedback mechanisms came into operation to reinforce the warming. The intervention of feedback processes introduces what is termed 'non-linearity' into the cause-and-effect sequence. Many ideas about the nature of the feedback processes have been put forward, and we should consider the most important ones.

The sea-level rise which is caused by ice melting has the potential to lift grounded marine ice sheets, allowing them to break up and drift offshore. A most important consequence of this is that the buttressing support provided by the grounded ice to the land ice lying upslope of it is removed; massive accelerations of the ice flow from the inland domes might then result. Fast-moving or 'surging' ice streams would then deliver ice to the coast very rapidly; by draining the interior parts of the terrestrial ice sheets they could cause just the kind of ice-sheet thinning referred to above as being required by the otherwise inconsistent record of terrestrial moraines and marine oxygen isotopes. This process of ice-sheet collapse has been termed 'downdraw'. The most important part of this mechanism for reinforcement of deglaciation is that it is not the slight increase in solar heating which melts the ice, but rather heat taken from sea water as the calved ice floats away to melt. Chilling of the oceans should result: it has been estimated that this effect could cool the upper 100 m of the North Atlantic at a rate of 1°C per annum. This indeed appears to be confirmed by the isotope record of the mid-latitude North Atlantic, where lowest water temperatures appear to have been reached at about 13–9 ka BP, well after the terrestrial glacial maximum, and during the time of maximum solar heating at high latitudes: this is taken to reflect the chilling produced by the influx of icebergs and meltwater (Ruddiman and McIntyre 1981). A mechanism related to this has been envisaged along the terrestrial margins of the ice sheets, but involving proglacial meltwater lakes rather than the sea. Accelerated ice loss into the darker waters of the proglacial lakes, where the icebergs would melt more readily, could accelerate frontal ice melting in a way similar to that just described.

Another feedback mechanism that has been considered is moisture starvation. Early melting would return fresh water to the North Atlantic. If this water formed a stratified layer overlying the salt, at least two effects might follow. Firstly, since fresh water freezes at 0°C (rather than the −1.9°C of sea water), sea ice cover might become more extensive, cutting off a source for the evaporation of moisture to sustain the ice sheets. Secondly, since the layer of fresh water would not readily mix with the underlying sea water, it would store most of the heat from summer, leaving the water below cooler than normal, and again restricting evaporation in the autumn and winter seasons. The more the ice melted, the more the remaining ice sheets would be starved of moisture, ensuring their rapid collapse.

A role for isostatic rebound behaviour of the crust under the ice sheets has also been envisaged (Peltier 1987). Initial ice loss would lower the ice cap surface, with the weight on the under-lying crust thus being reduced. Because the mantle is viscous, and only responds slowly, isostatic recovery would be delayed. Thus, instead of the ice cap rebounding, and so re-elevating the surface to colder heights, the melting would be accelerated by the lower elevation of the ice surface. Further melting would simply lower the surface even more, which in turn would bring it down to yet warmer elevations, and so forth.

Potentially the most important feedback mechanism to promote deglaciation involves the atmosphere, and particularly its green-house gas concentrations. The ice-core evidence presented earlier shows that in glacial times, levels of both methane and CO_2 are reduced. Some possible mechanisms for this were outlined in Chapter 2. By whatever mechanism in fact controls the glacial-interglacial changes in greenhouse gas concentrations, initial deglaciation would evidently be associated with an increase in concentrations of these gases. This would result in additional warming beyond that provided by the Milankovitch mechanism, and hence reinforce the warming.

A particularly interesting aspect of some of these feedback mechanisms is the way in which they relate to deglaciation in the Southern Hemisphere. Because of the largely oceanic nature of the high latitudes in this region, and the high heat capacity (i.e., thermal inertia) of water, the Milankovitch mechanism must have had a much reduced impact in the Southern Hemisphere. Nonetheless, deglaciation occurred just as rapidly here too. It might be imagined that the loss of the Northern Hemisphere ice sheets and their albedo-induced cooling would simply warm the global atmosphere and hence induce worldwide deglaciation. However, climatic modelling (e.g., Manabe and Broccoli 1985) suggests that ice loss only resulted in a small increase in the net amount of warmth available, rather too little to result in sufficient inter-hemisphere heat transport through the atmosphere to prom-ote deglaciation in the south. The explanation is as follows: though the high albedo of the ice sheets did reduce solar heating, the colder surface also emitted correspondingly less terrestrial long-wave radiation, so that the change in net radiation was much less than might be expected. It has thus been argued that one of the feedback processes needs to be involved to transmit the deglaciation trigger into the Southern Hemisphere. This might have involved the greenhouse gases in the atmosphere, heat transport by ocean currents, or destabilization of grounded ice sheets around Antarctica by the global sea-level rise triggered by ice loss in the Northern Hemisphere. These matters remain to be resolved.

Insights from global climate models

A final issue to which we should return is the way in which the terrestrial ice sheets fared once the North Atlantic began to cool rapidly after 74 ka BP. First, let us consider the cause of this oceanic cooling. The best explanation in fact comes from global climate models seeking to reconstruct atmospheric conditions as they were during the last glacial (e.g., Manabe and Broccoli 1985). These models in turn rely on the reconstructions of the Quaternary ice sheets based upon the field evidence of tills, moraines, and periglacial features reviewed earlier. The models show that the huge Laurentide ice sheet, and to a lesser extent the Scandinavian ice sheet, acted as barriers to the atmospheric circulation. The Laurentide ice caused the westerly jet stream of the middle troposphere to split into two sub-streams, one of which ran around the south of the ice and the other across its northern margin. Subsiding air over the ice resulted in a northward surface flow which adopted a course parallel to the northern ice margin also. Air following such a route would be cooled substantially by the chilled surface. The cold air mass then flowed between the Laurentide and Greenland ice sheets, across the Labrador Sea, and out over the North Atlantic. The movements of this air over the warmer sea surface would have extracted heat from it, resulting in the oceanic cooling revealed by the foram $\partial^{18}O$ analyses, and producing more extensive sea ice. According to the models, sea ice may have existed seasonally south to 46° or so. This sea ice in turn would have restricted evaporation and moisture levels in the atmosphere. Hence, according to the models, the ice sheets at glacial maximum were suffering net ablation rather than accumulation. The presence of meltwater ponds on the ice surface would have lowered its albedo, and further enhanced ablation. Thus, the small additional trigger provided by the Milankovitch maximum in summer heating may have been all that was required to trigger a small further ice retreat and then catastrophic deglaciation through the intervention of one or more of the feedback processes already described. The rapid terminations seen in the marine isotope record would then reflect the re-establishment of a single jet stream, and rapid warming of the North Atlantic.

Pattern of glacial–interglacial cycles

Our discussion of the glacial phenomena of the Quaternary so far has been biased, of necessity, toward the last glacial stage. This has meant that we have neglected one of the major issues which must be tackled if we are fully to understand the controls on Quaternary environments, namely, the longer-term rhythm of the glacial cycles.

As has been indicated in earlier chapters, the most prominent features of the Quaternary environmental instability after 0.9 Ma BP are the dramatic glacial terminations, in which a volume of ice which took perhaps 90 ka to accumulate is returned to the oceans in only 8 ka. Some of the environmental changes taking place during the terminations are truly remarkably rapid: at the end of the Younger Dryas stadial, southern Greenland warmed by 7°C in 50 years, according to the records from the Dye-3 ice core. Similarly, in only a few decades, the climate of the North Atlantic became milder and less stormy, following the retreat of sea ice there (Dansgaard *et al.* 1989). The rapidity of these events confirms the intervention of one or more feedback processes in reinforcing an initial tendency toward ice ablation produced by a very slight increase in summer solar heating in the high northern latitudes.

The oceanic record makes it clear that at least the last eight or nine such terminations have occurred about 100 ka apart, a frequency which matches that of the orbital change in eccentricity. This is yet another indication that Milankovitch rhythms underlie environmental change. However, it is problematic because there is no direct forcing resulting from changes in eccentricity: these only have an effect by modulating the amplitude of the 23 ka precessional variation. That is, solar heating of the earth does not vary significantly with a period of 100 ka. Therefore it is necessary to look for a separate mechanism which could be responsible for the striking 100 ka periodicity of glacial terminations.

Several hypotheses have been advanced seeking such a mechanism. In some, the explanation is sought in the interference of the several separate periodicities present in the orbital geometry of the earth. In others, the explanation involves the behaviour of the solid earth itself, through the mechanism of isostatic sinking of the crust under the load of glacial ice. This would lower the ice sheet such that ablation became dominant over accumulation. Once deglaciation had occurred, slow isostatic rebound of the unloaded crust would once more elevate the terrain to the point at which ice accumulation could again begin.

The conclusion must be reached that at present, no completely acceptable explanation for the 100 ka periodicity exists. This remains one of the major puzzles to be resolved in the continuing investigation of the Quaternary cryosphere. Clear evidence of the impact of the rapid terminations is to be found along the present coastlines, which have existed in their present configuration only since the marine transgression which resulted from the latest return of meltwater to the oceans. At full glacial times, sea levels were about 120 m below their present position. This lowering, and the very rapid Holocene marine transgression, are among the effects of continental glaciation and deglaciation felt in the oceans, and to which we turn our attention in the next chapter.

4 Quaternary Sea-Level Changes

I know not what I may appear to the world, but to myself I seem to have been only like a boy playing on the sea-shore, and diverting myself in now and then finding a smoother pebble or a prettier shell than ordinary, whilst the great ocean of truth lay all undiscovered before me.

D. Brewster,
Memoirs of Newton, Vol. 2, 1855.

Introduction

In this chapter we will consider one of the few environmental parameters which might in principle be expected to show some uniformity in its Quaternary variation: sea level. The oceans are all interconnected at present, so that any fluctuation in sea level must be transmitted throughout the oceans of the world and be felt everywhere. While this is indeed the case as a general rule, some rather complex effects influence the behaviour of sea level and make global fluctuations less regular than we might expect. The surface of the sea is a 'potentiometric surface', and lies horizontally with respect to the local net gravity field except where it is temporarily distorted by winds and currents. The local gravity field at any point, however, consists principally of the field produced by the earth itself, but modified by the particular water depth and by the properties of nearby rock masses or ice sheets. Consequently, the sea surface actually lies in a rather complex and peculiar configuration of highs (e.g., over the North Atlantic) and lows (e.g., over the equatorial Indian Ocean), the maximum elevation difference among which amounts to about 200 m. The three-dimensional form of this uneven surface is termed the 'geoid'. It is very similar to an 'ellipsoid', which is the regular geometric figure often used to represent the shape of the solid earth. The terrestrial ellipsoid is a sphere flattened by about 1/300, or about 40 km out-of-round. Global sea-level fluctuations are thus actually changes in the detailed form of the geoid, and the magnitude of the change experienced at any point depends partly on whether it lies relatively high or low on the geoid surface, as we shall see below.

Water added to the oceans by melting land ice or taken from them in phases of global cooling is nonetheless reflected in worldwide sea-level fluctuation, which, referred to the land, produces either transgression or regression. When a sea-level change results from changing water volume in the ocean basins in this way, it is described as 'eustatic'. The amount of rise or fall would not be exactly equal everywhere because the changing distribution of water mass alters the local gravity field simultaneously, so that the geoid readjusts. The record left by past sea levels around the margins of the continents is considerably blurred, in addition, by the tectonic movement of the land. Sea level as we ordinarily measure it is, of course, just the elevation of the line along which water and land meet; an apparent fall of sea level can thus in reality be caused by the land rising, and vice versa. A common cause of changed elevation of the land is 'isostatic' readjustment: this is produced by a change in the load borne by the crust, and is a consequence of the fact that the crust in turn is supported by the highly viscous but deformable mantle below. An increase in the load on the crust results in slow subsidence; removal of the load results in a similarly slow 'isostatic rebound' or recovery (Figure 4.1). This problem is particularly pertinent to the study of Quaternary sea-level changes, which were caused by the growth and melting of land ice; the changing ice mass borne by the continents resulted in isostatic subsidence or uplift and hence the evidence of old shorelines and other coastal markers has inevitably been moved. Considerable ingenuity often needs to be applied to unravel the resulting distorted record of sea-level change.

In working through the complex record of sea level, is it helpful to recognize 'gauge' and 'relative' sea levels (Chappell 1983). 'Gauge' sea level is the absolute level of the water surface which might be measured against some stable datum, unaffected by tectonics or isostasy. The centre of the earth would provide such a datum. The second way to consider sea level is in terms of 'relative' sea level, the position as we record it against the land. Evidence from the field (ancient shorelines, coral reefs, wave-cut notches, mangroves, etc.) generally only indicates relative sea levels, and in the discussion of Quaternary sea levels that follows, this convention is adopted.

Causes of sea-level fluctuation

The gauge level of the sea can be made to vary by a number of different processes (all of which also affect relative sea level, of course).

The largest fluctuations are caused by changes in the volume or holding capacity of the ocean basins themselves. A principal factor causing such change is the nature of activity along the mid-ocean ridge system which spans the globe. Along these ridges, volcanic activity results in warmth and youthfulness of the

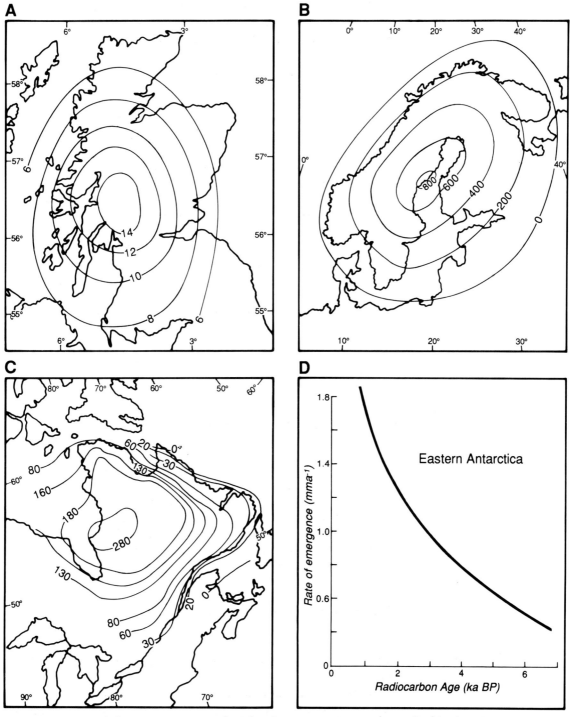

Fig. 4.1 Patterns of glacio-
isostatic rebound resulting
from the disappearance of
Wisconsin–Weichsel ice.
Emergence in m for:
(A) Scotland during the past

6.8 ka (after Sissons 1983)
(B) Fennoscandia during the
past 13 ka (after Mörner 1980)
(C) Northeastern North
America over the last 7.5 ka
(after Hillaire-Marcel and

Occhietti 1980
(D) Antarctica over the last
7 ka (after Adamson and
Pickard 1986)

oceanic crust, and hence the rocks stand isostatically high. Sea-floor spreading carries the new rock materials away from the ridges at rates measured in centimetres per year, and as this takes place, slow cooling results in subsidence of the sea floor. The broad profile of the oceanic ridge is thus defined by thermal isostatic processes.

Acceleration in the rate of sea-floor spreading, reflecting increased delivery of heat to the ridge system, carries the new crustal rocks away from the ridge more quickly, and hence they remain isostatically high over greater distances. This alters the shape and dimensions of the ridge system so that it occupies a greater volume within the ocean basin, and sea level must be displaced upwards. In a similar way, if the rate of sea-floor spreading slows, oceanic ridges become narrower and occupy smaller volumes in the ocean basins, and sea level is lowered. Such fluctuations in oceanic ridge volume are capable of producing the largest shifts in sea level, perhaps spanning ± 500 m; the fluctuations are also the slowest, typically occurring over tens of millions of years. Quaternary sea-level fluctuations, then, must relate to other causes that we shall shortly discuss.

Major marine transgressions in the Cretaceous, in contrast, were produced in the way just described by an episode of rapid sea-floor spreading, leading to 'tectono-eustatic' transgression (a gauge sea-level rise as well as a relative one). Sediments of this age occur over large areas of Australia, for example, and are important aquifers in the artesian groundwater system. The magnitude of the tectono-eustatic rise involved in the Cretaceous is considered to have been around 350 m (Mörner 1987).

The growth and melting of land ice provides the second major cause of sea-level change, and the one dominant in the Quaternary. Evaporation from the oceans occurs principally in the low latitudes where the water is warm, but the resulting fall in sea level when the water is stored as land ice is felt globally. As glacial conditions develop, sea level falls very slowly (perhaps averaging 1 m per thousand years, but reaching 5 m per thousand years during short periods of rapid ice growth). The lowering reached at the last glacial maximum amounted to 120–150 m below present level. Glacial 'terminations' are more rapid, with sea level rising again over no more than 10 000–20 000 years (i.e., at an average rate of 5–10 m per thousand years).

These 'glacio-eustatic' fluctuations are associated with a redistribution of mass among the ocean basins and the continents. When loaded with ice, the continents subside (and the oceanic crust in the partly emptied ocean basins rises); deglaciation results in isostatic uplift of the unloaded land, and subsidence of the ocean floors. Here we see some of the complications which arise in unravelling the history of sea level: the ice-loaded continents occupy a smaller area than the ocean basins, and the ice only occupies part of the most northerly and southerly continents (or land at high elevations elsewhere). Thus, the amount by which the continents subside or rise must be greater than the amount by which the oceanic crust moves. Further, in subsiding or rising, the crust must displace material in the underlying mantle, which is enormously viscous. It is probable that the time scale over which the continents subside or rise is different from that over which the oceanic crust does, since different amounts of vertical motion are involved, and because the mantle below continents has different thermal characteristics to that below the oceanic crust. Hence the trace of sea level left on the land will be a complex encoding of a host of processes acting at different rates.

This leads us to a third, related, process which can affect relative sea level. All subsidence and uplift of continents and ocean basins involves the compensating movement of displaced material in the mantle, just as the water in a bath must flow and redistribute itself when a block of wood is floated in the water, and causes displacement. Land ice grows and melts, and produces subsidence and uplift, too rapidly for the viscous flow of the mantle to keep pace. Thus, the subsidence of an ice-loaded continent produces deformation in the mantle, and the disturbance is slowly transmitted by viscous flow. The mantle deformation is complex, because the earth is essentially spherical, and depressing the crust in one place causes upward displacement in a surrounding zone (called a 'forebulge'). This disturbance spreads, like a ripple, around the globe to affect even distant areas which bore no ice. Changes in mantle configuration continue to occur in distant locations long after the ice or water load has stabilized, so that not even the timing of sea-level fluctutations can clearly be seen in the field evidence. Sites located on forebulges produced by ice loading (such as Holland and other areas facing the North Sea, on the forebulge of the Fennoscandian ice sheet, or the eastern seaboard of the USA, affected by the Laurentide ice sheet) experience a very large and prolonged relative sea-level rise during deglaciation, as the glacial forebulge subsides. Simultaneously, the deglaciated land areas experience falling relative sea levels, as isostatic uplift occurs in response to the removal of the ice load. There were of course multiple forebulges during the Quaternary, including one produced by the subsidence of Antarctica under its enlarged ice load. The interaction of multiple collapsing forebulges across the surface of the globe must have contributed greatly to the apparently different sea-level histories revealed at study sites worldwide. The magnitudes of the isostatic effects described here are considerable: collapse of the North Sea forebulge after the last glacial period is estimated to have lowered the affected areas by about 170 m; drowned coastal features consequently occur seaward of the present coast. Isostatic uplift of the glaciated areas of Scandinavia has left shorelines standing up to almost 300 m above the present coast; the uplift is continuing at rates of up to 9 mm per year (Devoy 1987b).

The final factor responsible for changes in relative sea level that must be mentioned here is 'hydro-isostasy'. Marine transgressions lead to an additional load of water being placed on the continental shelves, which are exposed during glacial stages. Underlying the shelves is continental crust; being more rigid than mantle material (elastic rather than viscous), the continental crust flexes under the changing load. This flexing leads to relatively small changes in relative sea level along the continental margins, amounting to perhaps a few metres. Clearly, the amount of flexure will relate to the rigidity of the rocks underlying individual continental margins, and the load of water covering the continental shelf (which will depend primarily on the width of the shelf). Geographical variability in these parameters will once again ensure that not even worldwide sea-level changes are indicated by field evidence at the same elevation at different locations.

In summary, the major factors influencing sea level and its fluctuation may be placed in three groups:

(i)   Those producing truly eustatic changes:
  • 'glacial eustasy' (loss or gain of water from ice growth or melting);
  • 'accretion of new water' (juvenile water, from igneous activity; discussed below).

(ii) Tectonic and isostatic processes which cause relative level change essentially worldwide without actually altering the oceanic water volume (so that the changes are not strictly eustatic):
  • 'geoidal eustasy' (changed configuration of the sea surface as a result of relocations of rock, ice or water masses, or astronomical effects);
  • 'tectono eustasy' (oceanic ridge or tectonic uplift or subsidence effects, including the subsidence associated with the fragmentation of supercontinents discussed below);
  • 'sedimento eustasy' (the filling of parts of the ocean basins by marine or terrestrial debris and sediments).

(iii) Local isostatic and tectonic effects which primarily affect a restricted area (and hence are in no sense eustatic) but whose effect may be transmitted to all parts of the globe with reduced amplitude via deformation of the mantle:
  • 'glacio isostasy' (ice loading and unloading effects);
  • 'hydro isostasy' (water loading and unloading effects).

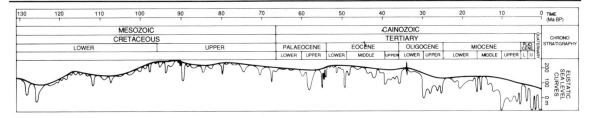

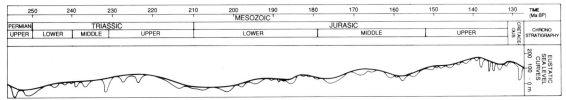

The timing of sea-level fluctuations

Over the enormous period of time during which the earth has possessed oceans, their depth may have varied in a systematic way as the volume of water at the surface changed. Little firm evidence on the volume of ocean water far back in geologic time exists, but it is often considered that this may have increased through time at a very slow rate. Initial condensation of water is likely to have proceeded in the early history of the earth, and it seems reasonable to suppose that in the Phanerozoic at least, the total oceanic water volume has been more or less fixed. Certainly, over Quaternary time scales, the addition of further 'juvenile' water must have been so slight that it may be omitted from consideration as a factor significantly influencing sea level.

Tertiary and earlier sea levels

The dominant influence of tectono-eustasy can be seen in inferred sea-level histories spanning the Phanerozoic (Figure 4.2), and based upon the sedimentary sequences around the continental margins. These show broad fluctuations of 300 m and more occurring over perhaps 50–100 Ma, with higher frequency (more rapid) fluctuations superimposed.

Essentially, two periods of high sea level can be recognized in such reconstructions: one occurred in the early-middle Palaeozoic, at about 450 Ma BP, the other at around 100–150 Ma BP, in the late Mesozoic. These high stands are paralleled by peaks in records of global igneous activity (Fischer 1984). The most probable explanation is that these 'supercycles', as they have been called, relate to the repeated fragmentation and aggregation of supercontinents. Pangea, the most recent supercontinent, began to break up in the Jurassic (during the interval of low sea level separating the two supercycle high stands). Supercontinents act to

Fig. 4.2 Long-term trends in sea level over the Phanerozoic period (after Haq *et al.* 1987). Note the rise in relative sea level in the Mesozoic, associated with subsidence of the dispersing fragments of Pangea

trap geothermal heat in the mantle; eventually, convectional overturning commences, and continental fragmentation begins. This sets off two processes which cause relative sea level to rise. First, the moving continents, being small, allow mantle heat to escape more readily: material deep in the crust and in the upper mantle thus cools, becomes more dense, and subsides isostatically. Second, as new oceans are produced along the fractures separating the continental fragments, a growing proportion of the sea floor is young, thermally buoyant, and standing isostatically high, displacing sea level upwards. Not surprisingly, apparent worldwide transgression follows, persisting over a period of time which appears to be about 75 Ma. As the continental fragments converge once more, the sea floor becomes old and subsides, and the new supercontinent once more traps mantle heat and begins to ride high isostatically: thus, regression commences. This cycle evidently takes about 300 Ma in total, so that only two supercycles are revealed in the Phanerozoic record. It thus appears likely that repeated fragmentation and aggregation of the continents, occurring in a broadly cyclic manner, may exert the fundamental, long-term control on the global sea level (Worsley *et al.* 1984).

During the Tertiary, sea level underwent a progressive but irregular decline from perhaps +300 m to present level. The decline was apparently punctuated by a series of major very rapid regressions and transgressions, perhaps of as much as 100 m in only 1–2 Ma; these remain to be fully documented. The overall falling trend presumably relates to the mechanism just outlined: the rate of sea-floor spreading declines as continental fragmentation proceeds, and the mean age of the sea floor consequently increases, producing steadily larger capacity in the ocean basins, and a gauge fall in sea level. Thus, we see that the Cainozoic trend of declining sea level is just a late stage of the second of the Phanerozoic supercycles.

Quaternary sea levels

The record of sea level is only really acceptably known for part of the upper Quaternary, perhaps the last 400 ka. The older evidence is dated by uranium disequilibrium techniques, with ^{14}C employed in the most recent 40 ka. Almost everywhere, tectonics or isostasy of some form has affected the field evidence. Surprisingly, the best records come from some of these unstable areas, including uplifted sites in Barbados and Papua New Guinea.

An excellent example is the extraordinary record derived from fossil coral reefs found in a staircase-like array above the present coastline of the Huon Peninsula in Papua New Guinea (see Figure 4.3). This area lies in the collision zone between the advancing Pacific and Indo-Australian lithospheric plates, and is consequently subject to continual tectonic influences. The result has

Fig. 4.3 Oblique aerial photograph of the coral terraces of the Huon Peninsula in Papua New Guinea. The prominent terrace in the middle of the photo dates from the last interglacial high sea-level stand

been fairly steady uplift of a fault-bounded block of land at a mean rate of 0.5–3 mm per year (varying systematically along the coast). Because this uplift relates to major ongoing plate-tectonic processes, it has not fluctuated widely during the late Quaternary.

The present Huon coastline, located at about 6°S, supports fringing coral reef communities. These do not flourish, however, because coral communities are continually elevated and suffer exposure. Consider, however, the situation during a glacial termination: meltwater, returning to the ocean basins, produces a worldwide transgression. The sea-level rise, occurring at a typical rate of some millimetres per year, then effectively keeps pace with the rising Huon Peninsula, and the coral communities are able to grow larger in size. Eventual stabilization of sea level brings conditions back to something like the present. The flight of fossil coral reefs running up to 700 m and more (Figure 4.3) thus represents the sequence of glacial terminations, the crest of each reef dating from the peak of a postglacial transgression. Falling sea levels are excessively hostile to coral growth, and the Huon Peninsula does not provide a record of the elevation of the lowest stands of the oceans during glacial times.

To extract a sea-level history from these reefs involves subtracting from their present elevation above sea level the part due to the steady uplift. The remainder represents the sea level at the time of

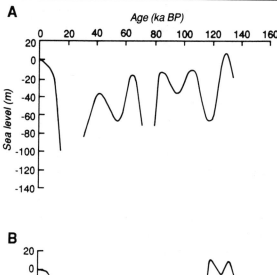

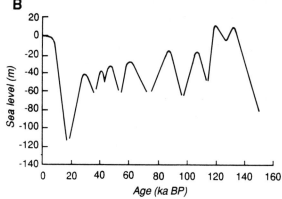

Fig. 4.4 Sea-level records for the late Quaternary derived from coral terraces at (A) Barbados (after Steinen *et al.* 1973; and Bard *et al.* 1990) (B) Huon Peninsula, Papua New Guinea (Aharon 1984). Both records lack evidence of the lowest levels reached by sea level during stadial and glacial phases, as explained in the text

coral growth. Dating the coral remnants then allows a full chronology to be established (see Aharon and Chappell 1986).

The record created in this way is shown in Figures 4.4 and 4.5, together with an oxygen isotope palaeotemperature record derived from marine microfossils: the similarity between the two records reinforces the conclusions drawn from them.

What does the record tell us about the history of sea level at this site?

Major reef complexes were produced by rapidly rising seas (rates of up to 8 mm per year) at 8.2 ka and 118–138 ka. These represent the Flandrian and previous Sangamon postglacial transgressions. The last interglacial sea level stood about 6 m higher than the present. After the last interglacial high stand, sea level trended downward, but with a series of reversals during short warming episodes ('interstadials') when smaller reefs were built. These have ages and relative sea levels of 107 ka (−12 m), 85 ka (−19 m), 60 ka (−28 m), 45 ka (−32 m), 45 ka (−38 m) and 40 ka (−42 m). The whole episode of falling sea level in the period 80–20 ka BP represents the last major Quaternary glaciation (the

Wisconsin of America and the Würm of Europe). The fact that even the interstadial high sea levels recorded at the Huon Peninsula were 12–42 m below present indicates that there was significant (and growing) land ice through the whole period.

The Holocene transgression

The termination of the last glacial phase poured vast quantities of water into the ocean basins. Ice melting took place at rates which varied from site to site. In north-east Canada, ice sheets lasted until at least 7 ka BP, but were completely gone from north-west Europe 1.5–3.5 ka earlier (Devoy 1987b). Reduction in the volume of Antarctic ice is estimated to have contributed 25 m to the transgression (Clark and Lingle 1979), but the bulk of the water was released from the Laurentide ice sheet.

Detailed stratigraphy and dating of the Holocene reef show that the maximum rate of sea-level rise in the Flandrian occurred at the Huon coast at about 9–10 ka BP (Chappell and Polach 1991).

The same transgression affected coastal areas worldwide, but there is a great diversity of apparent Holocene sea-level records from sites around the world (Figure 4.6). Some of this results from tectonic movement of the sites, from hydro-isostatic warping, and from dating problems. Taking an envelope which includes the bulk of the data (corrected for known major isostatic effects) reveals a generally clear picture of the behaviour of sea level. It is important to remember, however, that the range of effects described earlier, such as local geoid perturbations and isostatic warping, mean that in reality different histories of sea-level rise against the land did occur at various locations; at some sites, sea level might still have been at −30 m at 10 ka BP, while at others at the same time it had reached higher, say to −25 m. No single history of sea level necessarily applies in exact detail to any other

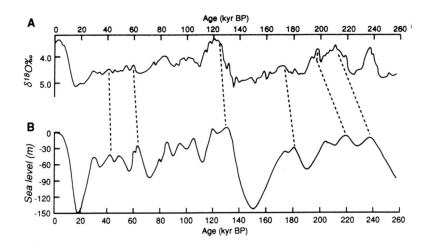

Fig. 4.5 Late Quaternary sea-level and associated oxygen isotope record derived from coral terraces, Huon Peninsula, Papua New Guinea (Aharon and Chappell 1986). (A) oxygen isotope record (B) sea-level record

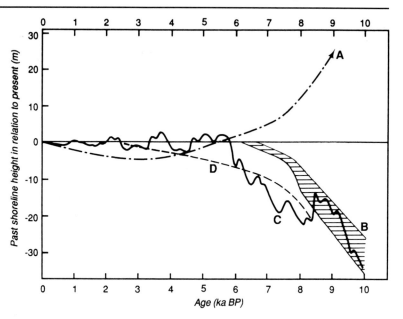

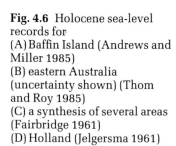

Fig. 4.6 Holocene sea-level records for
(A) Baffin Island (Andrews and Miller 1985)
(B) eastern Australia (uncertainty shown) (Thom and Roy 1985)
(C) a synthesis of several areas (Fairbridge 1961)
(D) Holland (Jelgersma 1961)

place. What follows is thus a generalized account.

After the commencement of deglaciation at around 17 ka BP, sea level rose rapidly, relative rates of rise of up to 30 mm per year being indicated at some sites. Present sea level was reached at around 6 ka BP. Fast though the rate of rise was, it is small in comparison with the rate at which the shoreline must have migrated landwards over the gently sloping continental shelves, which were progressively inundated as the transgression proceeded. For example, taking the width of the shelf off the north-west coast of Australia as 200 km, and the duration of the transgression as 10 ka, the rate of shoreline advance comes to 20 m per year, or about 40 cm per week! This is a phenomenally rapid and sustained process, which, through the continual migration of habitat, must have been particularly taxing for coastal flora and fauna.

Land bridges: a product of Quaternary low sea levels

During Quaternary low sea levels, the geography of the planet was considerably altered. The shelves were exposed around the major continents, with new river courses being cut across them as the enlarged land areas drained to the more distant sea. Additionally, dry 'land bridges' connected many areas now separated by straits (Figure 4.7). The Australian mainland was connected by such bridges both to Papua New Guinea in the north and to Tasmania. Britain and Ireland were connected to the European mainland, Alaska and Siberia were joined by the Bering land bridge, and

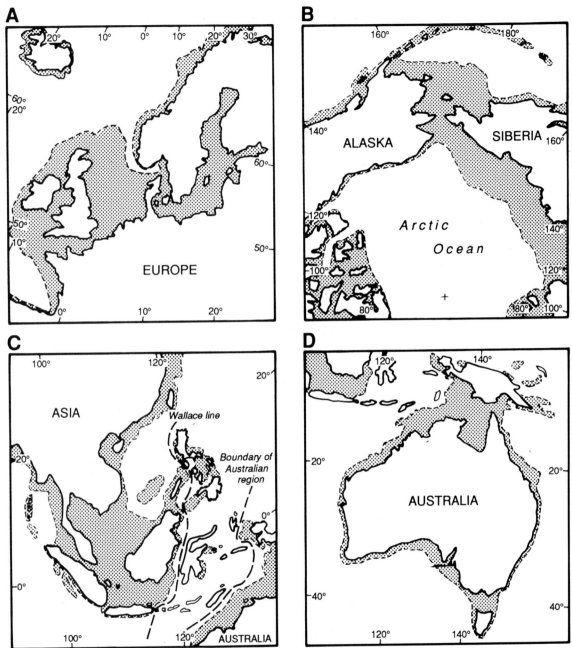

Fig. 4.7 Approximate extent of selected land bridges at the last glacial maximum (21–17 ka BP) drawn at the 200 m isobath.
(A) Europe
(B) Beringia
(C) south-east Asia
(D) Australia and Papua New Guinea

many islands of south-east Asia were linked and formed an extension of the Asian mainland. These transitory land bridges were exploited by the peoples of the time to journey among these landmasses. Many sites recording the occupation of the land bridges must now lie submerged and inaccessible.

The major coral reef communities of the present day would also have been exposed during glacial phases. Thus, the crest of the Great Barrier Reef of north-eastern Australia would have stood about 150 m above sea-level (Carter and Johnson 1986). Acid rainwater falling on the exposed limestone materials would have attacked them readily, forming a rugged landscape of jagged pinnacles with cave development occurring below the surface as water drained downward. Some of the caves which evolved at these times experienced roof collapse, a process common in present terrestrial cave environments; now drowned, these caves appear as 'Blue Holes', deep openings into the ancient reef below the contemporary reef crest, which can easily be seen from the air. Torres Strait, once part of the land bridge there, displays similar cave openings which are now drowned, as do other present-day reef environments, including the Caribbean.

Verifying the sea-level record from ice-volume studies

The fundamental control on sea level in the Quaternary was the advance and retreat of land ice. Thus, a record of land ice volume ought to provide a valuable check on the inferences made from coastal markers such as coral reefs, buried mangrove wood or shell beds.

Such a record can be obtained from the analysis of stable isotopes of oxygen in marine fossils and microfossils (see Chapter Five). Water evaporated from the oceans is enriched in the light isotopes, leaving the glacial seas isotopically heavy; deglaciation returns the light isotopes, making the oceans lighter isotopically. Organisms building their calcareous tests from oxygen extracted from the sea water reflect its isotopic composition at the time. The processes of isotope incorporation are also influenced by water temperature, so that in fact the composition of the tests reflects primarily ice-volume control on water composition, but also incorporates a temperature effect which must be removed. Indeed, revealing the temperature component (as a tool in 'palaeothermometry') is often the goal of isotope analysis.

Analysis of the isotopic composition of the shells of the giant clam *Tridacna gigas* collected from the coral terraces of the Huon Peninsula provides such data (see Aharon and Chappell 1986). These organisms only grow at depths less than 10 m, and may be used to infer the temperature of the surface layers of the tropical oceans; microfossils in deep-sea cores reflect more the temperature of the deeper waters. The results from the Huon are shown in Figure 4.5.

Clearly, the curves of isotopic composition and sea level undergo synchronous changes, and the trend of falling sea level through the period 80–20 ka BP is paralleled by a trend of decreasingly negative $\delta^{18}O$ values (i.e., progressive ^{18}O enrichment).

Analysis of the isotopic composition of land ice shows values much more negative (-37‰ in glacial periods and -31‰ in interglacials) as would be anticipated.

The Huon data can be used to attempt to unravel the temperature and ice-volume effects. Knowing the isotopic composition of the clams, it is possible to calculate by how much sea level must have been lowered through evaporation for the appropriate level of ^{18}O enrichment to develop. Comparing this with the known sea level from the reef analysis described earlier provides an estimate of the amount of isotopic compositional change which is likely to be the effect of changed water temperature. The Huon data suggest that the tropical ocean temperatures were similar to the present in the early part of the last glacial, but cooled to perhaps 3°C less than present toward the end of the glacial phase. This cooling was evidently insufficient to prevent continued coral growth.

Recent and historic changes in sea level

There is good reason to consider the contemporary trend of sea level, in view of the size and important trading role of coastal cities, and the survival of the habitats provided in low-lying coastal wetlands.

It appears from historical and contemporary records that sea level is rising steadily but slowly: a rise of about 10–15 cm over the past century is indicated (Gornitz *et al.* 1982), and the present rate of rise is taken to be 1–3 mm per year. These figures refer to the mean gauge rise, with relative sea levels still of course falling in areas experiencing rapid tectonic uplift.

The exact causes of the sea-level rise are not known. Mantle readjustments following the last deglaciation are undoubtedly still taking place, but it is possible that a major cause is global warming related to human activity. The significantly increased (and growing) levels of atmospheric carbon dioxide (CO_2) which have resulted from the Industrial Revolution and the widespread use of fossil fuels seem to be warming the atmosphere and oceans. Global warming leads to warming of the surface layers of the ocean, and the resulting volumetric expansion produces sea-level rise.

Another consequence of warming is that mountain glaciers undergo retreat, feeding additional meltwater into the rivers and hence to the sea. It has been estimated that this process may have raised sea level by 2–7 cm this century, and that a further 30 cm rise is possible in the next century (Meier 1984). Larger volumes of meltwater may in the much longer term be contributed from the massive Greenland and Antarctic ice caps.

Antarctic ice has the potential, however, to produce a significant sea-level rise in the relatively short term. The main east

Antarctic ice sheet lies on bedrock above sea level. The 10 per cent of Antarctic ice forming the west Antarctic ice sheet, however, is grounded below sea level. It is stabilized by adjacent ice shelves, such as the Ross ice shelf. Relatively little warming and sea-level rise would be required to destabilize this arrangement, and permit melting of the west Antarctic ice. The ice shelves could collapse in about a century, with complete disintegration of the west Antarctic ice sheet following over a further few centuries (Mercer 1978). The water so released could lift sea level by about 5 m. Interestingly, sea level during the last interglacial was about 6 m higher than present (as indicated earlier in the discussion of the Huon Peninsula record). Elevated shorelines at about +6 m are found in Antarctic, and are about 100 ka in age. It thus seems possible that the west Antarctic ice sheet did not survive the warming of the last interglacial; this adds weight to the idea that it might not survive the present one either, given the extra anthropogenic warmth!

The time required for this melting to occur gives us the opportunity to plan ahead, and to adapt our coastal settlements accordingly (and to moderate our production of greenhouse gases!). Deliberate control of sea level is possible: it has been estimated that water storage in dams has prevented 1–2 cm of sea-level rise this century (Gornitz *et al.* 1982). Water storage on a much larger scale would be required to halt a continuing rise, and the deliberate flooding of low-lying areas such as the rift valley between Israel and Jordan has been contemplated, but would bring with it severe environmental costs. Whether or not we seek deliberately to intervene in this way, the evident contemporary sea-level rise, perhaps the precursor of the larger change which may follow, obliges us to continue our observation of the sea, and to refine our knowledge of its past behaviour.

As we shall see in the next chapter, the sedimentary record of the ocean basins provides a great deal of information not only about sea levels, but on the temperature, salinity, and circulation of the oceans. When combined with the kinds of shoreline records discussed in this chapter, these data yield many additional insights into the behaviour of marine environments during the Quaternary, and their role in the diverse mechanisms of environmental change.

5 | Evidence from the Oceans

I should have been a pair of ragged claws scuttling across the floors of silent seas.

T. S. Eliot (1888–1965),
Morning at the Window.

Introduction

Oceans cover 71 per cent of the surface of the planet and contain the largest component of the planet's biosphere. It is therefore essential to understand the interactions within the oceans in addition to the influence they have on the climate of our planet. Two-thirds (67 per cent) of the land area of the globe is located in the Northern Hemisphere. Oceans occupy 61 per cent of the Northern Hemisphere, and 81 per cent of the Southern Hemisphere. The Southern Hemisphere is often termed 'maritime' in contrast to the Northern Hemisphere which is described as the 'continental hemisphere'. The implications of this asymmetry need to be examined in palaeoclimatic and palaeoenvironmental studies (Tchernia 1980; Emiliani 1981).

Oceans cover a very uneven topography ranging from usually narrow, gently sloping continental shelves to deep and narrow, elongated trenches, some as deep as 10 km or more. Three-quarters of the oceans (77 per cent) are deeper than 3 km, so that we need to consider the influence exerted by deep oceanic water masses on global climate (Labeyrie *et al.* 1987). A third of the oceanic water mass by volume (30 per cent) has a temperature between 1 and 2°C. Consequently, it is important to investigate the impact of this cold water mass on overall global temperature. We need to know how the oceans have changed during the Quaternary before trying to assess the effects they may have and may have had on global change (Open University Course Team 1989a,b,c). In addition, sea levels have fluctuated throughout Quaternary time (Chapter 4), sometimes influencing the chemical and physical attributes of the oceans. The purpose of the present chapter is to examine the changing nature of the ocean environment during the course of the Quaternary.

Properties of sea water and of the different water masses

Sea water is salty because it contains dissolved salts (solutes) which are the weathering products of continental and oceanic rocks and derive in part from sea-floor gases. Sodium and chloride are the principal ions found in sea water because they are among the most soluble constituents of weathering. Salinity is a measure of the total dissolved salts in sea water and the units commonly used are expressed as grammes of dissolved salts per litre of water or parts per thousand (‰). It is important to recognize that salinity values differ between oceans, or parts of them, and between different water depths. Although salinity differences may be small, they play a significant role when water densities are also taken into account. Surface-water salinities range from 33.5‰ in the Antarctic Weddell Sea to 37.5‰ in the North Atlantic. The latter value is indicative of high evaporation whereas the Weddell Sea is diluted by meltwater. A value of around 35‰ is found throughout most oceans, and is thus accepted as average for oceanic water. Several dissolved gases in ocean water are important owing to their abundance, and because they affect biological activity. These gases are oxygen, carbon dioxide, and nitrogen. Their significance is discussed later.

As mentioned earlier, a third of the oceans' water has a temperature between 1 and 2°C, rendering this water rather dense, especially when its salinity is high. For dense water to form in an ocean, it has to originate near the surface either through cooling or through evaporation, or both. Hence, cold water masses must originate near the poles, and this explains why the North Atlantic Deep Water (NADW) originates in the Norwegian and Greenland Seas, and the Antarctic Bottom Water (AABW), which is colder (usually cooler than 0°C), originates all around the Antarctic continent, especially in the Weddell Sea, Ross Sea and Prydz Bay areas (Bleil and Thiede 1990). This water mass is the most widepread in the world and is denser than the NADW. On approaching the Antarctic continent (Figure 5.1), the NADW rises progressively closer to the surface to become the Antarctic Circumpolar Water (Corliss 1983). This water reaches the surface at the Antarctic Divergence, where surface waters diverge (Figure 5.1). It is in this area of divergent water masses that biological activity is among the most prolific in the world (Hedgpeth 1969).

Above the deep cold water masses is a series of 'intermediate' water masses such as the Antarctic Intermediate Water (AAIW) (Figure 5.1). The AAIW, which in the Atlantic travels past the equator at a depth of about 1–2 km, is in fact the most widespread intermediate water mass in the world. Its temperature is usually between 2 and 4°C, and it has salinities lower than the AABW.

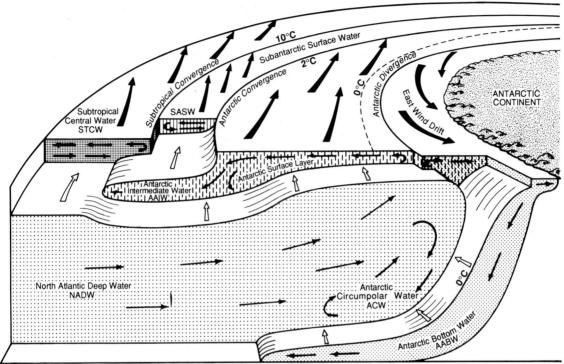

Fig. 5.1 Schematic diagram showing the various water masses in the Southern Ocean adjacent to Antarctica. Note the temperature of the different water masses which influence their respective densities. Also observe the direction of movement of the different currents, bearing in mind that where two water masses diverge from one another, there is upwelling of water to the surface; converging water masses induce downwelling. These phenomena are important when considering the movement of water masses in the oceans. Eventually, the AABW will continue flowing into the Atlantic Ocean (adapted from Hedgpeth 1969)

Characteristically, the 'upper water masses' are more localized because they are under the influence of surface currents so that temperature and evaporation/precipitation ratios may differ markedly between seasons. Consequently, the lateral and vertical extent of these water masses will also vary quite substantially. The upper water masses consist of waters which have a steep vertical gradient in temperature, salinity and dissolved oxygen content. There is thus a mixed surface layer which sits above a zone called the thermocline within which temperature decreases significantly with depth, and density increases. Below the thermocline are the 'central waters' which usually form the subtropical belts or gyres. The thickness of these central waters depends on their location. For example, where two water masses converge at the surface, as is the case at the subtropical convergence, warm water sinks so that the thermocline is lowered, and the upper layer increases in thickness.

Nevertheless, since water density is controlled by temperature as well as by salinity, some very cold water (under 1°C), which is less saline than the average ocean water, may sink down to the abyss simply because of its low temperature. Hence, several water masses, each with different densities, are superimposed upon one another in the oceans. These different water masses are not only characterized by different densities but also have other properties of importance to physicochemical processes in the oceans and to biological activity.

Deep, cold water is characteristically rich in nutrients and dissolved CO_2 and if such water is brought to the surface, as would occur in a zone of divergence, an 'upwelling' current occurs that is beneficial to fisheries. The best documented regions where this phenomenon occurs are along the south-western coast of Africa, off Namibia, and along the west coast of South America, off Peru. Upwelling of cold water, rich in nutrients, causes high organic productivity at the surface and results in rich fishery catches. Nevertheless, climatic links with areas of upwelling in the oceans and adjacent land surfaces are significant as cold upwelling offshore is often associated with aridity onshore (see Chapter 7).

The processes which operate in the oceans with respect to different water masses are basic to deciphering past changes. Surface temperatures, controlled by ocean currents and other processes below the surface, form part of the driving forces controlling the earth's climate. In addition, biological productivity in the oceans is determined by the physicochemical interactions which occur throughout the ocean's water column, especially carbon dioxide and oxygen uptake or discharge. There is 50 times more CO_2 stored in the oceans than in the atmosphere and the biosphere, so that the storage of CO_2 dissolved in the oceans or used by organisms secreting $CaCO_3$ skeletons (C and O are part of the skeletal framework), is vital to an understanding of carbon dioxide fluxes relevant to global climate (Broecker and Takahashi 1984; Broecker and Peng 1984). It is principally in high latitudes that CO_2 is taken up more effectively by the very cold, dense surface waters. Through the sinking of dense water, CO_2 will eventually end up in the deep part of the oceans where it is stored on a long-term basis (Anderson and Malahoff 1977). It is mainly in the tropics that CO_2 is vented through the surface to the atmosphere, especially where water masses diverge.

The extent of sea ice in polar regions also plays a significant role in controlling water density since ice helps induce the water below it to become more saline as a result of ice formation at the surface. In addition, the presence of sea ice, even if it is for the winter season only, helps prevent a CO_2 sink from occurring, in comparison with the period during which there is no ice cover (Martinson 1990).

Sea water composition is kept nearly constant because of the conservative nature of its major elements, and because biological activity and other physicochemical processes do not affect it significantly. Other elements present in smaller amounts (minor or trace elements) display what is called non-conservative behaviour because they are frequently influenced by biological processes (such as when trapped in the skeleton of organisms), or by physical processes such as where two water masses mix, or by dissolution of organically or inorganically produced precipitates. Changes in these 'unstable' elements will reflect changes in physicochemical processes within the water column or at the sediment–water interface. Dissolved gases also play an important part in oceanic processes. Dissolved oxygen, for example, is strongly affected by biological activity in the oceans, although temperature and salinity play an integral role in controlling the amount of dissolved oxygen in sea water. Dissolved oxygen levels increase with a decrease in temperature as does the amount of dissolved CO_2. The latter, of course, is of great importance for the preservation of calcium carbonate skeletons of organisms, and its effect needs to be taken into account when interpreting the fossil record, simply because species indicative of particular environmental conditions may have been destroyed through dissolution.

The composition of sea water is such that with a change in water temperature and pressure, and also in the partial pressure of dissolved CO_2, calcium carbonate in contact with that water may dissolve, more particularly in the deeper parts of the oceans. Calcium carbonate, being the principal component of planktonic organisms, dissolves through the water column as water depth and pressure increase and temperature decreases. It is principally the carbonate ion concentration in sea water, determined by these three parameters, which controls the preservation of calcium carbonate shells or tests. Consequently, it is necessary to determine the depth in the ocean where calcium carbonate returns to solution. This level is called the 'calcite compensation depth' (CCD), and varies between and within oceans (Takahashi 1975). Figure 5.2 shows depth contours which delineate the levels in the

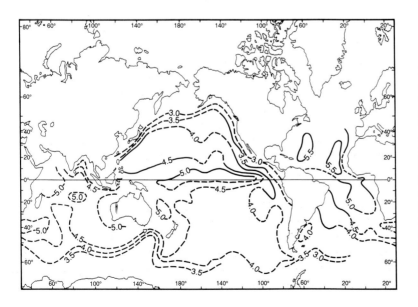

Fig. 5.2 Map showing contours (km) in the oceans defining the calcite compensation depth (CCD). This was obtained by examining sediment on the sea floor that contained calcite; below these depths, there should be little or no calcium carbonate preserved on the sea floor. Solid contours represent more than 20 samples per 10° square and dotted contours represent fewer than 20 samples (from Open University Course Team 1989a)

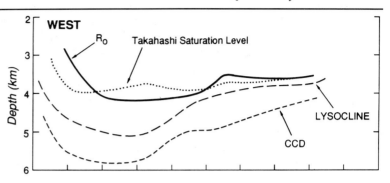

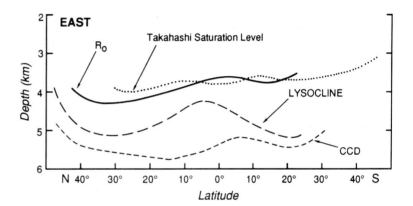

Fig. 5.3 North–South transects along both sides of the Atlantic Ocean showing the saturation level of carbonate in sea water as calculated by Takahashi (1975), the Ro level or depth in the ocean where calcium carbonate shells start to dissolve, the lysocline which corresponds to the level at which significant dissolution of calcitic shells is registered, and the CCD level below which no carbonate tests/shells are found. The different gradients along both sides of the Atlantic are principally caused by variations in the supply of carbon of terrigenous origin (from Vincent and Berger 1981)

oceans where calcite (the most stable form of calcium carbonate) becomes soluble. It is below these depths that under normal circumstances no calcium carbonate is found in sediment on the sea floor. The CCD levels are much deeper in the Atlantic Ocean because the total amount of dissolved CO_2 is greater there. A large supply of terrigenous material brought by rivers into the oceans will cause an increase in the amount of total organic carbon in sea water causing the CCD level to rise near the edge of continents (Figure 5.3).

The calcite compensation depth is important for ocean studies because it affects the preservation and/or dissolution of organisms that secrete a calcitic shell or test. The principal organisms of concern are the microscopic unicellular foraminifera (usually under 1 mm) and the much smaller organisms of algal affinity called coccoliths (of the order of a few micrometres in diameter).

So far, we have discussed the level in the ocean where calcium carbonate (principally calcite) is totally dissolved. We also need to take into account the level in the ocean where calcium carbonate starts to dissolve. Some organisms have a thinner and smaller skeleton than others, and so are more prone to dissolution. Recognition of this phenomenon is important when using fossil taxa to interpret palaeoenvironments as some taxa are 'missing' from the record because of dissolution processes which operated while the remains of the organisms descended through the water column or were lying on the sea floor. The level in the ocean where calcium carbonate starts dissolving preferentially for thinner and fragile tests is labelled Ro in Figure 5.3. However, it is the level below this, called the lysocline, where only the resistant forms of organisms are found. The lysocline level usually parallels the CCD level (see Figure 5.3). There is another type of calcium carbonate, called aragonite, which is secreted by organisms and precipitated directly from the water. Corals are among the best known organisms with an aragonitic skeleton. Most corals are found in the tropics because it is in those regions that water temperature is sufficiently high to allow precipitation of aragonite.

Aragonite is a less stable form of calcium carbonate and thus dissolves much more readily than calcite. Consequently, the aragonite compensation depth (ACD) in the oceans is much closer to the surface than the CCD, and can be used as a more sensitive recorder of CO_2 in the oceans through time, especially during periods of oceanic changes when water masses alter their proportion significantly, or when upwelling occurs. Identification of the location and timing of calcite and aragonite dissolution through the study of deep-sea cores is indicative of processes occurring in the oceans, and can help to define their relationship with climatic change. Microplankton, especially foraminifera and coccoliths, are so abundant in some parts of the oceans, that they constitute a large proportion of the biomass and so help to regulate the amount of carbon produced by the biosphere.

Silica forms the lattice of several important groups of organisms, and is another significant component of the oceanic chemical budget. Among the algae are organisms called diatoms which secrete a variety of siliceous pill-box shaped tests called 'frustules'. Radiolarians comprise another group of unicellular organisms related to the foraminifera. Silica, frequently labelled as opaline silica because of its amorphous and porous texture, is secreted by these micro-organisms (less than 1 mm in diameter) which commonly inhabit zones of high productivity. The remains of siliceous organisms are found in the oceans where carbonate concentration is low (as in the polar regions where CO_2 is high), or where carbonate is absent or below the CCD. The study of biogenic siliceous remains thus becomes important for the reconstruction of past conditions in these parts of the oceans. Silica dissolves in the oceans, especially in the upper 50 m of the water column (the solubility of silica increases as the temperature decreases), but also within the sediment as well, especially at the sediment–water interface, and this will affect the fossil record and its interpretation.

Dissolution of biogenic silica is highest within the waters at the surface and progressively decreases down to approximately

500–1000 m depth, depending on the ocean, before reaching a steady state. For small diatoms, and for the coccolith plates which are so minute, sedimentation down to the deep sea may take up to a century. This gives plenty of time for dissolution to occur. However, a rapid sedimentation rate may prevent the slow process of dissolution from occurring at the sediment–water interface.

Microfossils as tools for palaeoenvironmental reconstruction

Foraminifera (also called forams) are ubiquitous marine organisms which secrete a test consisting of a series of small chambers. Most forams have a calcitic test. There are two main types of foraminifera, characterized by their mode of life (Funnell and Riedel 1971). One is labelled planktonic because the organism is able to control its position in the water column, although most individuals occur near the surface and migrate up and down during their life cycle, sometimes as much as several hundred metres. The other group of foraminifera, called benthic, lives on the sea floor or sometimes within the upper few centimetres of the sediment at the bottom of the ocean. Numerous benthic forams have a test made of an agglutination of debris from the sea floor, and thus differ from those in the other benthic group which have a calcareous test. A wealth of information has been obtained on the ecology of planktonic forams, especially about their modern distribution in the oceans (Bé and Tolderlund 1971). Since there are only about 40 species of planktonic forams, it is possible to define their respective biogeographical boundaries. Figure 5.4 shows the distribution of these organisms in the present-day oceans.

Study of the fossil remains of the organisms on the sea floor can greatly elucidate the characteristics of ocean surface waters for the entire Quaternary during which most of these species existed. Several planktonic species of forams are characterized by a different direction of coiling for the chambers made by the individual throughout its ontogeny. Coiling is either sinistral (to the left) or dextral (to the right) and the direction of coiling in some taxa seems to be broadly related to a particular temperature regime. Different coiling types for distinct species are now used to relate to portions of the oceans with specific surface temperature regimes. The best known example is the foram *Neogloboquadrina pachyderma* which characteristically has a sinistral coiling direction in cold water. A change to dextral coiling in this species seems to occur in waters with a mean annual temperature greater than 9°C. Several other species listed in Figure 5.4 have different coiling directions in regions with different thermal regimes. Study

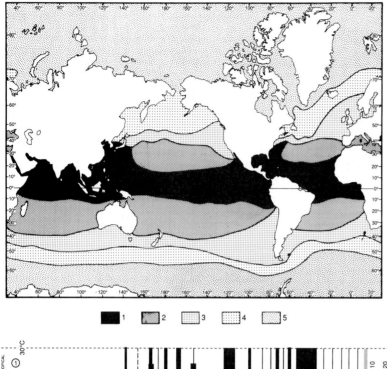

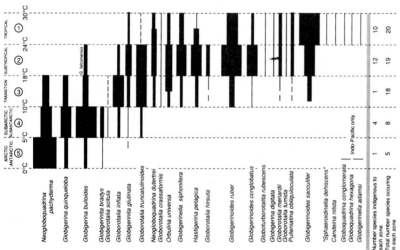

Fig. 5.4 Map showing zoogeographical zones in the oceans defined by planktonic foram assemblages. Taxa belonging to the five zones are shown below. The thickness of the bars in the lower figure represents the relative abundance of each taxon in each zone (adapted from Bé and Tolderlund 1971)

of chamber arrangement in planktonic forams thus allows us to detect a shift in surface water masses during the Quaternary. Nevertheless, one should be aware that coiling direction is not always entirely consistent with temperature signals.

Several authors have tried to relate foram size, shape, surficial texture and other architectural factors to particular temperature regimes and/or geographical zonations. All of these features may be related to the capability of individual forams to secrete a test (so that for some species it is easier for crystallization to occur at a certain 'optimum' temperature), or to cope with variations in water density (controlled by salinity and temperature). Temperature should also control the rate of calcification and the chemical composition of the foraminiferal test.

Benthic foram morphology reflects its mode of life (Corliss and Fois 1990). Narrow, elongated and cone-shaped forams are often burrowers (infauna), whereas flat and broad ones live on the surface (epifauna). Observations of benthic foram morphology and diversity are starting to provide information on organic productivity at the surface above the site where the forams are living since the amount of organic matter supply, through its 'showering', can affect foram diversity and rate of growth. It now appears possible to relate the infaunal/epifaunal ratio to the interaction between phenomena which occur at the surface and at the bottom of the ocean.

Oxygen isotopes

Foraminifera have also been successfully used in palaeoceanography by studying their isotopic composition with respect to the stable isotopes of oxygen and carbon (Savin and Yeh 1981; Vincent and Berger 1981; Bradley 1985). The principle behind the use of stable isotopes in foraminifera is that the ratio between the two isotopes of oxygen ^{16}O and ^{18}O taken up by the organism during test formation is controlled by temperature and the isotopic composition of the ambient water. With knowledge of the latter two variables, the isotopic composition of foraminifera can be used to reconstruct palaeotemperature. The formula commonly used to determine the isotopic difference ($\delta^{18}O$) between these isotopes of oxygen is:

$$\delta^{18}O = \frac{(^{18}O/^{16}O)\text{sample} - (^{18}O/^{16}O) \text{ standard}}{(^{18}O/^{16}O) \text{ standard}} \times 1000$$

The units are in parts per thousand (‰). The standard commonly used for forams is PDB, a Cretaceous belemnite from the Pee Dee Formation in North Carolina. Standard Mean Ocean Water (SMOW) is the standard for present-day water and is given a nil value (0‰).

Fractionation occurs between the two isotopes of oxygen of interest here. The lighter isotope ^{16}O preferentially escapes as vapour during water evaporation so that rain water is isotopically lighter than the ocean water from which it originated and the $\delta^{18}O$ of carbonate shells decreases as water temperature increases (see figures 5.5 and 5.6).

During glacial times, when sea level had dropped by approximately 120 m (Chappell 1987; also Chapter 4), much water was locked up in ice caps and mountain glaciers (Chapter 3). This water, in the form of ice, was isotopically enriched in ^{16}O, so that ocean water became proportionately enriched in [18]O (Shackleton 1987). About 0.11‰ change in the $\delta^{18}O$ in planktonic carbonate fossils, such as forams, represents a 10 m change of sea level, although estimates vary somewhat between authors. The difference between isotopic composition of sea water during a glacial period and today is approximately 1.20‰, with the ocean water being isotopically heavier during glacial times. Measurements of the isotopic composition of foraminifera from Quaternary cores are represented as curves of the isotopic composition of the water and its inferred temperature. Many workers have opted to study cores taken in tropical regions where it has been postulated that very little temperature change occurred between glacial and interglacial episodes. The record of isotopic change in the tropics has been interpreted as relating almost entirely to the effect of global ice volume and hence sea-level change, amounting to approximately 120 m, equivalent to an isotopic shift of 1.20‰. However, elsewhere in the oceans an isotopic change in the forams found in cores results from both temperature and ice-volume changes (Mix and Ruddiman 1985). Nevertheless, the amplitude of changes recognized in cores from nearly anywhere in the oceans between glacial and interglacial periods is sufficient to allow the patterns of changes through time to be used for correlation between different cores at different locations.

If one examines a much longer record spanning the entire Quaternary, it becomes obvious that fluctuations in the isotopic values found in planktonic foraminifera are somewhat cyclical. Figure 5.6a displays such a pattern which is repeated about every 125 000 years. In the core V28-238 taken in the west equatorial Pacific (Shackleton and Opdyke 1973), compaction of the sediment in the core is such that the upper portion of the core appears thicker than an equivalent time period lower down the core. The difference in isotopic composition between the benthic and planktonic forams analysed from the same core is used to identify the differences between both the surface temperature as well as the smaller differences in $\delta^{18}O$ between bottom and surface water (Birchfield 1987). At present, there is still disagreement about these differences because the isotopic composition of deep ocean water during glacial times is not precisely known. We still do not know how much the bottom water, which has a slightly different isotopic composition, mixes with surface water and thus alters the isotopic value at the surface (Shackleton 1987).

Several other problems also occur because it is necessary to calibrate the isotopic data against other factors which influence the isotopic composition of foraminifera. For example, one has to determine the fractionation between the two isotopes of oxygen for different taxa, and in addition, to determine the size fraction of forams because different sizes grown in the same water apparently have different isotopic signatures (Vincent and Berger 1988).

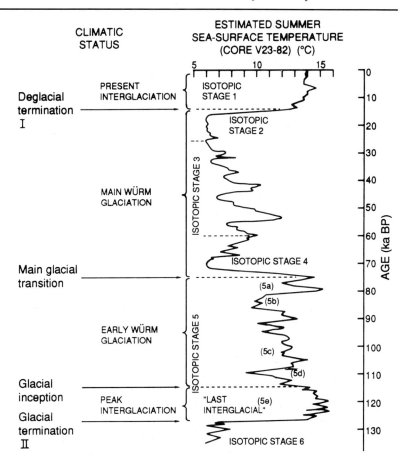

Fig. 5.5 Diagram showing the estimated sea-surface temperature (SST) for the North Atlantic Ocean based on foraminiferal assemblages of core V23-82 in the north-west of the Atlantic. Dating is obtained through correlation with other cores (based primarily on tephra layers and sea-level curves from Barbados), and the various, commonly used, oxygen isotope stages are also indicated on this figure (modified from Bradley 1985)

Some foram species are in closer 'isotopic equilibrium' with sea water compared to others which diverge from the expected value (see below), and which consequently register an enrichment in one of the two isotopes. In addition, some planktonic foram taxa have a mode of life which permits them to live at different water depths (where temperature may differ markedly, and water isotopic composition less so), so that different chambers of the one specimen may have different isotopic values. There is some evidence that it is the rate of calcification which controls the oxygen isotopic fractionation in forams. If correct, this would explain the different isotopic values for the different foram sizes.

The life style of the forams can also influence the isotopic composition of their tests. To allow gametogenesis to occur, some forams add an extra calcitic layer which is often in isotopic equilibrium with the ambient water, itself at a very different temperature to that of the surface. (Gametogenesis is the period when gametes are released in the deeper parts of the water column where the forams live prior to dying and sinking to the sea floor.)

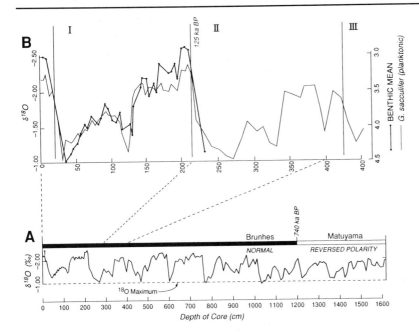

Fig. 5.6 Oxygen isotope stratigraphy of core V28-238 from the western equatorial Pacific.
(A) $\delta^{18}O$ values for the planktonic foraminifer *Globigerinoides sacculifer* in parallel with the palaeomagnetic stratigraphy (the Brunhes-Matuyama polarity boundary is placed around 740 000 years BP). Note the cyclicity of the record with the most negative values returning to the same points. (B) is an enlarged portion of (A), covering approximately the last 250 000 years of the record to show variations in the $\delta^{18}O$ content of *G. sacculifer* and the mean values of benthic forams. Note the transition between isotope stages 6 and 5e placed around 127 000 years BP and the sharp nature of this transition. Note also the difference in $\delta^{18}O$ values between the benthic and planktonic forams. Roman numerals represent the timing of the last three glacial maxima (slightly modified from Vincent and Berger 1981)

We need to know more about these organisms' mode of life before interpreting their isotopic signatures.

Isotope stratigraphy, as it is called, is used to provide a basic age for cores taken from anywhere in the world's oceans (Jansen 1989). For example, the last major change in $\delta^{18}O$ in a core, with values becoming more enriched in ^{16}O, is interpreted as a result of the melting of the ice caps after about 20 ka BP. This phase is labelled as the transition between isotope stages 1 and 2 (see Figure 5.5). The previous similar pattern in cores occurs at the transition between isotopic stages 5 and 6. Such cyclicity has now been recognized in all deep-sea cores and it is accepted that the cycles relate to astronomical cycles which govern the position of the earth with respect to the sun, the tilt of the earth and the wobble of the earth along its axis (see Chapter 3). The importance and amplitude of some of these cycles have been examined for several marine cores and it has been recognized that three main cycles lasting about 100 000, 41 000 and 23 000/19 000 years consistently occur (McIntyre 1989). The cycles correlate fairly well with astronomical forcing, but it is important to realize that the effect of the forcing may be felt differently at different latitudes. For example, the 41 ka cycle has a more pronounced effect in high latitudes, compared to the 19 ka cycle which is more prevalent in middle to low latitudes. The 41 ka cycle has been the main driving force behind the changes in glaciated areas, and so has the 100 ka cycle for the North Atlantic. However, the amplitude of some of these cycles has been such that their influence on climatic variation during the entire Quaternary has not always been the same. For example, the 100 ka cycle is considered to have only played a dominant role during about the last 650 000 years of the Pleistocene, whereas the 41 ka cycle had a significant amplitude during the entire Matuyama Chron, spanning 2.5–0.7 Ma. These cycles have been deciphered mainly from isotopic changes in foraminifera and from variations in the total $CaCO_3$ content in marine cores.

Carbon isotopes (^{13}C, ^{12}C) are usually analysed in conjunction with oxygen isotopes, and have been used to obtain information on the origin of organic matter used by planktonic organisms since there is a marked difference between the carbon isotopic ratio of ocean water compared to organic matter of continental origin. There is also some dispute as to whether temperature causes fractionation between the two isotopes of carbon in calcareous microplankton. In addition, methanogenesis, which occurs within the ocean and more significantly within the sediment, may also alter the carbon isotopic ratio.

It has not been possible to use carbon isotopes in foraminifera to correlate global events, as has been done with the oxygen isotopic record. On the other hand, isotopic shifts in carbon isotopes have been used to infer substantial changes in planktonic blooms at the ocean's surface, as well as changes in primary productivity (causing large amounts of $CaCO_3$ to precipitate), and changes in the supply of usually fresh water of continental origin to the ocean. For example, an isotopic shift was detected in the latter part of the Pleistocene in the Gulf of Mexico, and was interpreted as a sudden influx of glacial meltwater carried into the Gulf by the Mississippi River (Broecker *et al.* 1989).

Nitrogen isotopes (^{15}N, ^{14}N) have also been applied in combination with carbon isotopes to determine the origin of the organic matter in deep-sea cores. Upwelling of cold, dense and nutrient-rich water near the ocean surface can be detected from the carbon and nitrogen isotopic signature of organisms which live near the surface, such as forams and nannoplankton.

Transfer functions

Another technique used to reconstruct conditions in the oceans through time has been to establish the relationships which exist between assemblages of species and the ecological conditions which control them, especially sea surface temperature (SST). Thus, faunal association and composition have been used to compute a palaeoclimatic index. In a sense, it is like plotting a ratio of selected warm-versus-cold species to obtain an indication of the temperature in which the faunal assemblage lived. Multivariate statistical analysis has been successfully used to quantify the past conditions in the oceans through correlation of modern-day species with established oceanic conditions. Consequently, several research groups have come up with 'transfer functions' for different oceans after having related SST to the presence of faunal assemblages recovered from the top of cores collected in different oceans. Naturally, an important uncertainty pertains with regard to interpreting the information as the remains of organisms found on the sea floor are not necessarily modern. Phenomena such as reworking by bioturbation or bottom-current activity, or slow sedimentation, may affect the accuracy of the correlation.

Four major faunal assemblages (from polar, subpolar, subtropical and tropical regions) have been commonly chosen to establish the equations or transfer functions used to calculate sea surface temperature, and to distinguish between summer and winter values. In fact, information obtained from these transfer functions has been used in parallel with oxygen isotope curves, and a good correlation between the two was established, confirming the usefulness of transfer functions despite the uncertainties they contain. Fossil groups used for the transfer functions are principally foraminifera, but also include radiolarians, diatoms and coccoliths. The use of transfer functions climaxed during the CLIMAP (Climate: Long-range Investigation Mapping and Prediction) project (McIntyre 1981; CLIMAP Project Members 1981). One important objective of the CLIMAP project was to establish the conditions on the globe during the last glacial maximum, nominated as 18 000 years ago. For this project, sea surface temperatures were reconstructed and are presented for August in Figure 5.7. This figure represents the state of knowledge in 1981 when the CLIMAP maps were first published. Some of the temperature reconstructions presented by CLIMAP are still being disputed and are now being checked by other means.

This approach has also been applied to other periods of the Quaternary. For instance, the ocean during the last interglacial has been compared with the present day, because sea level is considered to have been approximately the same. A map representing the differences in SST for February and August was compiled (see Figure 5.8) in order to determine past conditions and to validate the use of transfer functions using faunal/floral assemblages and stable isotopes (Ruddiman 1984).

The information about past conditions in the oceans, especially for surface water, can be used in combination with continental reconstructions to model past climates. One aim is to model oceanic processes through time in order to detect how ocean currents interacted with the atmosphere (Broecker 1987; Broecker and Denton 1990). This type of work is still in its infancy.

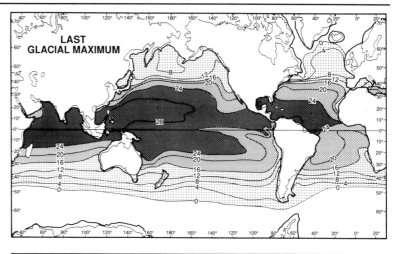

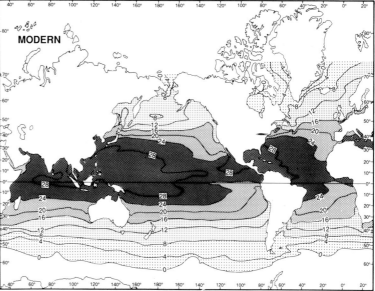

Fig. 5.7 Reconstruction of surface ocean temperatures in August for the glacial maximum 18 000 years BP, and comparison with the present day. This reconstruction by the CLIMAP group (see text) was based on transfer functions from several planktonic organisms and on foraminifera oxygen isotopic composition (modified from McIntyre 1981)

Cadmium, barium and germanium analyses

Several other chemical analyses of marine organisms have recently been used to supplement the isotopic record of the oceans. The trace element cadmium, and the more recently studied barium, have been recognized as important indicators of nutrient levels in the oceans. Cadmium, being directly linked with the amount of phosphorous in the oceans, has been used through the study of the Cd/Ca ratio in foraminifera to detect the nutrient content in the oceans, especially for different water masses when cores are taken at different water depths. This exciting new field of research will help us better understand past

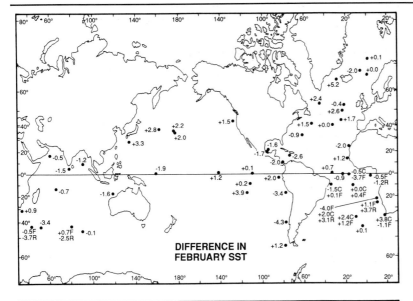

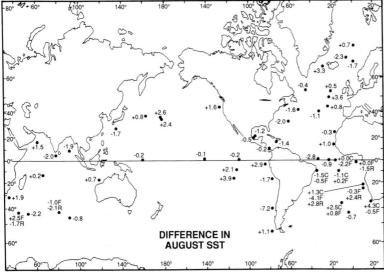

Fig. 5.8 Difference between the estimated sea-surface temperature (SST) in °C of the oceans for two different seasons between the peak of the last interglacial (isotope stage 5e) and the present, based on oxygen isotopic values of foraminifera and transfer functions of different planktonic organisms (after Ruddiman 1984)

and present processes in the oceans, including the location and timing of ocean ventilation which controls CO_2 release into the atmosphere, and is also relevant to calibrating the radiocarbon record (see Appendix). This record will eventually be used for correlation with that of the air bubbles found in ice cores (see Chapter 3). Cadmium is thus a useful tracer in palaeoceanography, especially when used in combination with $\delta^{13}C$ in foraminifera (Boyle 1990).

Barium has been used in parallel with cadmium analyses in forams to reconstruct the nutrient levels in the deep water (Lea *et al.* 1989). For example, it has been postulated that circulation in the deep oceans differed during parts of the Pleistocene, and that nutrient levels were higher during glacial times in the North Atlantic in comparison with today. Barium has also been analysed in corals to determine a change in nutrient level in the surface ambient water. A change in nutrient level could be explained by the shoaling of the thermocline that enabled deeper water, richer in nutrients and also in barium, to emerge at the surface where corals grow (Lea *et al.* 1989). This preliminary work was on modern corals from the Galapagos Islands which come under the occasional influence of upwelling induced by SST anomalies caused by El Niño–Southern Oscillation (ENSO) events (see Chapter 6). Cadmium, on the other hand, was analysed in the same project to demonstrate the effect of upwelling, but the results were not as obvious, probably because cadmium is more readily scavenged by organisms. Techniques using trace elements have tremendous potential for improving our understanding of oceanic processes such as the behaviour of different water masses and currents through time.

The content of germanium in biogenic silica (diatoms and radiolarians) has been used to determine the amount of silica present in ocean water. In a sense, germanium behaves like a heavy isotope of silica, and although only a small proportion of germanium takes the place of silicon atoms (one in a million), it is possible to reconstruct the Ge/Si ratio of sea water through time. This ratio has been used to show that productivity had apparently dropped during glacial periods on the basis that silica uptake, and consequently accumulation of biogenic silica on the sea floor was lower then.

Aeolian dust and pollen

A substantial programme of research aimed at correlating oceanic and continental events has been accomplished in the Atlantic Ocean to analyse the record of airborne dust originally deflated from Africa, and deposited out at sea (Hooghiemstra *et al.* 1987). This work sought to establish a record of arid events on the continent, and to correlate this record with oceanic conditions. The principle has been to distinguish between the supply of quartz grains coloured red by iron oxides and considered typical of a desert origin, and to compare it with white quartz usually originating primarily from fluvial material deposited near the continental edge. It has also been possible to establish the prominence of trade winds for different latitudes through time from the study of those terrigenous remains originating from the deserts (Chapter 7). Similarly, the study of pollen originating from the continents and recovered from marine cores has enabled palaeoecologists to reconstruct the vegetation record spanning long periods of the Quaternary record which are frequently not found on land because the relevant deposits have been eroded by glacial activity. The study of marine planktonic organisms (through faunal/floral associations and their chemical composition) and of marine sediment composition, in combination with the study of aeolian dust and pollen, provides a means of establishing past environmental conditions of global importance (Prell 1984). This kind of multidisciplinary study is essential to provide the necessary correlation between the global climatic fluctuations and changes in the biosphere on land and at sea.

Ice-rafted material

Ice-rafted debris transported by icebergs and sea ice can clarify the extent of ice transport, and the direction of oceanic currents (Keany 1976). It is now well known that particular clay minerals, such as chlorite, can originate principally from the weathering of rocks in the polar regions. The presence of such a clay may once again be indicative of the extent of ice-rafted material. Similarly, studies were made also of the distribution of diatoms which grow on sea ice in order to determine, from the distribution of the diatom frustule remains in cores, the extent of sea ice through time.

The Atlantic and its significance

To describe in detail events which occurred in individual oceans and at particular periods of time is beyond the scope of this chapter. Because of its proximity to the largest concentration of research institutes, the Atlantic Ocean has been more extensively studied than any other ocean. However, caution is needed in extrapolating from the Atlantic to the other oceans, since the Atlantic is a somewhat unusual ocean in several respects. It is fairly narrow and is bounded by large landmasses which set a particular pattern of ocean circulation. Atlantic water is usually saline in comparison with the other larger oceans. Deep water in the Atlantic is under the influence of cold water originating in the Norwegian and Greenland Seas to the north and in the Antarctic Seas to the south. The Pacific and Indian Oceans have no such deep cold water in their northern region. Nevertheless, this saline water which originates in the Atlantic may have a profound effect once it enters other oceans via the deeper waters. The Atlantic may have a significant effect on global climate because saline water can help generate more water vapour to the atmosphere. The Atlantic is a driving force for numerous other processes such as the control of deep cold water. There is still a great need to

determine the links which may have occurred between the
different oceans with respect to different water masses (for
nutrient levels, salinity and total dissolved CO_2 levels) before we
can establish the precise impact oceans have had upon climate on
land and upon the biosphere in general.

6 | Rivers, Lakes and Groundwater

*Suppose, now, that the Nile should change its course and flow
into this gulf – the Red Sea – what is to prevent it from being
silted up by the stream within, say, twenty thousand years?
Personally, I think even ten thousand would be enough.*

Herodotus (c. 485–425 BC),
The Histories. Book Two.
(Trans. Aubrey de Sélincourt, 1954).

Introduction

Rivers, lakes and groundwater are treated together in this chapter
for a number of reasons. Clearly, all represent forms of surface and
near-surface water, and must have experienced changed environmental controls as continental water balances shifted during the
Quaternary. Rivers, lakes and groundwater are, however, linked
in more direct physical ways. Where the water balance is normally positive (i.e., where a surplus of precipitation over the
amount lost to evaporation is available) rivers and lakes are often
fed by slow seepage from the groundwater stored in higher parts
of the landscape. Such seepage sustains river flow during rainless
periods and similarly helps maintain lake levels. In arid regions,
the reverse may be the case: river flow is lost through the porous
stream-bed into the underlying alluvium, where it recharges the
groundwater store. Elsewhere, lake behaviour is closely related to
the state of the regional groundwater store. For instance, near-
surface, saline groundwater may crop out at the surface in shallow
saline lakes which episodically dry out in the summer or during a
series of drier years. Such saline lakes affect the surrounding
vegetation and may be associated, for example, with the deflation
of salts from the dry lake bed onto the surrounding landscape. A
final reason which may be mentioned for linking the subjects of
this chapter is that, especially in low-gradient alluvial plains such
as the Riverine Plain of south-eastern Australia or the Gezira Plain
of the central Sudan, infilled river channels left by previous
episodes of river incision act in a number of ways to guide the
movement of groundwater in the sedimentary basin. Rivers, lakes
and groundwater then are merely the major manifestations of an
integrated system of pathways through which surface and subsurface water move through the landscape. An important aspect of
this network of water flow pathways is that it also constitutes a

major avenue through which dissolved rock materials, plant
nutrients and other solutes are transported. This aspect, however,
will not be pursued in detail here.

Rivers of the present day

Riverine environments globally constitute a vast, complex, and
incompletely understood set of landscapes. The range of simple
physical characteristics spans the permanently flowing 'perennial' streams, such as those of the tropical rainforests and other
humid areas, through 'intermittent' streams whose seasonally
varying flow relates to regular monsoonal rain or spring snow-
melt, to the shortlived 'ephemeral' streams of the desert areas
whose brief floods after local rain rapidly give way to dry, sandy
channels. In any of these environments, floods of widely varying
size may occur, resulting in potentially major changes in channel
size, in bank scour and collapse, and in alteration to floodplains,
bars, and islands. Across this range of environments, the stream
channel may adopt one of a number of recognizable basic forms,
including the 'braided' form which displays multiple small inter-
twining channels and which is characteristic of streams supplied
with large quantities of sandy and coarser debris. Streams carrying finer silty materials often display the smoothly sinuous
'meandering' form which is very characteristic of lowland floodplain sites. Streams in the high latitudes are greatly affected by
ice, which forces banks to deform during the winter; glacial
meltwater streams draining major mountain belts are fed enormous loads of sediment as moraine is dumped at the glacier snout,
and proceed in turn to set down great depths of alluvium, often
rapidly filling lowland valleys. Steep mountain streams carry very
coarse sediment particles, while those on the lowland alluvial
plains may be able to move only silts and clays. Sediment
particles must thus be set down, according to size, as streams flow
from upland areas onto lowland plains. Rivers of macrotidal
coasts may experience tidal influence 100 km inland, and may
undergo daily tidal depth fluctuations in their lower reaches of
5 m and more. Rivers of tectonically active areas in many cases
erode their courses downward as the terrain is elevated, leaving
suites of terraces and the remains of old stream beds high above
the modern channel; elsewhere, tectonic stability and low gradients result in streams whose rate of change is almost immeasur-
ably slow. In some areas, streamflow removes water to the oceans
('exorheic drainage'), while in others, such as the enormous Lake
Eyre basin of Australia, flow is directed toward an inland lake
('endorheic drainage'). Large areas exist which possess no surface
streams at all ('arheic') although this may not always have been
the case (Issawi 1983), while in others the 'drainage density' may

be very high. The course followed by streams may be determined largely by the underlying rocks and their structural arrangement, or it may be essentially random in nature. The course of many streams has nowadays been set deliberately by engineering works designed to straighten or alter their course or depth for purposes such as navigation and flood control.

All of these influences contribute to the vast diversity of riverine landscapes. However, there are certain environmental factors which exert a similar influence on all rivers and streams. These include the nature of the drainage basin which delivers water and sediment into the river, the external climate experienced by the drainage basin, and the tectonic stability of both the landscape which is drained by the stream and the lake or ocean level (the 'baselevel' of the drainage system) to which it drains. Major Quaternary sea-level fluctuations, for example, affected the baselevel of all exorheic streams worldwide, although the effect on an individual river would be moderated by other local factors such as the resistance of the rocks into which the stream was incising its valley, and whether the catchment area was ice-bound or escaped Quaternary glaciation.

Characteristics of some contemporary rivers

The principal characteristics of some modern river systems are listed in Table 6.1. This tabulation reveals the wide range of sizes, sediment loads and discharges which exists.

Table 6.1 Physical characteristics of selected major river systems

River	Location	Annual discharge (km^3/a)	Dissolved load ($t \times 10^6$) a^{-1}	Suspended load ($t \times 10^6$) a^{-1}	Mainstream length (km)	Basin area ($km^2 \times 10^6$)
Amazon	South America	6300	223	900	6300	7.18
Zaire (Congo)	Africa	1250	36	43	4700	3.82
Orinoco	South America	1100	39	210	2500	0.99
Yangtze	China	900	226	478	5000	1.94
Ganges– Brahmaputra	Asia	970	136	1670	2900	1.48
Nile	Africa	90	17	57	6700	3.349
Murray–Darling	Australia	12	2	6	3770	1.072
Fly	Papua New Guinea	190	—	115	1120	0.076

Australian rivers are relatively minor in most aspects by world standards, reflecting the overall dryness and flatness of the continent. Global average figures for river flow are always distorted by the characteristics of the Amazon, which is so large in flow volume that this single stream carries about 20 per cent of all the water flowing off the land surfaces of the globe. The characteristics of many Asian rivers (such as the Yellow River) reflect their peculiar environmental history combined with human use and modification of the catchment area.

Factors influencing river environments during the Quaternary

We will now consider the principal factors which must have affected riverine environments during the Quaternary. The factors considered are:

(i)   baselevel change and tectonic effects;
(ii) catchment water balance and erosional processes;
(iii) catchment fluvial (river channel) processes.

Baselevel change

For exorheic streams, a major series of disturbances during the Quaternary was caused by repeated shifts in baselevel as sea level rose and fell with the glacial–interglacial cycles. The magnitude of the fluctuation exceeded 120 m (see Chapter 4). During glacial periods, the course of rivers would have been lengthened by the distance across the continental shelves. In falling through the additional 120 m or so, significant additional erosional work would be possible, so that canyons were cut in many places through the sediments lying on the exposed shelves, such as the narrow gorge of the Mississippi River (called the Mississippi Trench) running across the shelf in the Gulf of Mexico (Bloom 1983). The disturbance would progressively have been felt through much of the inland drainage basin also, with all points along the course of the stream lying higher above baselevel when sea level was lower. Incision of channels was the common (but not invariable) consequence. Such incision leaves remnants of the original valley floor standing above the newly incised streams to form 'river terraces'. Many streams display multiple river terraces which trace an earlier path of the stream. Terraces may also, of course, be produced by river incision which results from tectonic uplift; this 'rejuvenates' the lower reaches of the stream which formerly were nearly at baselevel, and leads to an upstream progression of erosion. Terraces may be produced in other ways as well, as we shall shortly see.

Effects almost opposite in tendency are produced as sea level rises and the continental shelves are once more inundated during deglaciation. River canyons on the shelves are flooded, and the seas occupy the lower reaches of the stream valleys. Regions formerly well above baselevel then are located almost at sea level, and the ability of the stream to carry sediment is reduced. River-mouth regions would thus slowly fill with deposited sediment at times of high sea level (like the present day). These materials will be cut into during the next time of low sea level, with events of this kind repeating themselves every 100 ka or so during the upper Quaternary. The stratigraphy of the lower reaches of rivers must therefore be expected to show a confusing array of sediment bodies of varying age. Eroded materials will be laid down in a similarly complex sequence on the continental shelves and below the canyons on the shelf slope and rise.

A schematic diagram showing the formation of terraces produced by low sea-level stands is included in Figure 6.1.

Catchment water balance and erosional processes

During the Quaternary, as temperature and precipitation fluctuated, the water balance of stream catchments worldwide underwent repeated change. Undoubtedly, complex patterns of change were involved, with reduced evaporation combining with rainfalls regionally shifted by varying amounts to produce unevenly distributed water surpluses. Seasonal distribution of rainfall must have altered in some areas; lower temperatures also affected the vegetation cover, and hence the protection offered to the soil against erosion. Sediment washing into streams must thus have been altered both in quantity and kind, with perhaps more sand and coarser materials being stripped from relatively unprotected slopes during glacial phases.

Stream channels, which were adjusted to carry the sediment loads fed to them during interglacials, underwent changes in form and function as a consequence of the onset of cold conditions. These changes of river form and process are grouped under the title 'river metamorphosis' (see Schumm 1977). Streams carrying quantities of coarse debris probably became wider and shallower, with a somewhat less sinuous course being slowly developed. Streams which perhaps had a meandering form (and which had been carrying fine silts or clay materials) would become more like contemporary braided channels during glacial times, and undergo the reverse form change as interglacial conditions evolved. Where the river flow was strongly seasonal, the exposed bed sediments in some cases were blown by the wind ('deflated') to form sand dunes in areas adjacent to the channel. Once again, changes of these types happened repeatedly during the Quaternary so that we must expect the present landscape to be composed of a rather fragmented assemblage of deposits of varying age and character.

Those streams which drained the melting margins of the great Quaternary ice sheets must all have experienced significant and repeated changes in discharge and sediment load through the glacial–interglacial cycles. In particular, great volumes of sediment must have been contributed to them in the early stages of deglaciation. The Mississippi River system of North America

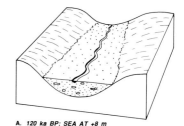

A. *120 ka BP: SEA AT +8 m*

B. *20 ka BP: SEA AT -150 m*

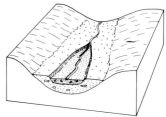

C. *TODAY: RIVERS NOW GRADED TO 0 m*

Fig. 6.1 Schematic representation of valley aggradation and subsequent incision and terrace formation which result from baselevel change

experienced such 'proglacial' conditions as the Laurentide ice retreated (see Chapter 3 for the chronology of this retreat). River sediments preserved in various areas show that during this phase, the central Mississippi became a large braided channel because of the load of sands and gravels fed to it (Baker 1983).

Catchment fluvial processes

The changed conditions over the catchment area produced by Quaternary environmental change had more involved consequences for river channels than those just mentioned. Imagine, for example, a catchment whose less protected slopes shed into nearby streams, in the early part of a glacial phase, much increased sediment loads as former soils and weathered materials were rapidly eroded. This increased sediment load would overtax the stream's carrying capacity, and some materials would be set down (the channels downstream would be forced to 'aggrade', or build up their beds). Valleys lower down in the drainage system might be 'alluviated' in this way to substantial depths as the excess materials were deposited. Eventually, however, the available weathered materials on the catchment slopes would be exhausted. In this circumstance, relatively sediment-free water would begin to be shed from the bare upland catchments and collect an appropriate load of material by scouring or incising the valley floors at lower elevations which were previously experiencing the aggradation just described. The overall landscape effect, then, would be terrace formation as small remnants of the aggraded valley floors were left standing on the valley sides above the newly incised streams. This terrace formation would happen some time after the climatic swing into glacial cold conditions, and would in fact occur at a time when no external or 'extrinsic' environmental change was taking place. Landscape changes of this sort, occurring at a time when there is no evident external trigger for the change, represent examples of what is termed 'complex response' in the riverine environment (Schumm and Parker 1973). This complex response (involving unknown time-lags as weathered materials are stripped from the uplands) makes the interpretation of river terraces and sediment bodies one of the most problematic areas of Quaternary environmental reconstruction.

Examples of Quaternary riverine environmental change

In order to illustrate some of the processes explained in general terms above, and to see how they are manifested in contrasting environments, we will examine the Quaternary record of a sample

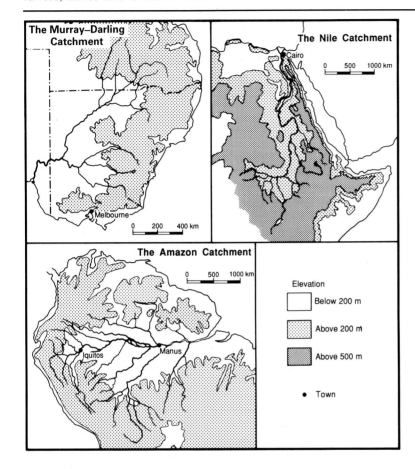

Fig. 6.2 Location and catchment relief of the Nile, Amazon, and Murray–Darling river systems

of drainage basins spanning the globe. River systems which provide suitable examples for our purposes are:

(i)   the Nile system in northern Africa;
(ii) the Amazon system of South America.
(iii) the Murray–Darling system of south-eastern Australia;

The locations of these river systems and their catchment areas is shown in Figure 6.2.

The Nile river system in northern Africa

The Nile river system is presently an exotic or 'exogenous' one for much of its course: that is, it flows through dry areas which would not normally be expected to display such streams. The Murray–Darling system in Australia, which we shall consider shortly (page 116), is also of this kind.

The Nile system consists of three major branches: the White Nile, which drains large lakes in the southern headwater areas of Uganda; the Blue Nile, which drains the high northern and central plateau areas of Ethiopia; and the Atbara, which also drains northerly areas of Ethiopia. The Blue Nile and the Atbara supply the bulk of the sediment carried by the lower Nile, but the White Nile is critically important in sustaining the Nile flow during the dry winter of Ethiopia. From the headwaters of the White Nile just south of the equator, the system flows nearly 7000 km north to its delta in the eastern Mediterranean, crossing 35° of latitude (Adamson *et al.* 1980). Clearly, therefore, the Quaternary fluctuations in the behaviour of this river must have related to the altered environments of areas whose climates ranged widely (presently, equatorial to hyperarid). River metamorphosis of the middle and lower Nile in Egypt can be expected to relate to the altered environments of the distant Ethiopian uplands, in particular, as these supply much of the water and sediment load of the present-day river, and thus probably also its Quaternary predecessors.

In fact, the middle and lower Nile display just such a complex record of river sediments as we might expect. Sediments covering parts of the modern landscape, or partly filling valleys in the system, have been much studied for the information that they contain relating to the development of human settlement along the Nile, and dated using isotope techniques and thermoluminescence (Schild and Wendorf 1989).

Two major periods of alluviation along the Nile are recognized from the late Pleistocene, but more are likely to have occurred and remain unidentified because of burial or removal of the sedimentary evidence by later erosion. The episodes of alluviation are taken to reflect altered conditions in the Ethiopian uplands during Quaternary cold phases. During these times, the highest areas were transformed from alpine grasslands to bare soil and rock with ground ice and periglacial processes; at lower elevations, tree and shrub cover were replaced by grassland. The treeline was probably lowered by 1000 m during the coldest times (Adamson *et al.* 1980). In association with the colder conditions, rainfall was also lower than that of the present, but still delivered in a fairly intense wet season. Smaller volumes of water would have been fed into the rivers, but with a larger sediment load (and a coarser one also) because of the reduced ground cover. A lowering of the stream competence or carrying capacity in this way with a simultaneous increase in the load of a sediment fed from the catchment area not surprisingly resulted in alluviation along the Nile.

In the best dated of the periods of late Pleistocene alluviation, spanning the period 20–12.5 ka BP (and thus coinciding partly with the coldest part of the last glacial), the sediments along the Nile reveal that the river underwent metamorphosis to adopt essentially the braided form, with a sandy floodplain occupied by a water flow much smaller than that of the modern river, perhaps only 10–20 per cent as large. Continued deposition of sediments scoured from the upland catchment built up the floodplain progressively during the period of alluviation. The distant Mediterranean would have been about 130 m lower but would have had no influence on river incision or deposition upstream of the most northerly cataract. Outside the Nile valley, the environment was hyperarid during the cold conditions, and the valley became an especially important habitat for people and animals (Butzer and Hansen 1968; Clark 1980). Even within the valley, the dry conditions permitted extensive dune development and it must still have been a relatively harsh environment.

As the climate warmed in the period 15–5 ka BP, the vegetation in the upland catchments recovered, and despite the increased rainfall associated with the warming, loss of sediments was reduced, and the eroded materials were restricted to finer grain sizes. Deposits of these materials left along the Nile by floods reflect a metamorphosis to a more typical suspended-load stream, with higher sinuosity.

This picture of the late Quaternary behaviour of the Nile is supported by other essentially independent methods of environmental reconstruction. For example, during the glacial cold conditions, the headwater lakes which feed the White Nile were much shallower, and did not overflow as they presently do (Livingstone 1980). The White Nile itself may have been blocked by dune sands. In this situation, the lower Nile would not only have a much lower discharge; the sediments it carried would no longer reflect the geology of the White Nile source areas. In interglacial conditions, with the White Nile flowing, the composition of the materials carried by the lower Nile would shift accordingly.

Analysis of materials drilled from the Nile delta (Foucault and Stanley 1989) shows a composition of heavy materials (used as fingerprints of the geology of the Blue and White Nile catchments) which fluctuates through the last 40 ka BP. In the period 20–10 ka BP the mineral assemblage is consistent with the major mineral source area being the Ethiopian uplands; following this, materials consistent with an increased White Nile flow appear. The White Nile also appears to have been active in the period 40–20 ka BP. These dates are in good accord with those based on fossil materials within the Nile valley itself, and with the model of Nile behaviour already outlined.

A final confirmation of the scenario comes from the sediments of the eastern Mediterranean. Drill cores of sediment here display layers of black mud rich in organic matter, called 'sapropel'. These layers reflect conditions in which the ocean becomes layered or 'stratified', so that vertical mixing is restricted and organic materials do not oxidize. Two sapropel layers are found in the eastern Mediterranean; their accumulation is placed in the intervals 11.8–10.4 and 9–8 ka BP (Rossignol-Strick *et al.* 1982). The formation of these sapropels may be related to the occurrence of major flooding along the Nile as the climate warmed after the last glacial phase. Such floods would deliver large volumes of fresh water which, being less dense than sea water, floats as an upper layer and so stratifies the ocean. The floods in the periods when these sapropels were formed would have been very large indeed; the rainfall at the time was higher than that of the present, and there may have been great releases of water as the lakes in the White Nile headwaters began to overflow once more. Indeed, the high rainfalls may have been a phenomenon of the equatorial areas especially, so that it is only the extraordinary Nile, crossing such a wide latitudinal belt, which carried the influence of these rains as Nile floods all the way to the distant Mediterranean. It has been calculated that the Nile, in the late glacial phase when these floods occurred, may have carried a discharge 250 per cent larger than that of today, and could have delivered in 15 years sufficient fresh water to cover the whole Mediterranean to a depth of 25 m (Rossignol-Strick *et al.* 1982).

A brief period of aridity over equatorial Africa interrupted the period of flooding and permitted the development of two separate sapropel layers.

In summary, the late Quaternary history of the Nile river system includes a dry phase at 20–12.5 ka BP associated with cold conditions, rapid erosion of the Ethiopian uplands, and valley alluviation, and a terminal Pleistocene to mid-Holocene moister interval at 12.5–5 ka BP associated with stabilization of upland slopes and river metamorphosis to the large single channel associated with the modern Nile (Adamson *et al.* 1980). Since about 5 ka BP, conditions have become increasingly arid, and today no water flows into the Nile at all during its 1000 km path through Egypt. There is of course a more extensive history of valley incision and alluviation related to tectonic events and to the 3000 m drop in the level of the Mediterranean during the Messinian salinity crisis, but the influence of these more remote events on the Nile is at present less well documented.

The Amazon river system

The Amazon river system of South America drains an enormous catchment area spanning Peru, Guiana and Brazil, which amounts to more than 7×10^6 km². The river system runs approximately west–east just south of the equator, spanning 33° of longitude but only 12° of latitude, draining the Peruvian mountains and flowing into the Atlantic Ocean. Because all of this area lies within the tropics, and is situated so as the receive high rainfalls averaging 2300 mm per year, the Amazon carries an astonishing volume of water: the mean discharge is 175 000 m³/s (Sioli 1984). This amounts to nearly 20 per cent of the entire flow of fresh water carried by the combined rivers of the world. The Amazon is also exceeded in total length only by the Nile system, and through much of its middle and lower sections runs on a very low gradient. Despite this, rapid erosion in its rugged Andean headwaters combined with the enormous water flow result in the Amazon carrying the third largest tonnage of suspended sediments globally (exceeded only by the Huang Ho river and the Ganges–Brahmaputra). It is estimated that the Amazon carries 9×10^8 tonnes of suspended sediment per year (Sioli 1984).

The history and Quaternary behaviour of this enormous river system are not well known. Fractures or depressions in the underlying Precambrian crystalline basement (perhaps related to stress patterns associated with the break-up of Africa and South America) are the fundamental control on the orientation of the Amazon valley. This depression contains sediments varying widely in age, including large deposits of Tertiary materials.

Some sediments from the Amazon are carried out across the continental shelf at the Atlantic seaboard and spill down the continental slope to form a large sediment accumulation known as the Amazon cone. Sampling of the materials composing the cone has revealed that it has only been accumulating there for a few million years, since about the end of the Miocene (Damuth and Kumar 1975). This is interpreted to mean that prior to this, at least part of the present Amazon basin drained westward into the Pacific. Subsequent uplift of the Andean mountains must have diverted the system into its present valley, guided by the underlying structural features.

The present-day Amazon is associated with a wide variety of environments, which include the low-lying forest areas which are flooded annually, known as the *'várzea'*. Higher areas which lie above flood level are known as *'terra firme'*. In the *várzea* areas, the Amazon and its tributaries meander in complex loops through young sediments, and are flanked in many areas by flights of terraces ranging up to 90 m or so above river level. Even higher level surfaces, up to 180 m above river level, are formed in eroded Tertiary materials (Klammer 1984).

The exact origin of the *várzea* materials remains to be discovered. Much of this material must be Holocene in age, laid down during and since the Flandrian marine transgression. During periods of glacial low sea level, the Amazon must have incised its course, cutting a channel out across the continental shelf. The rate at which this took place may, however, have been exceptionally slow in the case of the Amazon because of its enormous sediment load. Postglacial transgression would have inundated the deep valley system, causing aggradation to begin. The enormous volume of the system again probably means that infilling would be very slow to complete, and there is evidence that inland, perhaps 1000 km from the river mouth, infilling of the pre-Holocene valley is still actively proceeding. Tributaries in some sub-catchments of the Amazon which lack high ground carry only very small suspended sediment loads (these are termed the 'clear water' rivers in contrast to the turbid, sediment-laden 'white water' rivers like the main Amazon itself) (Irion 1984). Even slower valley aggradation could be anticipated in these systems, and many show very clear flights of river terraces.

The terraces of the Amazon system indeed pose unanswered questions (Bigarella and Ferreira 1985). It may be that they are related to baselevel fluctuations during the Quaternary glacial–interglacial alternation. This would be particularly probable because of the very low elevation of so much of the Amazon catchment, which lies less than 100 m above sea level over vast areas. At the mouth of the Rio Negro, 1500 km from the sea, low water level is only 15 m above sea level; even at Iquitos, low-water level is only about 100 m above sea level. The gradient of the valley over great distances could thus have been doubled when sea level stood 100 m lower than today, because the length of the river would only change by the width of the continental shelf, resulting in considerable impetus to incision. However, the formation of terrace systems in such an enormous river basin requires that there be substantial time for migration of the effects from the coast inland, and it may be that the repeated baselevel swings occurred too rapidly for this process to be completed.

It is also possible to imagine that the Amazon terraces relate to climate change affecting the catchment in just the way already described for the Nile, with drier conditions causing overloading of rivers with sediment, and forcing aggradation downstream. A subsequent return to wetter conditions could then promote renewed incision and terrace development. There is some palynological evidence suggesting that drier conditions have periodically affected parts of the Amazon basin (Colinvaux 1989), and Damuth and Fairbridge (1970), who studied the sedimentary record of the Amazon cone and adjacent areas, conclude that the mineralogy of the sediments is consistent with the existence of arid to semi-arid climates over much of equatorial South America during the last glacial. However, the true relative importance of baselevel fluctuation and catchment climate change remain to be resolved in the case of the Amazon, as does the part played by tectonic activity, especially uplift.

The Murray–Murrumbidgee–Darling river system of south-eastern Australia

The rivers which flow westwards from the uplands of south-eastern Australia (such as the Murray, Murrumbidgee and Darling) descend onto a vast alluvial plain which is partly composed of sediments laid down by marine sedimentation during a major transgression into the Murray basin during the Tertiary. Stream gradients are extremely low, partly because the plains are flat and additionally because these rivers have sinuous, meandering courses which require the water to travel much more than the direct distance to the sea.

During the Quaternary, the course of these rivers changed repeatedly as bank erosion occurred or as flooding resulted in new channels being cut on the floodplains. The plains now preserve evidence of these former channels as infilled 'palaeochannels' which are not easy to see at ground level but which are visible in aerial photographs. Some of the palaeochannels relate to glacial times, when rainfall and erosion processes in the highlands were different to those of today, and others to interglacial times, which might have been not unlike the present.

In fact, three different major kinds of channel are visible on these plains, which are collectively known as the 'Riverine Plain' (Bowler 1978). The present-day channels represent one kind. They are generally very sinuous, and carry dominantly fine suspended-sediment loads. In cross-section, the channels are broadly dished, and in the case of the Murray, for example, about 6 m deep in the middle section of the river course.

Palaeochannels of two distinct forms can also be identified (Schumm 1977). These include a set of channel remnants which resemble the present river (i.e., having a highly sinuous meandering course) but which are much larger; these often form large oxbow lakes on the floodplain of the present river. These old channels were about twice as wide as the modern one, as well as deeper, and their meanders were correspondingly larger in size. They must have carried a considerably greater discharge, and been able to shift more and coarser sediment as well because of the higher flow velocities which must have been generated in them.

A second variety of palaeochannel of very different form can also be identified. These display a much less sinuous channel, and hence flowed on a steeper gradient (moving more directly down the slope of the alluvial plains). On the basis of the channel remnants, they must have been nearly three times as wide as the modern rivers, but much shallower. From this and the preserved channel sediments it can be concluded that these were essentially 'bedload' channels, moving vast quantities of sand.

The palaeochannels similar to, but larger than, the modern rivers must relate to times when the catchment shed more water than at present. Their form is essentially that of the present channels, but scaled-up. The absence of coarse sediments in these channels suggests that the highlands, as the source area of the bulk of both the water and the sediment load, must have been well protected by vegetation as a result of humid conditions.

In contrast, the wide low-sinuosity palaeochannels are likely to be the product of a drier environment. Under such conditions, the highlands would have been less well protected by plant cover, and storms would have been able to generate both large floods (from the barer surface) and strip the larger quantities of coarser sand which now fill these channels. The straighter course must have evolved in response to the greater sediment load, and provides another clear example of river metamorphosis resulting from environmental change. The change from a wide, shallow, sand-transporting stream to a narrower, more sinuous, suspended-load stream as the climate moderated provides an important example of a constraint on metamorphosis. Because of the great length of the river system, incision or aggradation as a consequence of a change in incoming sediment load becomes a less probable outcome for the stream than does gradient change. In these systems gradient was apparently adjusted by a change in planform sinuosity. Thus, halving the sinuosity doubles the channel gradient, and vice versa. River metamorphosis must thus have been associated with dramatic bank erosion but relative stability of bed elevation. In the case of shorter rivers, the same catchment change might well be accommodated more completely by bed scour or aggradation and less by planform change. The particular kind of metamorphosis experienced will therefore relate to aspects of the catchment geomorphology, as well as to the nature of the external environmental change itself. Once again we see, therefore, that the morphologic response of rivers to environmental change can be complex indeed. The link between the form and sediments of palaeochannels and the regional environment at the time they were active is an indirect one, modified by parameters which we may as yet not fully understand.

Metamorphosis is revealed in the alluvial history of other river systems in Australia, many of which remain to be dated and explored in detail, and similar large alluvial plains elsewhere. For example, the multiple anastomosing channels of the Cooper Creek system in south-western Queensland presently carry dominantly fine suspended sediments and deposit very fine alluvium as floods wane. Underlying the present channels, however, is a system of sand-rich, sinuous, meandering palaeochannels (Rust and Nanson 1986). The age of these is not known with any certainty, but it has been suggested that repeated change from sandy to muddy channel systems through the Quaternary may reflect dominantly the effects of climatic change.

River metamorphosis and climatic change

The actual fluvial processes involved in the meandering-to-anastomosing channel metamorphosis remain to be unravelled.

The Gulf coastal plain in North America shows sets of low-sinuosity, gravel-transporting channels as well as high-sinuosity channels which carried sands and silts. The metamorphosis reflected in these different channel forms is taken to reflect the change in sediment and water loads resulting from the change from relatively arid to more humid conditions of the late Quaternary (Baker 1983). A history of changing fluvial environments has also been described for the Son River in north-central India (Williams and Royce 1982; Williams and Clarke 1984). Here again the sedimentary record, with a chronology based upon ^{14}C dating of shells, charcoal and carbonate, shows a relatively wide, seasonal, and low-sinuosity channel in the last glacial transformed to a narrower, less seasonal and more sinuous channel, carrying and depositing a finer sediment load. In response to the wetter climate of the Holocene, the Son has incised its Pleistocene floodplain to a depth of about 30 m. Once again, the transformation of the river can be interpreted in terms of the change from a sparsely vegetated watershed during the cold and dry glacial conditions, to a better vegetated one which consequently sheds reduced amounts of sediment during the milder and wetter Holocene. An understanding of the mechanisms behind such transformations can usefully be applied to the development of forecasts of the effects of future environmental changes (see Chapter 11; Knox 1984).

Evaluating river hydrology from palaeochannel evidence

A most attractive possibility is raised by the preservation of Quaternary channels, terraces, and riverine sediments. This is to attempt to reconstruct, on the basis of these kinds of evidence, the former flow conditions of the rivers and hence to make inferences about the environments of their drainage basins in a quantitative way. Certainly, it is possible to infer from preserved river sediments something of the speed of the flowing water responsible for carrying them. If suitable cross-sections of the old channels can be measured, it is then possible in principle to estimate the volumetric water flow that formerly existed. In a similar way, by using relationships established between channel geometry (say, the form of meander bends) and modern discharge, and extrapolating these to palaeochannels, it is possible to say something of the former flow conditions. A range of methods for this kind of study is reviewed by Williams (1984). However, the relationships involved in such palaeohydrologic inference are complex and only incompletely understood. The complex response of river systems outlined earlier makes the interpretation of river terraces equally involved. As our knowledge of river behaviour develops, it will be possible to make further use of the evidence of palaeochannels. At present, however, the difficulties in making firm quantitative inferences from them confound most attempts. Lakes, however, offer more readily available sources of data on the environmental history of their surroundings, and we turn now to consider some of these.

Lake morphology and origin

Lakes and other non-flowing waterbodies owe their origin to a great variety of geological circumstances. Some of the largest lakes in the world occur in depressions left in the landscape as a result of glacial erosion. Commonly, these lakes are deep (with depths of more than 100 m) and are found in glacial valleys in mountainous regions. Numerous lakes in the Alps belong to this category. Other large lakes are relict depressions left in the landscape by a retreating ice sheet, such as the Laurentide ice sheet in North America (see Chapter 3 for more details). The Great Lakes are the best example. The sedimentary record in those lakes formed by glacial ice begins from the time of the most recent retreat of ice from the lake basin. Hence, most of these lakes have a record of sedimentation that spans fewer than 18 000 years. More discussion on the record of these lakes will be given later in this chapter.

There are several other types of large lakes which owe their origin to phenomena other than those associated with glaciation. Principally, those lakes occur in large and sometimes blocked depressions such as Lake Eyre, which occurs in the lowest portion of the Australian continent; Qinghai Lake, which again occurs in a vast depression on the plateau in northern China behind the Himalayas; the Aral Sea which occupies an extensive sedimentary basin once connected to a large seaway that became closed as a result of tectonic activity; or in large tectonic depressions like the very long and deep (over 1000 m) Lake Baikal in the USSR and Tanganyika in East Africa. Because these latter two depressions in which the lakes occur were not formed as a result of glacial activity, the sedimentary record in the lakes extends much further back in time than that of the glacial lakes. In fact, the sedimentary sequences below Baikal and Tanganyika extend to several kilometres, as recently recognized from seismic profiles run over both lakes.

Among other types of lakes, it is necessary to distinguish those that receive most of their water from surface drainage and directly from rainfall, from those that are principally fed by groundwater. The former type encompasses crater lakes which are perhaps the best type of lake for the reconstruction of climatic records. These lakes can either be formed as a result of the impact of extraterrestrial material, for example, meteorite impact, or as the result of volcanic activity, either by a volcanic eruption or the result of an explosion below ground caused by basalt coming in contact with an aquifer or by a pocket of gas (e.g., CO_2 originating from the mantle) expanding on its way to the surface. In both cases, a crater is formed at the surface. Under normal circumstances, crater lakes act like gigantic rain gauges because most of the water that enters the lakes results from precipitation directly over the lakes. There is usually no river and little groundwater input into those lakes.

The second type of lake is usually referred to as a 'groundwater window' because it is most often found in arid and semi-arid regions where evaporation is greater than precipitation, and this causes the groundwater to rise and then to seep into the lake if the former is near the landscape surface (see Chapter 7). In fact, the level of these lakes can also be a good indicator of changes that occur on a regional basis, reflecting for example the amount of water entering the lake's catchment. A great proportion of the groundwater-window lakes yields saline water simply because of the excess of evaporation in the area, and the solutes (soluble mineral components) that are picked up and transported by the water within the sediment interstices before emerging above the floor of the lake.

Other small lakes, sometimes with a long record of deposition, are karstic lakes which occur in regions where dissolution of the local lithology (e.g., limestone, gypsum, laterite) operates substantially. Although these lakes are usually characterized by a small area of catchment compared to the others mentioned above, there is a frequent occurrence that the lakes' records are disturbed by further dissolution of the lake margins, thus causing slumping of the sedimentary sequences along the margins. Similarly, some lakes that are found in shallow (under 5 m) depressions behind dunes along the coastline, or behind aeolian deposits deflated from lake floors, characteristically have a small catchment and also a short-lived record due to the frequent migration of coastlines and associated dunes. This is in contrast with the other types of lakes with a large catchment, which have the potential to inform on hydrological, and thus climatic, regimes for usually extensive regions, and which provide information of greater relevance on a continental scale. Nevertheless, only major events are recorded accurately in large lake basins whereas minor events/changes are usually more faithfully recorded in the small lakes which readily respond to (and thus register) small hydrological changes.

There are also less significant lacustrine systems such as small pans, oxbow lakes (occurring in cut-off meanders of rivers), and even recently established artificial lakes, dams and farm dams which usually yield a short record of sedimentation, but which are of relevance to changes occurring nearby and commonly resulting from anthropogenic interaction with the environment (especially for dams). Nevertheless, some of these small aquatic systems, and their deposits, can provide relevant information about short-term climatic change and provide a high-resolution record of change. High-resolution implies that it is possible to decipher a record of change from the sedimentary record of a waterbody that represents short-spaced events, such as on a seasonal level or even less, compared to the other records recognized in lakes or the oceans which cannot, under normal circumstances, provide records spanning episodes of less than centuries.

Associated features

In order to reconstruct the history of lakes it is necessary to obtain indications of lake-level changes. The latter is of importance because a change in hydrology in a particular region, often best represented by an alteration of the ratio of precipitation to evaporation, will cause the lake volume to change, sometimes dramatically (Forester 1987). For example, a lake may have retained water for centuries (and hence be called a 'permanent' lake) and change status to become an 'ephemeral' lake, commonly dry but filling up occasionally, as a result of a change in the climatic/hydrologic regime. However, before discussing geomorphological features indicative of lake-level change, it is necessary to point out that lake water budgets also cause water chemistry and aquatic biota to change too. These features are discussed later in this chapter because they occupy an important place in the reconstruction of palaeoenvironments.

The best way to describe the various features characteristic of different lake phases is to examine a full, deep lake with permanent water which progressively changes to a shallow, ephemeral lake as a result of a change of hydrological regime (De Deckker 1988). Emphasis is placed on features which are distinctive of the different lake phases, but which can also frequently be recognized in ancient lake deposits so as to permit reconstruction of lake histories (Figure 6.3).

Fig. 6.3 Geomorphological, sedimentological and chemical features associated with different types of water budget affecting a lake. The diagram is arranged to distinguish the characteristics of a large lake during a wet period and evolving progressively to become a dry playa partly covered by a salt crust. Note that in this particular system where a lake ends up with NaCl-dominant water, a change in sediment mineralogy is also registered with a dominance of gypsum ($CaSO_4$) crystals, and finally with a halite (NaCl) crust. Note also that the deep-lake phase (called '*megalake*') is characterized by a stratified water column that does not mix for at least substantial periods of time, and that leads to the preservation of laminated sediments because of the absence (due to the lack of oxygen) of burrowing organisms that would disturb the layering (after De Deckker 1988)

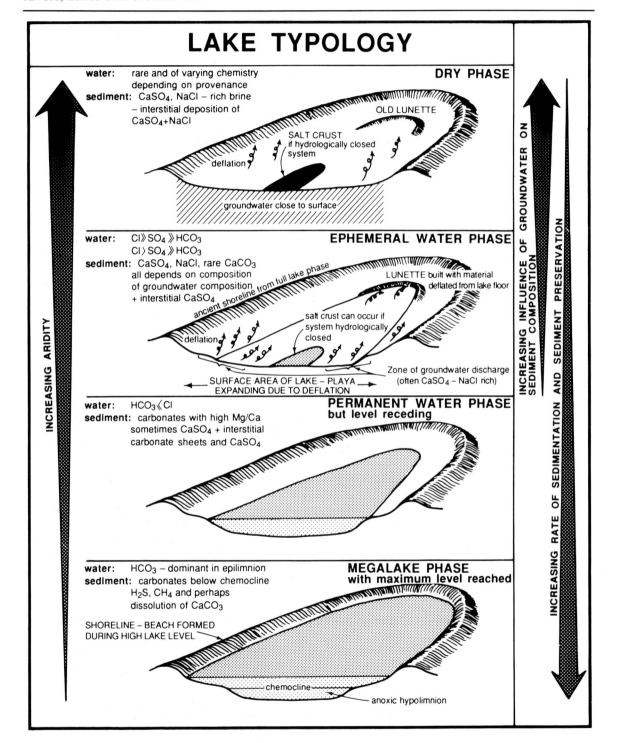

LAKE TYPOLOGY

DRY PHASE

water: rare and of varying chemistry depending on provenance
sediment: $CaSO_4$, $NaCl$ – rich brine – interstitial deposition of $CaSO_4 + NaCl$

OLD LUNETTE

SALT CRUST
if hydrologically closed system

deflation

groundwater close to surface

EPHEMERAL WATER PHASE

water: $Cl \gg SO_4 \gg HCO_3$
$Cl > SO_4 \gg HCO_3$
sediment: $CaSO_4$, $NaCl$, rare $CaCO_3$ all depends on composition of groundwater composition + interstitial $CaSO_4$

ancient shoreline from full lake phase

LUNETTE built with material deflated from lake floor

salt crust can occur if system hydrologically closed

deflation

Zone of groundwater discharge (often $CaSO_4$ – $NaCl$ rich)

← SURFACE AREA OF LAKE – PLAYA →
EXPANDING DUE TO DEFLATION

PERMANENT WATER PHASE
but level receding

water: $HCO_3 \ll Cl$
sediment: carbonates with high Mg/Ca sometimes $CaSO_4$ + interstitial carbonate sheets and $CaSO_4$

MEGALAKE PHASE
with maximum level reached

water: HCO_3 – dominant in epilimnion
sediment: carbonates below chemocline H_2S, CH_4 and perhaps dissolution of $CaCO_3$

SHORELINE – BEACH FORMED DURING HIGH LAKE LEVEL

chemocline

anoxic hypolimnion

INCREASING ARIDITY

INCREASING INFLUENCE OF GROUNDWATER ON SEDIMENT COMPOSITION

INCREASING RATE OF SEDIMENTATION AND SEDIMENT PRESERVATION

The large, deep lakes are characterized by a substantial, deep water column which has an important effect on the occurrence of the biota in the lake and the preservation of their skeletal remains on the lake floor, as well as the composition and preservation of the inorganic sediment. Several of the large lakes are so deep that the amount of dissolved oxygen is so low that common invertebrates and vertebrate organisms cannot live below a certain water depth. The consequence of this 'de-oxygenation' of the water is that a layering of the water column occurs, thus causing different layers of the water to become stratified. Those lakes become stratified as a result of chemical changes (e.g., changes in oxygen, pH and compositional levels) or physical changes (e.g., changes in density and temperature with depth [note that these two features interact]); thus, the bottoms of these lakes are devoid of organisms which otherwise would disturb the sedimentary record by burrowing and other forms of bioturbation. Consequently, lakes which are stratified retain an undisturbed sedimentary record, which may sometimes be laminated. The latter consists of a thinly layered sequence caused by colour or compositional change in the sediment that usually results from alternation of biological remains (e.g., siliceous skeletons of algae) falling to the bottom of the lake. An algal bloom at the lake's surface can in turn engender a change in the chemical composition of the surface water which may force the precipitation of some inorganic minerals (the commonest being tiny crystals of calcium carbonate such as calcite and aragonite, typically a few μm long) which subsequently settle on the lake floor above the remains of siliceous algae, forming a layered sedimentary sequence. Frequently, such a laminated sequence is characterized by an alternation of pale and dark layers and it has the potential to provide a high-resolution record if the laminations can be assigned to particular phenomena such as seasonal/annual changes.

The shorelines of large lakes (and not necessarily deep ones) are characterized by beach deposits (Figure 6.3) consisting of coarse material that frequently shows signs of reworking. During a phase of lake-level recession (regression), material previously reworked and deposited at a shoreline will remain as part of the landscape, and thus leave evidence of a former lake level. On the other hand, during a lake-level rise (transgression) across previous levels, former shorelines/beach deposits will be eroded away, and thus a loss of previous records occurs. Commonly, remains of aquatic organisms mixed with those of terrestrial material (e.g., plants) are found in shoreline deposits.

As water level recedes, the dissolved salts in the water become concentrated, and a chemical evolution of the water ensues (see Figure 6.3). Depending on the nature of the rocks surrounding the lake and in the catchment, the rocks' weathered components which are soluble end up in the lake and their variety will control the chemical composition of the lake water. Of the two principal types which are commonly found in lakes, one at the earliest stages of solute concentration (in other words as salinity progressively increases) has bicarbonate as the dominant ion, whereas in the other type of water, the bivalent cations, calcium and/or magnesium, are dominant. The progressive evaporation of the lake water will force some precipitation of minerals such as the ubiquitous calcium carbonate which is also one of the least soluble. Calcite, which is a common form of $CaCO_3$, will frequently precipitate first at the expense of the minor components, forcing a further dominance of others. In other words, once calcite has precipitated (this is called the 'calcite branch point'), the chemistry of the water will follow one of two pathways: one evolves from a water enriched in bicarbonate to the detriment of the bivalent cations (Mg and Ca), whereas the other is bicarbonate depleted and enriched in either, or both, of the above-mentioned cations. The bicarbonate-rich waters will see a further enrichment, and finally precipitation of carbonate minerals, with an increase in salinity compared to the other type of chemical pathway which eventually will register the precipitation of gypsum ($CaSO_4 \cdot 2H_2O$ [labelled only as $CaSO_4$ for simplicity in Figure 6.3]), and eventually halite (NaCl, which is the same composition as table salt) with a progressive salinity increase. (This second type of chemical pathway has a dominance of some ions in similar proportions to that of oceanic water, especially if it were to be evaporated.) It is important to distinguish between the two types of chemical pathways, water chemistries and salinities as they both have a significant effect on the presence and absence of aquatic organisms. By reversing roles, the remains of aquatic organisms recovered for the sediments of a lake can yield information on the water quality of the lake, and a relationship between water quality and salinity can be used to reconstruct climate. For example, the evaporation/precipitation ratio can be estimated depending on salinity levels if there is good control of salt budgets, and also the amount and source of weathering (thus leading to discovery of a change in lake catchment) can be postulated from a change in water chemistry.

Once a lake level has receded sufficiently to uncover a large portion of the lake floor, deflation of surficial material (namely, sediments as well as remains of organisms) can occur during dry and windy periods (see Chapter 7). Hence, this deflated material may be transported for long distances, and even be finally deposited in the oceans some thousands of kilometres away. A large portion of this deflated material eventually accumulates to form dune deposits. The best example of this phenomenon is represented by the crescent-shaped dunes formed downwind of lakes. With a further lake-level drop, salinity continues to increase and diversity of aquatic life (plants and animals) decreases, and the final step will generally see a salt crust forming on the lake floor. This salt crust usually prevents further deflation of lake-floor material. It is during low-level lake phases or when a lake is dry that several of the sedimentary features that originated during previous high-level lake phases are completely or partially destroyed (through deflation, bioturbation or non-biological, mechanical processes such as the formation of mud cracks), thus blurring the record of lake history.

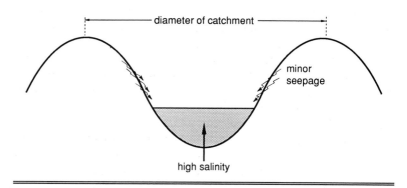

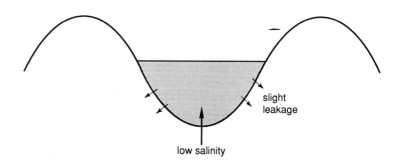

Fig. 6.4 Schematic diagrams showing characteristic features associated with crater lakes under two specific hydrological regimes.
Top: When evaporation is greater than precipitation, the water level remains low and solute concentration is high assuming that most of the water in the lake originates mostly from within the lake's small catchment, as defined by the periphery of the crater rim.
Bottom: Represents the situation where the evaporation to precipitation ratio is reversed, the water level is then high and solute concentration low. Consequently, crater lakes can be considered as gigantic rain gauges, and thus either lake-level or salinity changes can be used to monitor evaporation–precipitation changes

Lake status

Lake status, as previously defined, provides the best possible indicator of the climatic history of a region. Nevertheless, it is also necessary to recognize that large lakes with a large catchment have the potential to provide information on the evaporation to precipitation ratio (E/P) over a very large area. In fact, a slight change of this E/P ratio may not necessarily be recognized easily by features such as shoreline or water salinity changes. On the other hand, it is principally the dramatic changes in lake levels that are to be recognized, and to be correlated with a change in the hydrological regime affecting either the entire region, or at least part of it.

It is the small lakes, with a small catchment, that provide a more accurate account of slight changes of the E/P budget over the lake's catchment (Figure 6.4). The disadvantage of these small lakes, on the other hand, is that they do not usually exist for a long period of the geological time scale, and often if there is a dramatic change in the E/P ratio, such as that due to an excess of water, the lake may overflow and a loss of information relevant to the climatic change may be lost if solutes are flushed out of the catchment. Similarly, no shoreline features that would accurately record the extent of the lake could form. An excess of evaporation would cause the lake to dry and loss of sediment caused by deflation would again suppress some of the information of value for a reconstruction of the lake history.

Aquatic organisms

Aquatic organisms are numerous and diverse. They inhabit almost any type of aquatic environment, from a shallow ephemeral pool to a hypersaline lake (Figure 6.5). Organisms also can survive drought by burrowing in the mud, or survive a long desiccation phase by producing drought-resistant eggs. Other organisms only survive and reproduce in permanent lakes. The recognition of species which have very specific ecological requirements enables the palaeoecologist to use the fossil remains of organisms for the reconstruction of ecological changes in lakes, and thus infer physico-chemical changes that directly translate to hydrological and climatic changes (Frey 1969; Meriläinen *et al.* 1983; Löffler 1987; Gray 1988). It is also important to recognize that numerous species of aquatic organisms are to be found in very specific water chemistries, and that it is therefore possible from the identification of fossil remains to reconstruct changes through time in chemical pathways in a lacustrine system. The latter can give information on the particular types of water chemistry in which the organisms lived.

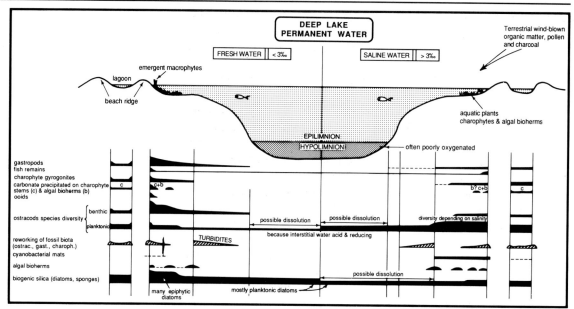

Fig. 6.5 Schematic diagram showing some of the organisms most commonly used to reconstruct the history of lakes. The profile shows a deep lake in the centre and shallow ones on either side. Two types of lakes are recognized, one fresh and the other saline. Note that most organisms can be found in both fresh and saline ecosystems, but diversity would naturally decrease with a salinity increase. Some organisms like gastropod molluscs are restricted to shallow portions of a lake as they are frequently herbivores and feed on aquatic plants (emergent macrophytes, some of which produce pollen and seeds), and algae fixed to the lake floor (e.g., charophytes which secrete calcareous egg cases called gyrogonites) that require light to grow. Ostracods are calcareous, bivalved microcrustaceans that are ubiquitous aquatic organisms with well-defined ecological requirements. Algal mats, made principally of blue-green algae (cyanobacteria) forming small mats, but occasionally large mounds (bioherms that occasionally are lithified) grow in the lakes at shallow depths. Siliceous remains of algae (principally diatoms and chrysophytes) and sponges can also be used to reconstruct the history of lakes since they are excellent ecological indicators. Diatoms especially have very particular salinity, chemical and nutrient requirements, and thus provide information on water quality. Care should be taken when assessing the lake's record from fossil remains as some may be reworked at shorelines or along steep flanks by turbidity flows, or may be partly dissolved by physico-chemical processes after death of the organisms (after De Deckker 1988)

The most commonly available organisms which enable us to reconstruct water chemistry are diatoms (algae with a siliceous skeleton), and invertebrates such as ostracods (microscopic crustaceans related to shrimps), gastropods (snails) and bivalve molluscs which all have a calcareous shell made of $CaCO_3$. All of these organisms are commonly found in fresh and saline lakes, and other physico-chemical parameters control or influence their presence and abundance in lakes. Some of the important parameters are: water temperature (which principally controls hatching and reproduction), pH, salinity, dissolved oxygen, and trophic/nutrient level.

Figure 6.5 shows the types of organisms, and their remains, that are to be found in lakes. This diagram places emphasis on the occurrence of organisms in lakes, be they saline or fresh. Discussion here only refers to the organisms which are commonly represented in lakes and which have remains that preserve readily. Not mentioned in this diagram are several other organisms which have been used by palaeoecologists to reconstruct lake histories, such as aquatic insects (remains of larvae and adults), vertebrate remains (e.g., fish, frogs), remains of unicellular organisms such as the cestate thecamoebians (which resemble the well-known marine foraminifers), some foraminifers brought into lakes by birds (and which can survive as well as reproduce in the lakes provided the salinity and water chemistry are similar to that of sea water), and remains of crustaceans among which are crabs and other decapods and isopods. More commonly used are the cladocerans (commonly known as water fleas), rotifer eggs and algal remains (some of which are even distinguishable by the different pigments recovered in the sediments).

Of importance too are pollen, spores and seeds of aquatic plants which may be used to help reconstruct conditions in a lake and define its status. Naturally, the pollen which is blown into lakes from elsewhere can also indicate the vegetation in the vicinity of the lakes (see Chapter 8 for more details), and thus complement the reconstruction of the history of a region through a definition of the hydrological budget of a lake and of the effect of the water regime on the vegetation in the area.

Quaternary lacustrine records

Features characteristic of large deep lakes have been recognized on all continents. In North America, the high lake-level phases are referred to as 'pluvial lake' phases whereas in Australia they are called 'megalake' phases. The latter term is preferentially used in Australia and should be worldwide because the fact that the lakes retained water may not necessarily be the result of more rainfall (as the word 'pluvial' would infer) in the lake catchment. Instead, a high lake-level stand may be more the result of less evaporation or more cloud cover, or even a change in runoff condition (e.g., due to a change in vegetation cover). The best documented example of large lakes having formed during the Quaternary in the United States is Lake Lahontan in the now arid south-west. The Great Salt Lake (GSL) (see Figure 6.6(a); Benson and Thompson 1987) in the state of Utah is now very saline (with a NaCl-

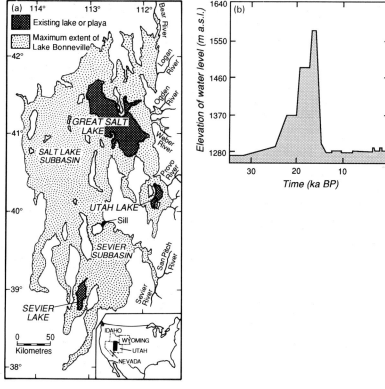

Fig. 6.6
(a) Map showing the maximum extent of the Palaeolake Bonneville
which existed 18 ka ago and the present configuration of the Great Salt
Lake, Sevier and Utah Lakes in the western United States. These three
lakes represent the shrunken remnants of Lake Bonneville under the
present climatic and hydrological conditions (after Benson and
Thompson 1987)
(b) Schematic reconstruction of lake-level fluctuations for Great Salt Lake
for the last 32 000 years derived from sedimentological, mineralogical
and microfossil studies carried out by Spencer *et al.* 1984. Note the
step-like rise in lake level prior to the glacial maximum registered at 18 ka
followed by the extremely rapid lake-level drop (after Spencer *et al.* 1984)

dominant chemical pathway) and is a relic of the once much
larger Lake Bonneville which during its high-lake phase reached
an altitude close to 1600 m compared to the present-day GSL level
of 1270 m (see Figure 6.6(b); Spencer *et al.* 1984). In fact a drop in
lake level of some 300 m occurred during approximately 2000
years, i.e., at a rate of 15 cm per year assuming a constant rate of
lake-level drop. The maximum high lake-level stand occurred
at 16 ka BP at a time when glaciers started melting in North
America (and when most of Australia was extremely dry).

There are now sufficient data available to be able to reconstruct
the history of lake-level changes for large North and South
American, African and Australian lakes covering the last 18 000

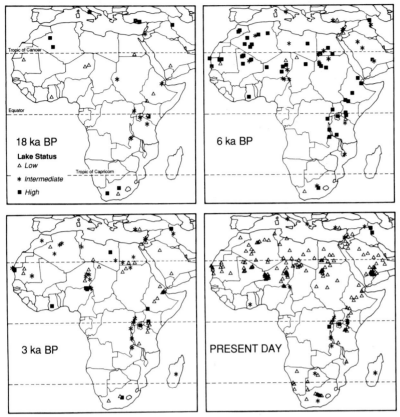

Fig. 6.7 Composite diagram showing the status of lake levels registered on the African continent for four specific time-frames: 18 ka, 6 ka, 3 ka and present (after Street-Perrott *et al.* 1989)

years since the last glacial maximum had a significant effect on the landscape in the Northern Hemisphere. Examination of maps of the world detailing lake status (low, intermediate and high lake level) indicates that not all lake levels were of the same status, even on a continental-wide basis at the same time. Figure 6.7 shows as an example lake-level status for the African continent for four selected time-frames that demonstrate the differences that exist on a continental basis between major lacustrine systems and the need to be cautious before attempting to infer broad climatic generalizations from lake-level changes (Street-Perrott *et al.* 1989). Different regions register a drop in lake level whereas others may have high lake levels. This phenomenon relates to changes in precipitation regimes that may be affected by large-scale intercontinental atmospheric trends (Open University 1989). For example, today when the South East Trade Winds are strong near the coast of South America (Figure 6.8), thus causing upwelling (see Chapter 5 for further details) along the Peruvian coast, rainfall is recorded in northern Australia, even affecting the

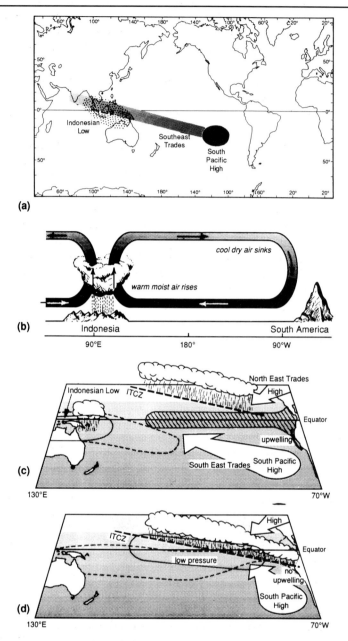

Fig. 6.8
(a) Map showing the position of a typical atmospheric high-pressure cell
(High) in the South Pacific above Tahiti that corresponds to a Low above
Indonesia and northern Australia and which both would correspond to
an excess of rainfall in the latter area as indicated in diagram (b).
(c) Schematic diagram to show the effect of the High in the South Pacific
under normal circumstances and the corresponding effect on ocean
temperature. In such a situation, upwelling of cold water and nutrients
occurs along the Peruvian coast, and a large portion of the eastern
equatorial Pacific, including the southwestern portion of the United

level of lakes in central Australia, whereas the western part of the
United States is usually undergoing drought conditions. This
phenomenon of intercontinental connection is now quite well
documented and warns against comparing lake-level status (e.g.,
Highs) from different sites on a global scale, or between two
continents, in order to demonstrate a globally wetter climate. In
fact, it is the prevailing winds over the equatorial Pacific that form
the link between the different rainfall patterns on either side of
the Pacific. This is defined as the Southern Oscillation Index
which relates to a difference in atmospheric pressure measured in
Tahiti and Darwin in northern Australia, and which defines an
overall set of atmospheric conditions across the Pacific. A high-
pressure cell in the South Pacific is paralleled by a low-pressure
cell over northern Australia that strongly affects summer rainfall
there. In the opposite case, drought conditions prevail over a large
portion of the Australian continent. Such a transcontinental
relationship in climate that affects lake level needs to be recognized to explain the changes that are registered in lake records
across the globe. More details of the El Niño and the Southern
Oscillation Index and global teleconnections affecting climate are
presented in Chapter 10.

Lake histories

The record of lake-level fluctuations in large lakes can be obtained
through several methods of investigation. The first is the study of
ancient shorelines which can help document past lake-level
highs. Such geomorphological features are sometimes so obvious
and extensive across the landscape that they are distinguishable
on aerial and even sometimes satellite photographs, especially in
extensive lacustrine systems. It must be remembered though that
only the regressive phases leave shorelines on the landscape. A
transgressive lake phase will tend to rework ancient shorelines
except for the one that is formed during the maximum extent of
the lake. In a sense, the way shoreline sequences can be interpreted with respect to lake-level fluctuations is the same as for
glacial moraines left on the landscape after a glacial retreat.

Another method for reconstructing lake-level histories is to
examine the sedimentary and palaeontological record of the large

States, remains dry (see cross-hatched area). Rainfall is substantial over
northern Australia under that scenario. Ocean surface temperatures
greater than 28°C are circled by a stippled line.
(d) During El Niño years, there is no upwelling, the South East Trades are
weaker, ocean temperature is lower and the Intertropical Convergence
Zone (ITCZ) is closer to the equator in the eastern Pacific. As a
consequence of the above conditions, there is a rainfall deficit over
northern Australia and most lakes are dry (after Open University 1989)

lakes. Different mineralogies, sediment chemistry and faunal/ floral composition of fossiliferous beds will relate to particular sedimentological and biological processes that directly relate to lake-level histories. In addition, recent research to decipher the relationship that exists between water chemistry (e.g., the particular chemical pathways discussed previously in this chapter) and several taxa of calcareous-shelled ostracods, and also the siliceous diatom frustules, has permitted the reconstruction of the chemical evolution of the lake waters through time, and has thus related chemical changes to climatic change (Forester 1987).

Naturally, one should also be aware of the potential reworking of sediment and fossil remains along lake margins where wave activity is predominant, and also of the possible slumping along lake flanks where turbidity flows are frequent (Figure 6.4). In addition, dissolution and diagenetic effects can alter the sedimentary and palaeontological record.

Several sophisticated techniques have recently been used to decipher the history of lakes. Geochemical analyses of microfossil remains of the sort now commonly applied to the marine record (see Chapter 5 for further details) have successfully been used on lacustrine material. Because of the almost complete absence of foraminifers in lakes, investigations have had to focus on different organisms, and more frequently on the calcareous-shelled ostracods and gastropod and bivalve molluscs. Stable isotopes of oxygen have been analysed from the calcareous remains of organisms mentioned above with the purpose of defining past physico-chemical parameters that existed in lakes, such as temperature and water composition, with respect to two isotopes of oxygen (^{16}O and ^{18}O) and carbon (^{12}C and ^{13}C) (for further details on the isotopes refer to Chapters 2 and 5).

The best documented example so far using the composition of stable isotopes of organisms, combining ostracods and bivalve (pelecypod) molluscs, was carried out by Lister (1988) from specimens extracted from a core from the large peri-Alpine Lake Zürich to document a change in mean annual air temperature and changing lacustrine productivity. Examination of Figure 6.9 shows that the sharp positive shift in the $\delta^{18}O$ registered in the pelecypod shells just after 12.8 ka BP is interpreted as a change of inflow into the lake probably caused by isotopically light water originating from the melting of glacial ice. The only change in $\delta^{18}O$ after that time occurs as a slight shift around 8.5 ka BP interpreted as a slightly warmer interval that is eventually followed by continuous climatic amelioration during the rest of the Holocene. The $\delta^{13}C$ of the biogenic carbonates, on the other hand, can be used to interpret the palaeoproductivity of the lake. For example, the shift in negative $\delta^{13}C$ values just prior to the onset of the Holocene is interpreted as a significant increase in surface productivity. Additional positive and negative shifts during the

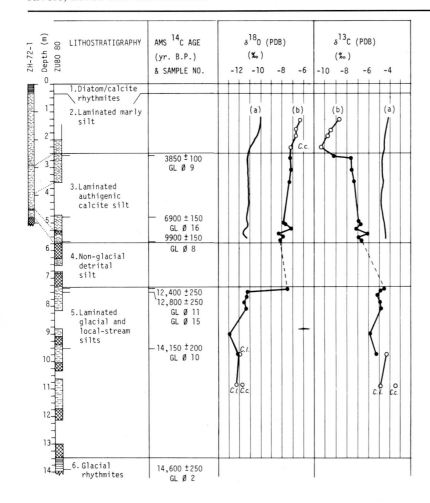

Fig. 6.9 $\delta^{18}O$ and $\delta^{13}C$ values (related to the PDB standard) of (a) inorganic carbonates precipitated in Lake Zürich during approximately the last 15 ka, and of (b) biogenic carbonates (open circle: the ostracods *Candona candida* [*C.c.*] and *Cytherissa lacustris* [*C.l.*]; and filled circle: the pelecypod mollusc *Pisidium conventus*) found in the sediment for which a lithological log is presented in the left margin (after Lister 1988)

rest of the Holocene document events of reduced and enhanced productivity. The significant shift in $\delta^{13}C$ around 3.85 ka BP is considered by Lister (1988) to result from a change in trophic level in the lake caused by a surplus of organic matter reaching the bottom of the lake which, in fact, may even have caused the disappearance of the pelecypods from that location.

Similar isotopic investigations were carried out by Fritz *et al.* (1975) on a core from the large Lake Erie in North America where fossil ostracods, pelecypod and gastropod molluscs were analysed for their isotopic composition to document climatic changes of the Great Lakes region and the evolution of the lake since deglaciation. The latter would have affected the hydrological regime of the lake and this can be detected through isotopic changes registered in the biogenic carbonates. Through that study, two major climatic improvements were found, one occurring between 13 and 12 ka BP and the other between 10 and 8 ka BP (for the significance of these events refer to the discussion in

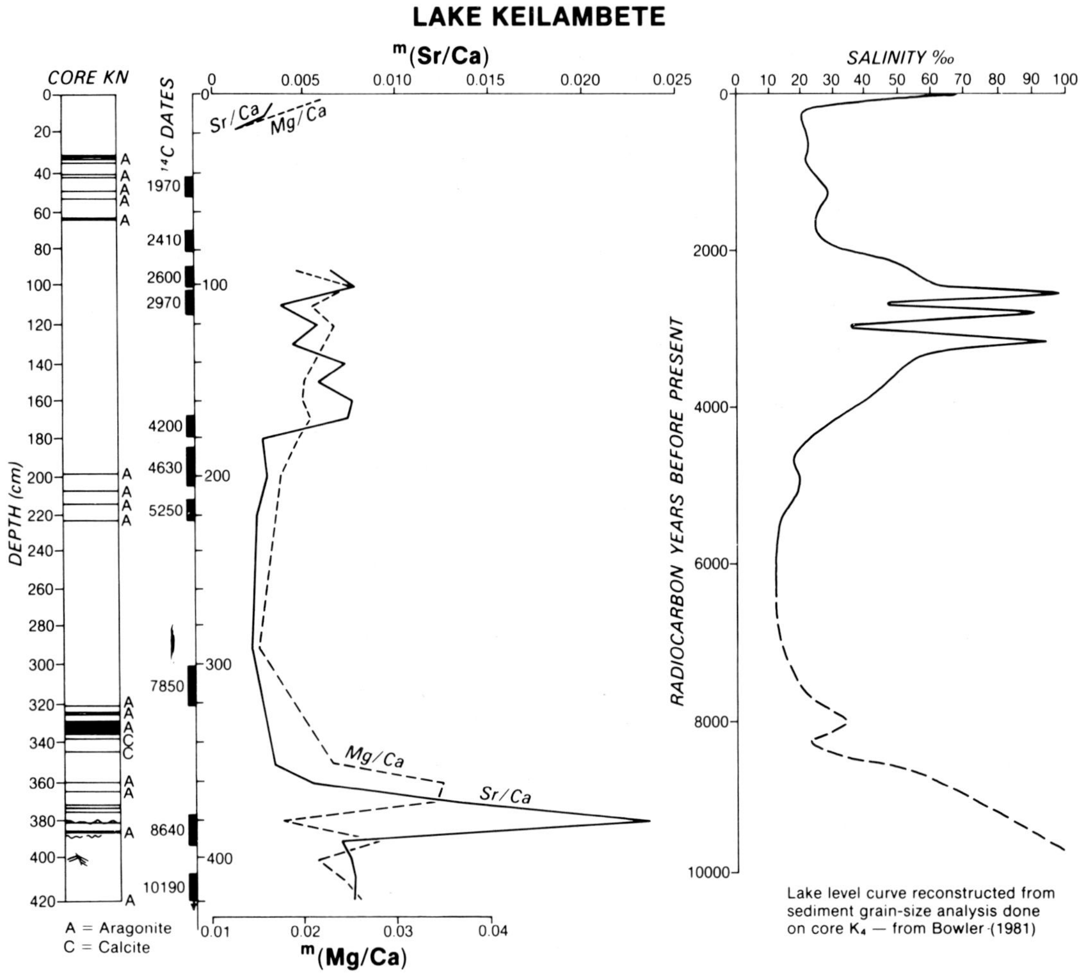

Fig. 6.10 Plots of the mean value for individual layers of the molar ratios of Sr/Ca and Mg/Ca measured on ostracod shells from a 4.2 m long core (representing the entire Holocene) from the maar Lake Keilambete (data from Chivas *et al*. 1986) compared to the smoothed palaeosalinity curve of Bowler (1981) based on sediment grain-size (which is related to water depth in the lake) from some 40 layers from an adjacent core. A schematic stratigraphic log on the left margin of the diagram shows the presence of calcitic and aragonitic layers that testify to the saturation nature of the lake waters through time with respect to carbonates. Sr/Ca values in the ostracod shells are related to salinity changes in the lake (see Chivas *et al*. 1986)

Chapters 3 and 5 of the isotopic record of the North Atlantic corresponding to the Younger Dryas event).

Recent investigations of trace elements, namely magnesium and strontium, which replace calcium atoms in the calcite lattice of ostracod shells from non-marine environments, have permitted a better definition of physico-chemical conditions in lakes, namely temperature, ionic composition and also salinity. In the best available example so far, Chivas *et al.* (1986) were able to relate a change in the atomic ratio Sr/Ca in the valves of single ostracods from a Holocene sequence from a crater lake, the volcanic maar Lake Keilambete in south-eastern Australia, to salinity changes in the lake (Bowler 1981; Chivas *et al.* 1986). An increase in salinity in a lake with a well-defined/constricted catchment, such as that of a crater lake (Figure 6.4), for example, as a result of an increase in evaporation, should only register an increase in solute concentration and not a chemical change such as modification in the Sr/Ca ratio of the water. However, in the case of Lake Keilambete, the Sr/Ca of the ostracods (and thus of the water too) did change and could be directly related to a change in salinity (Figure 6.10) since the lake water must have remained supersaturated with respect to the bicarbonate ions despite the fact that some carbonate precipitates frequently formed within the lake (A. R. Chivas, pers. comm.).

Techniques now used to reconstruct lake histories are becoming more sophisticated, but it is necessary to be aware that the selection of lacustrine sites is very important in order to obtain the relevant information. Two lakes adjacent to one another frequently can react differently to the same climatic conditions; for example, if their morphologies differ (e.g., a conical waterbody will evaporate at a different rate compared to a lake that is large and extremely shallow) or the quality of their water differs (e.g., water turbidity affects the trapping of solar heating, and the biota also can affect the water transparency).

In this chapter we have seen that different lake types provide diverse sorts of palaeoclimatic information, and different organisms, their chemical composition, and sediments yield information on various types of conditions and phenomena. A multidisciplinary approach to the study of lacustrine deposits will therefore provide the most complete array of information.

7 | Evidence from the Deserts

Ubi solitudinem faciunt, pacem appellant. (They create a desert and call it peace.)

Tacitus (c. 55–117 AD),
Agricola, 30.

Introduction

Deserts are remarkable repositories of palaeoclimatic information. Many deserts landforms reflect the operation of processes which are no longer active today. Every desert in the world has its legacy of dry or saline lakes, many of them part of once integrated but now defunct and segmented drainage systems. The fossil fauna of great deserts like the Sahara is an eloquent witness to a time when the presently parched wilderness was able to sustain an abundant plant and animal life, including such large tropical herbivores as elephants and giraffes, as well as a widespread aquatic fauna of turtles, hippos, crocodiles and Nile perch (Monod 1963). In sheltered valleys in the Aïr Mountains of Niger, in the southern Sahara, there are remnant populations of savanna primates: anubis baboons (*Papio anubis*) and patas monkeys (*Cercopithecus patas*), which reached the mountains during wetter times when the West African savanna woodland was considerably more extensive than it is today (Monod 1963; Newby 1984). The dwarf crocodiles which apparently still inhabited some of the permanent waterholes of the Tibesti Mountains of the south-central Sahara in the 1950s (A. T. Grove, pers. comm.), but which may well now be extinct (see Lambert 1984), are (or were) part of this palaeoclimatic heritage, prompting us to ask a number of questions. When and why were our present-day deserts once able to sustain such an abundance of plant and animal life, and why can they not do so today?

Many of the animals which once roamed the Sahara were observed by wandering bands of late Stone Age hunters, and they recorded what they had seen in a multitude of remarkable naturalistic rock engravings and rock paintings (Lhote 1959; Roset 1984). These upper Palaeolithic paintings and petroglyphs depict most of the larger savanna mammals and are scattered throughout the Sahara wherever suitable rock walls were available. With the advent of Neolithic animal domestication in the early to mid-

Holocene, the subject of the paintings changed, and great herds of domesticated cattle are depicted on smooth rock faces in mountains as far apart as Jebel 'Uweinat ring-complex in the far south-east of Libya, the famous Tassili sandstone plateau in southern Algeria (Lhote 1959), and the Aïr Mountains in Niger (Roset 1984). The great herds of brindled cattle depicted with their herdsmen in the Neolithic rock-art galleries of the Sahara, together with the occasional archaeological find of the bones and horn cores of prehistoric domesticated cattle (Smith 1980, 1984; Gautier 1988), raise a further interesting question. To what extent did Neolithic overgrazing by large herds of hard-hoofed cattle, sheep and goats accelerate soil erosion by wind and water, and initiate humanly induced processes of desertification, especially in the drier second half of the Holocene (Williams 1982, 1984a and b; Stiles 1988)? In this chapter we try to answer some of these questions, but we return to the subject of desertification in Chapter 11.

Present-day distribution of arid and semi-arid regions

Before attempting to identify the relative impact of former climatic and human influences upon desert environments, it is appropriate to begin with a brief analysis of the causes of aridity and the reasons for the present-day distribution of arid and semi-arid areas depicted in Figure 7.1.

Fig. 7.1 Present-day distribution of tropical and temperate deserts and semi-deserts (after Tricart and Cailleux 1972, Figure III; The Times Atlas 1980, p. xxiv)

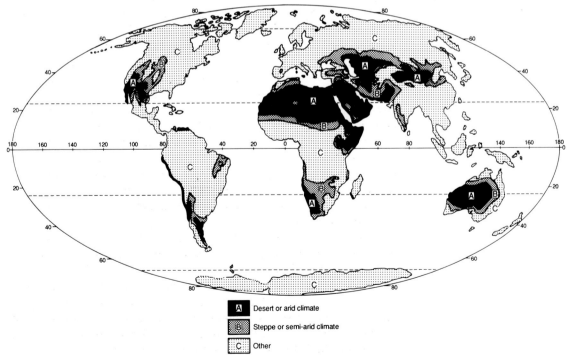

A Desert or arid climate

B Steppe or semi-arid climate

C Other

Deserts are regions where the rainfall is too low and erratic and the evaporation too high to enable many plants and animals to survive. Those species which do survive in arid areas are physiologically and behaviourally well adapted to use scarce water efficiently. Low rainfall is not in itself an adequate criterion of aridity. In many cold areas of the world where the rates of evapotranspiration are very low, a relatively dense vegetation cover may exist even when local precipitation is quite low. Another, more utilitarian definition of a desert is a region where sustainable agriculture is not possible without irrigation. If we accept this working definition, then roughly 36 per cent of the land area of the globe is either arid or semi-arid and provides a home for about 13 per cent of the world's present human population (Crabb 1982).

Figure 7.1 shows that the major deserts of the world today either lie astride (or very close to) the tropics of Cancer and Capricorn, or else are situated in the interior of mid-latitude continental regions, often in the rain-shadow of high mountain ranges such as the Rocky Mountains of North America, and the Altai, Tien Shan and Kunlun Mountains in western China. What factors are responsible for this very particular distribution pattern?

Causes of aridity

Global atmospheric circulation is primarily controlled by latitude. Insolation is at a maximum over the equator, so that the surfaces of both land and sea become warm and they in turn heat the air above them by convection. The warm air rises, expands and cools adiabatically. Since warm air can store more water vapour than an equivalent volume of cold air, the rising air soon becomes saturated with respect to water vapour and any excess water vapour condenses to form clouds. Convectional uplift induces further cooling, leading to precipitation of water droplets which eventually coalesce into larger drops and fall as rain. As the air aloft becomes colder and denser it begins to subside. The tropical latitudes are zones of atmospheric subsidence. By now the air has shed much of its excess water vapour so that the air which descends over the two tropics is habitually dry.

The hot tropical deserts are in latitudes where the atmospheric pressure is high for much of the year and sustained by dry subsiding air. As the air subsides it is compressed and becomes warmer so that its capacity to absorb additional water vapour is increased. The result is that the relative humidity of desert air is usually very low although the absolute amount of water vapour retained in the air may be quite large, becoming evident only when the night temperatures fall sufficiently for the evanescent desert dew to precipitate on chilled rock surfaces.

Since the tropical anticyclonic deserts are a direct result of global atmospheric circulation, their location is determined by latitude and has very little to do with the regional distribution of land and sea. The two polar deserts (not shown on Figure 7.1) are also under the influence of semi-permanent anticyclones and of cool, dry, subsiding air. As a result of this latitudinal distribution of high-pressure cells, the oceans in both polar and strictly tropical latitudes also receive very little precipitation, and are in effect the marine counterparts of the continental deserts.

The mid-latitude deserts, including those of western China and Uzbekistan are dry because of their geographical location in the centre of large continental land masses far removed from the influence of moist maritime air. This factor of 'continentality' or distance inland applies to all the larger deserts, including the Sahara and Australia. Indeed, rainfall tends to decrease rapidly away from the coast in all parts of the world except those close to the equator. In the case of the hot tropical deserts, the effects of continentality therefore serve to reinforce those of latitude.

Three additional factors may enhance the aridity resulting from latitude and continentality, or may themselves be direct causes of reduced precipitation. These factors are the presence of cold sea-surface water offshore, the rain-shadow effect, and low relief inland. They may operate individually or in concert.

The presence close offshore of cold upwelling water or a cold ocean current is an effective cause of coastal aridity in tropical latitudes. The Atacama desert in Chile and the Namib desert in southern Africa are flanked offshore by the cold Peru and Benguela currents, respectively. The western borders of all the tropical or 'Trade Wind' deserts are washed by cool ocean currents associated with the ocean gyres which flow clockwise in the Northern and anticlockwise in the Southern Hemisphere. When moist and relatively cool maritime air masses move onshore they encounter a land surface which is warmer than the adjacent ocean surface. The result is a reduction in the relative humidity of the former maritime air, so that it develops an increased capacity to absorb rather than to shed moisture. The major source of moisture in these narrow coastal deserts is the coastal fog which blows inland in winter and this effect is greater if there is high relief close to the shore.

The rain-shadow effect is a globally universal phenomenon and is not peculiar to deserts. Wherever there are hills or mountains close to the coast, the incoming moist maritime air will be forced upwards. As the moist air rises it is cooled adiabatically, attains vapour saturation, and sheds its precipitated water vapour as rain or snow. The air then flows over the coastal ranges and downhill, becoming warmer and drier for the reasons outlined earlier. The region inland of the coastal ranges is said to lie in their rain-shadow and will always remain drier than their coastal flanks.

Patagonia lies in the rain-shadow of the Andes. Other examples of rain-shadow deserts are the arid or semi-arid areas immediately leeward of the Rockies, the Himalayas, the Ethiopian uplands and the Eastern Highlands of Australia. Extreme examples are the Dead Sea Rift and the Afar Depression, both of which are flanked by high mountains while they themselves lie close to and in places well below sea level.

The rain-shadow effect is increased when the region inland of the humid coastal ranges is low-lying and devoid of any significant relief. Some of the driest deserts are those in which the landscape consists almost entirely of extensive plains or low plateaux, such as the stone-mantled 'gibber' plains of central Australia and the gravel-strewn 'serir' of southern Libya. The converse is also true and the high mountains of the central and southern Sahara, such as Tibesti (3415 m), the Hoggar (2918 m) and Jebel Marra (3042 m) receive sufficient rainfall for them to have served as refugia for plants, animals and humans throughout the Quaternary (Messerli *et al.* 1980; Maley 1980; Rognon 1967, 1989; Williams *et al.* 1980).

There is one additional factor which deserves mention, for it also provides a link between present and past. Late Cainozoic uplift of the Tibetan Plateau (see Chapter 2) resulted in the development of the easterly jet stream which now flows from Tibet across the Arabian peninsula towards Somalia, accentuating the aridity in those regions and helping to create a desert right on the equator (See Figure 7.1; Rognon and Williams 1977).

The distribution of our present-day deserts therefore reflects the lingering influence of past tectonic events as well as the combined effects of at least six other major factors. These factors are the prevalence of dry, subsiding, anticyclonic air masses over deserts (itself controlled by latitude and global atmospheric circulation), a vast land area, coastal ranges, low relief inland, cool ocean water close offshore, and a subtropical jet stream aloft. Within the time-frame of the Quaternary, these factors did not change very drastically, although the degree to which they influenced aridity did vary in response to changes in oceanic and atmospheric circulation associated with orbital perturbations and global temperature fluctuations (see Chapters 3, 5 and 10 for details).

The desert environment

Before embarking on the delicate task of trying to reconstruct the Quaternary environmental changes to which all the deserts were subjected, it is useful to describe some of the characteristics of existing desert environments. For clarity and simplicity we group desert landforms somewhat arbitrarily into erosional and depositional. We begin by considering the desert landscape as a whole.

Perhaps the single most characteristic attribute of deserts throughout the world is their lack of a perennial and integrated system of drainage. Desert streams are ephemeral. They flow episodically, for variable distances, depending upon the intensity and duration of sporadic rainstorms in their upper catchments (Walther 1900; Berkey and Morris 1927; Jutson 1934; Cooke and Warren 1973; Mabbutt 1978; Rognon 1989; Thomas 1989). Of course, allochthonous rivers like the Nile may flow through parts of a desert, but in their desert sojourn they do not gain any additional water from perennial tributaries. On the contrary, rivers which traverse deserts constantly lose water by seepage to the often deep-seated regional groundwater-table. Most desert rivers are endorheic (see Chapter 6) and flow into closed depressions like the Tarim basin in China or the Lake Eyre basin in Australia. The fossil river valleys of the Sahara, the Gobi and Western Australia have long interested geologists. Today they are broad linear depressions filled with Cainozoic alluvium which is often cemented with iron, silica or calcium carbonate to form low erosional remnants or sometimes extensive sheets of resistant ferricrete, silcrete or calcrete (Lamplugh 1902). In Mauritania, Namibia and Western Australia these valley-fill calcretes may also contain variable amounts of secondary uranium minerals precipitated out of slowly moving groundwater originating from the Precambrian host rocks which form the valley interfluves.

A further attribute of all desert landscapes is their clarity and starkness, for their erosional landforms invariably show a high degree of adjustment to rock type and geological structure. In more humid regions evidence of similar structural control is usually well camouflaged by deep soils and a more or less continuous cover of vegetation. Desert soils are usually skeletal lithosols on rocky hillslopes, or almost unweathered deposits of gravelly alluvium or wind-blown quartz sand. Nevertheless, in sheltered mountain valleys in the heart of even our greatest deserts, and especially along their semi-arid margins, relict deep-weathered mantles and well-developed palaeosols are often sufficiently well preserved to allow a sequence of formerly wetter climatic regimes to be reconstructed, sometimes in considerable detail (Williams *et al.* 1987).

Many desert landforms are exceedingly old. The vast desert plains of the central and western Sahara have been exposed to subaerial denudation for well over 500 million years (Williams 1984a), as have the Precambrian shield deserts of the Yilgarn Block and the Pilbara in Western Australia. It is misleading to consider such well-known desert monoliths as Ayers Rock (Uluru) in central Australia, or the granite inselbergs of the Sahara as diagnostic of aridity, for they owe their present morphology to prolonged and repeated phases of weathering and erosion under a succession of former climates, few of which were particularly arid.

Linked to this longevity of most erosional desert landforms is another somewhat paradoxical attribute of many desert landscapes: the close juxtaposition of very young depositional features with very ancient erosional landforms. It is these young landforms and sediments, whether aeolian or fluviatile or lacustrine, which best retain the imprint of past environmental changes, most notably the rapid climatic fluctuations of the late Tertiary and Quaternary, to which we now turn.

Late Cainozoic cooling and desiccation

The onset of late Cainozoic aridity and the slow emergence of the deserts portrayed in Figure 7.1 were associated with the global tectonic events discussed in Chapter 2. As a result of the lithospheric plate movements shown in Figure 2.1, a number of changes in global atmospheric circulation resulted from the changing horizontal and vertical distribution of land and sea, and certain continents or regions also moved into dry tropical latitudes, most notably North Africa and Australia.

The origin of the Sahara as a desert was associated with several independent tectonic events. Slow northward movement of the African plate during the late Mesozoic and Cainozoic saw the migration of much of North Africa from wet equatorial into dry tropical latitudes. A slight clockwise rotation of Africa during the Miocene and Pliocene brought Africa into contact with Europe, and was accompanied by crustal deformation and rapid uplift in the Atlas region, and by volcanism and updoming in Jebel Marra, Tibesti, the Hoggar and Aïr Mountains (Williams 1984a).

Two additional factors were responsible for accentuating the late Cainozoic desiccation of North Africa. One was the gradual expansion of continental ice in high latitudes associated with the post-Eocene cooling of the Southern Ocean and the North Atlantic (Schnitker 1980). This cooling was initiated by the break-up of Laurasia and by the separation of Australia from Antarctica (see Chapter 2). It culminated in the establishment of a large ice cap on Antarctica by 10 Ma ago, and in a sudden increase in the volume of Northern Hemisphere ice caps towards 2.5 Ma ago (Shackleton and Opdyke 1977; Shackleton *et al.* 1984). One effect of the progressive build-up of high-latitude ice sheets was to steepen the temperature and pressure gradients between the equator and the poles, resulting in increased Trade Wind velocities. Faster Trade Winds were better able to mobilize the alluvial sands of an increasingly dry Sahara and to fashion them into desert dunes. The first appearance of wind-blown quartz sands in the Chad basin, for example, is towards the end of the Tertiary, when they occur interstratified among Plio-Pleistocene fluvio-lacustrine sediments (Servant 1973). The associated lacustrine diatom flora indicates temperatures cooler than those now characteristic of this region (Servant and Servant-Vildary 1980), reinforcing the notion that the late Pliocene was both cooler and drier along the tropical borders of the Sahara. Pollen evidence from Pliocene Lake Gadeb in the south-eastern uplands of Ethiopia is also consistent with this conclusion (Bonnefille 1983), suggesting that intertropical cooling and desiccation may have been closely bound up with the expansion of Northern Hemisphere ice caps towards 2.5 Ma ago.

A second factor contributing to the late Cainozoic desiccation of the Sahara, and briefly alluded to earlier, was the Neogene uplift of the Tibetan plateau and the ensuing creation of the easterly jet stream which brought dry subsiding air to the incipient deserts of Pakistan, Arabia, Somalia, Ethiopia and the Sahara. In this context, it is interesting to note that carbon and oxygen isotopic analyses of Neogene palaeosols and fossil herbivore teeth collected from the Potwar Plateau of Pakistan reveal a dramatic change in flora and fauna between 7.3 and 7.0 Ma ago (Quade *et al.* 1989). Prior to 7.3 Ma, there was a dominance of C3 plants (see p. 217) indicative of forest and woodland. After 7.0 Ma, C4 plants were the most abundant, indicating a rapid expansion of tropical grassland at the expense of forest. Quade and his co-workers have interpreted this change as being consistent with a major strengthening of the Indian summer monsoon during the very late Miocene, if not indeed with the actual inception of the monsoon. Changes in the Cainozoic flora and fauna of the Sahara show a similar trend. During the Palaeocene and Eocene much of the southern Sahara was covered in equatorial rainforest, and there was widespread deep weathering at this time. During the Oligocene and Miocene much of what is now the Sahara was covered in woodland and savanna woodland, but by Pliocene times many elements of the present Saharan flora were already present. Maley (1980) has reviewed the evidence from pollen preserved in scattered localities in northern Africa, concluding that replacement of tropical woodland by plants adapted to aridity was already under way during the late Miocene and early Pliocene.

Late Cainozoic cooling and desiccation was not confined to the vast tropical arid zone which extends from the western Sahara across Arabia as far as north-western India (Figure 7.1). Nor was it peculiar to the Northern Hemisphere. The loess of north-western China first began to accumulate 2.4 Ma ago (Heller and Liu 1982; Kukla and Zhisheng 1989). Summer aridity first became apparent around the Mediterranean basin at about this time (Suc 1984). In the tropical Andes there was a major change in the flora towards 2.5 Ma (Hooghiemstra 1989). The onset of aridity in South America, South Africa and Australia is harder to pinpoint with precision, but the evidence from geomorphology and geochemistry and from the fossil flora and fauna is all indicative of progressive Neogene desiccation on land and of a post-Eocene cooling of the oceans to the south and west of all three southern land masses (Zinderen Bakker 1978; Barker and Greenslade 1982; Dingle, Siesser and Newton 1983; Williams 1984c; Fasano 1989; Zarate and Kukla 1989; Williams, De Deckker and Kershaw 1991).

Quaternary glacial aridity

From about late Pliocene times onwards, the great tropical inland lakes of the Sahara, Ethiopia and Arabia began to dry out. The formerly abundant tropical flora and fauna of the well-watered Saharan uplands became progressively impoverished as entire taxa became extinct, and a once integrated and efficient network of major rivers became increasingly obliterated by wind-blown sands. Allowing for local differences in timing linked to regional climatic and tectonic factors, a similar sequence of events was also true of the deserts of China, India, Australia and southern Africa. With the advent of the Quaternary, a global pattern of climatic oscillations now became established, apparently linked to and modulated by orbital perturbations, although other mechanisms may also have played a role (Berger 1981, 1989; Peltier 1982; Mörner 1989; Huggett 1991).

In Chapter 3 we saw that the astronomical cycles vary in amplitude and frequency, and are responsible for variations in the amount of solar energy received at different times in the past, and at different latitudes (see also Berger 1981). The 41 ka cycle is linked to changes in the obliquity of the earth's axis, which is now 23°27′ but oscillates between 22°30′ and 25°. The 100 ka cycle is controlled by variations in the eccentricity of the earth's orbit around the sun. The present eccentricity value is such that the earth now receives 6 per cent more energy when closest to the sun than when furthest from it; at maximum eccentricity the value is closer to 27 per cent. The precession of the equinoxes varies with the changing distance between earth and sun. The 21 ka precession cycle has two peaks, one of 19 ka, the other of 23 ka. The relative influence of each of these cycles has varied during the course of time, so that the duration of glacial–interglacial cycles has also varied, with concomitant repercussions for all of the earth's surface, including the deserts.

Prior to 2.4 Ma (and the rapid expansion of North American ice caps), the dominant cycles evident from magnetic susceptibility measurements of deep-sea cores from the Arabian Sea and the eastern tropical Atlantic were the 23 ka and the 19 ka precession cycles, but after 2.4 Ma the 41 ka obliquity cycle became dominant (Bloemendal and de Menocal 1989). Oxygen isotope records from deep-sea cores spanning the entire Quaternary reveal that the early Pleistocene from about 1.8 to 0.9 Ma was subject to frequent low-amplitude fluctuations in the oxygen isotope differences between glacial and interglacial maxima (Williams *et al.* 1981). Since changes in the oxygen isotopic composition of benthic foraminifera broadly reflect changes in global ice volume (see Chapter 5; and Shackleton 1977, 1987), the changing magnitude and frequency of Quaternary glacial cycles should also be reflected in the severity and frequency of former cycles of Quaternary aridity and of desert expansion and retreat. Let us now confront this hypothesis with evidence emanating from the deserts.

Recent evidence from North Atlantic cores shows that the 41 ka obliquity cycle was dominant during the Matuyama magnetic chron from 2.47 to 0.7 Ma (Ruddiman and Raymo 1988). During the last 735 ka of the Brunhes magnetic chron the 100 ka orbital eccentricity cycle was dominant. The present-day Atlantic exerts a powerful influence upon snow accumulation in North America and Europe as well as upon rainfall and drought in North Africa, and a similar influence is discernible in the geologically recent past.

During the last 600 ka at least, maximum concentrations of Saharan desert dust in equatorial Atlantic deep-sea cores coincide with times of glacial maxima and low sea-surface temperatures (Parkin and Shackleton 1973; Parmenter and Folger 1974; Bowles 1975; Sarnthein *et al.* 1981). A similar pattern of glacial aridity is evident also in the Gulf of Aden and the Red Sea. The isotopic composition of planktonic foraminifera from deep-sea cores in this region shows that during the last 250 ka at least, glacial maxima were times of extreme aridity, with increased sea-surface salinity reflecting even higher rates of evaporation than prevail there today (Deuser *et al.* 1976). What was the impact of these alternating wetter and drier Quaternary climatic phases upon the deserts?

Glacial and interglacial desert environments

To equate glaciations with aridity and interglacials with an increase in desert rainfall is to over-simplify. The reality is both more complex and more interesting. We have long known that during the last glacial maximum towards 18 ka, aridity was more widespread in the intertropical zone (Fairbridge 1970; Williams 1975), Trade Winds were stronger (Parkin 1974), dunes were active well beyond their present limits (Grove and Warren 1968; Sarnthein 1978; Talbot 1980) (compare Figures 7.2 and 7.3), and considerable volumes of desert dust and loess were deposited on land (and far out to sea, Parkin and Shackleton 1973), in regions

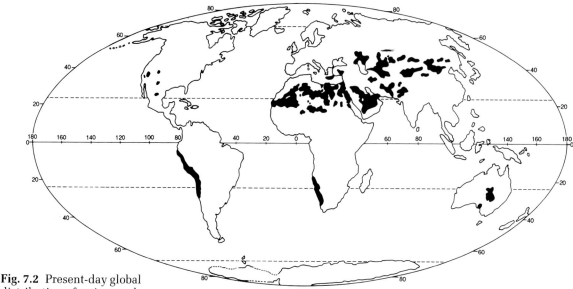

Present-day distribution of active sand dunes

Fig. 7.2 Present-day global distribution of active sand dunes (after Goudie 1983, Figure 3.1)

where such dust mantles are now vegetated and relatively stable (Figures 7.4 and 7.5; and Liu *et al.* 1987, 1989).

We noted earlier that maximum concentrations of desert dust in equatorial Atlantic deep-sea cores coincided with glacial maxima. Such dust is easily recognized by its high degree of sorting, shown in the very characteristic cumulative frequency curves of dust-particle size illustrated in Figure 7.6. Although dust mobilization presupposes aridity and appropriately strong seasonal winds to

Fig. 7.3 Global distribution of active sand dunes during last glacial maximum (18 ka) (after Sarnthein 1978, Figure 1)

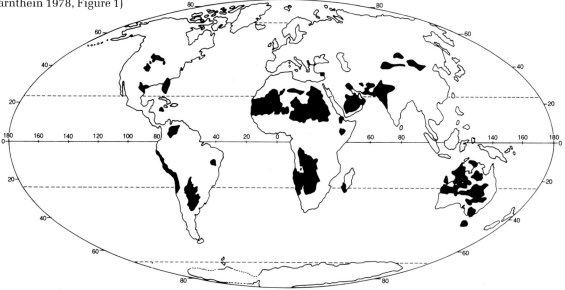

Distribution of active sand dunes at 18ka

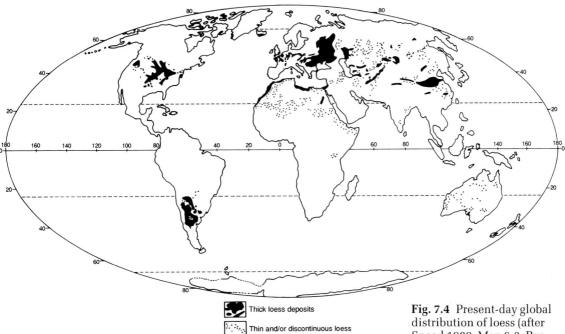

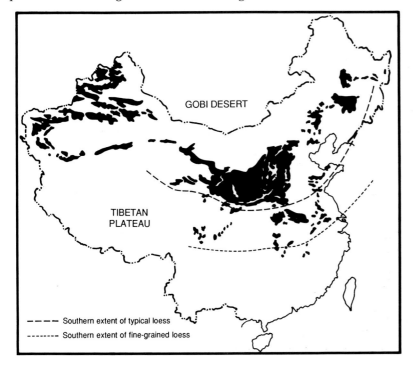

Fig. 7.4 Present-day global distribution of loess (after Snead 1980, Map 6-3; Pye 1984, Figure 1; Pye 1987, Figure 9.1; Thomas 1989, Figure 11.1)

entrain the dust plumes aloft and offshore (Morales 1979; McTainsh 1980, 1985; Pye 1987), the glacial loess deposits of Eurasia and North America (Figures 7.4 and 7.5) reflect the former presence of unvegetated Pleistocene glacifluvial outwash sedi-

Fig. 7.5 Present-day distribution of loess in China (after Liu *et al.* 1982, Figure 1; Pye 1984, Figure 2)

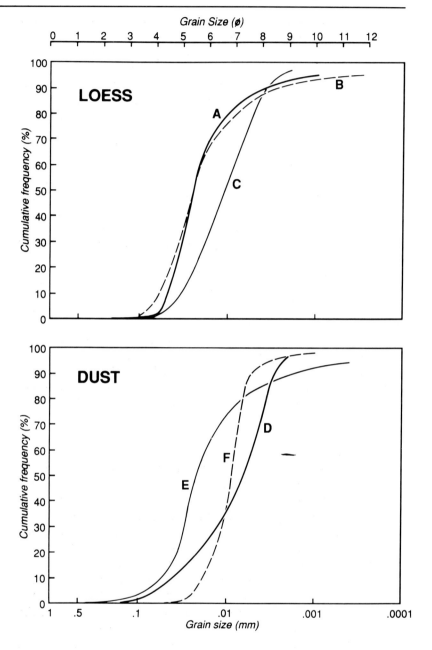

Fig. 7.6 Cumulative grain-size frequency curves of loess and desert dust. (A) Sanborn loess (after Swineford and Frye 1945); (B) Siberian loess (after Péwé 1981); (C) Malan loess (after Liu *et al.* 1981); (D) Mongolian dust recorded in Beijing (after Liu *et al.* 1981); (E) Arizona dust (after Péwé 1981); (F) Saharan dust recorded in London (after Wheeler 1985)

ments upwind of the loess mantles rather than simply aridity and strong winds. Although the parna sheets of Australia (Butler 1956) may resemble the loess deposits of north-central India (Williams and Clarke 1984) in terms of their physical characteristics, a fine well-sorted grain-size distribution is not in itself indicative of either a glacial outwash or a desert dust origin.

Dunes, alone, are not diagnostic of aridity, since dune mobilization depends upon such factors as sand supply, vegetation cover and surface roughness as well as wind velocity and turbulence (Wasson and Hyde 1983; Wasson and Nanninga 1986). The source-bordering dunes referred to in Chapter 6 are most frequently found in semi-arid regions. Prerequisites for their formation are a regular (usually seasonal) replenishment of river-channel sands or sandy beaches by longshore drift in deep lakes, as well as a strong seasonal unidirectional wind and a lack of riparian or lake-margin vegetation (Williams 1985). The first prerequisite – regular renewal of the sand supply from seasonally active rivers – effectively precludes a fully arid climate. It is also worth emphasizing that only a small proportion of all deserts are covered in sand dunes, only one-fifth in the case of the Sahara, and two-fifths of the much less arid Australian desert, and many of the Australian dunes are only active along their ridge crests (Ash and Wasson 1983). The pioneering and now classic empirical and experimental work of Bagnold (1941) demonstrated that the volume of desert sand transported increased exponentially with wind velocity above a certain threshold value (Figure 7.7). Where sand supply and wind speed are not limiting factors, dune mobilization will increase as vegetation cover decreases. Since plant cover in dry areas is governed primarily by rainfall and evapotranspiration, there is a close relationship between the amount of rainfall and the average outer limit of active dunes in such deserts as the Thar desert of Rajasthan (Goudie *et al.* 1973) or the Sahara, where the southern limit of presently mobile dunes coincides remarkably closely with the 150 mm isohyet (Figure 7.8).

When all the 'caveats' mentioned above are taken into account, it becomes very clear that in many of the world's hot deserts, including the Gobi and Uzbekistan deserts, the dominant climate during the last glacial maximum (18 ka ± 3 ka) was drier, windier and colder than today, although the summers may still have been very hot (Bowler 1978; Sarnthein 1978; Wasson *et al.* 1983; Bowler and Wasson 1984; Williams 1985; Thomas 1989; Dong *et al.* 1991). Previously fixed and vegetated dunes along what are now the semi-arid margins of these deserts became mobile, so that the effective range of the Sahara extended 400 to 600 km further south (Grove and Warren 1978; and Figure 7.8), and that of the Rajasthan desert some 350 km to the south-east (Goudie *et al.* 1973). Many of the desert lakes which immediately prior to about 18–20 ka had occupied deflational hollows or tectonic depressions and were full and fresh, now dried out or became hypersaline (Gasse *et al.* 1980; Harrison *et al.* 1984; Street-Perrott *et al.* 1985), as previously perennial desert-margin rivers became seasonal while seasonal rivers became, at best, highly intermittent or ephemeral streams (Rognon 1989).

With glacially lowered sea levels, the desiccating influence of greater continentality was also enhanced. Stronger Trade Winds

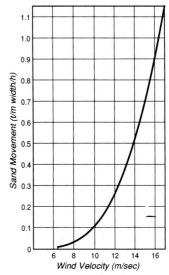

Fig. 7.7 Relationship between wind speed and volume of desert sand transported (after Bagnold 1941, Figure 22)

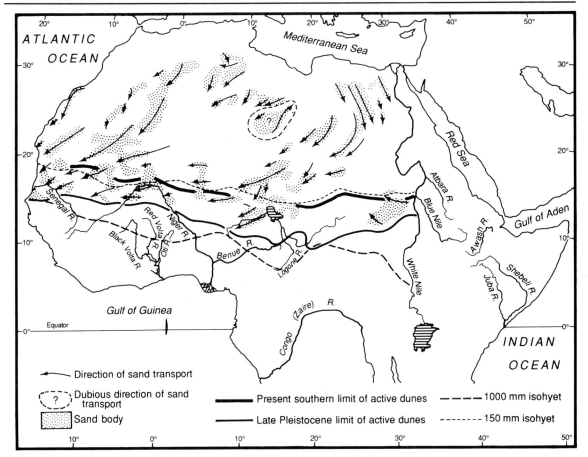

Fig. 7.8 Directions of sand transport in the Sahara and its southern margins based on alignments of present-day and later Pleistocene desert dunes (after Grove 1980; Mainguet and Cossus 1980; Mainguet, Canon and Chemin 1980)

associated with steeper pressure gradients between equator and pole caused increased upwelling of cold water close offshore, further accentuating the aridity of coastal deserts.

The contrast between the terminal Pleistocene aridity of the Sahara and its verdant early Holocene status is hard to imagine today, but it had an enormous impact on the late Stone Age and early Neolithic peoples who witnessed these changes (Chapter 9). As postglacial temperatures and sea levels rose around the world, evaporation from the intertropical oceans also increased, and the previously weakened summer monsoons of Northern Australia, India and West Africa once more became reliable sources of seasonal rainfall (Pastouret *et al.* 1978; Williams 1984b, 1985). Throughout the previously dry tropics, groundwater levels rose, aquifers were replenished, lakes refilled, hitherto mobile dunes became vegetated and stable, and savanna woodland and grass-land re-occupied what are today the semi-arid regions of the world. A remarkable and well-integrated drainage network became established in many parts of the Sahara, Arabia and Rajasthan, all of which were studded with innumerable freshwater lakes and ponds reaching a peak towards 9 ka (Faure 1966; Singh *et al.* 1974; McClure 1976). This was the time of 'le Sahara des Tchads' (Balout 1955) when 'aqualithic' upper Palaeolithic hunter-fisher-gatherer communities (Sutton 1977) used barbed bone harpoons to obtain Nile perch and other large fish from the Saharan lakes; hippos, crocodiles and turtles also featured in their diet (Clark 1980).

The climatic conditions of the warm and wet early Holocene were very similar to those of the last interglacial at c. 125 ka (oxygen isotope stage 5e of the deep-sea core record), as was the global distribution of land and sea. The desert environments no doubt oscillated between these two extremes, with the interglacials (125 and 9 ka) being mostly slightly warmer and very much wetter than today, and the glacial maxima (140 and 18 ka) colder and mostly drier. However, not all arid phases coincide with glacial maxima, any more than do all humid phases with interglacial times. For instance, Lake Chad in the southern Sahara (Servant 1973) and Lake Abhe in the Afar desert of Ethiopia (Gasse 1975) were both very high for at least 10 000 years before 18 ka, when they fell rapidly. They were then intermittently dry (Lake Chad) or dry (Lake Abhe) until 12 ka, rising rapidly thereafter to reach peak levels at 9 ka. Since about 4.5 ka both lakes have remained low apart from occasional brief transgressions. Very schematically, we could consider the c. 30 to 18 ka phase of high lake levels as a humid glacial phase, the 18 to 12 ka regression as an arid glacial phase; the early Holocene transgression as a humid interglacial phase; and the late Hololcene interval of low lake levels as a dry interglacial phase. This fourfold subdivision caricatures reality, and also ignores local hydrological and geomorphic controls over rainfall, runoff, evaporation, seepage losses and groundwater inflow (Fontes *et al.* 1985; Abell and Williams 1989).

Correlating the terrestrial and marine records

Earlier in this chapter we noted the influx of Quaternary desert dust into the oceans. Most Quaternary land records are discontinuous in space and time and many contain very big stratigraphic gaps, in contrast to the long, continuous records from deep-sea cores. An outstanding exception to the general rule that most Quaternary desert records are fragmentary is the unique Chinese loess deposits, which have been studied in great detail by Professor Liu Tungsheng and his Chinese colleagues (Heller and Liu 1982; Liu 1987; Kukla and An 1989). Figure 7.9 represents one attempt to correlate the palaeomagnetically dated loess deposits and associated palaeosols at Lochuan in north-central China with

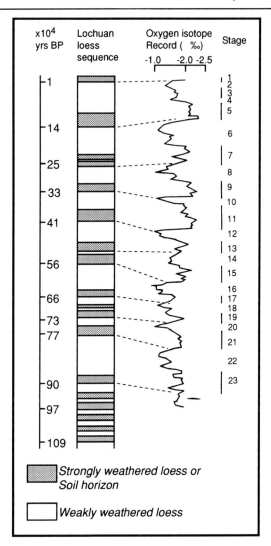

Fig. 7.9 Correlation between the Lochuan loess/palaeosol sequence of China and the North Pacific oxygen isotope record of core V28-238 during the last million years (core record stops at 970 ka). Soils formed or the loess became strongly weathered when the relative concentration of ^{18}O was high, indicating warmer and wetter climatic conditions (after An Zhinsheng and Liu Tungsheng 1987)

the North Pacific oxygen isotope record from core V28-238 for roughly the last one million years. Cold phases (even isotope stage numbers) discernible in the North Pacific generally coincide with times when loess accumulation was rapid and soil formation was at a minimum. Warm phases (odd numbers) mostly coincide with weathering of the loess mantles and with times of soil development. Ignoring the smaller fluctuations (0.5‰ or less) evident in the oxygen isotope record, it seems reasonable to accept An and Liu's conclusion that interglacials or interstadials evident in the North Pacific deep-sea core record were synchronous with intervals of warmer and wetter climate in north China. Conversely, glacials and stadials coincided with colder and drier intervals during which accumulation of unweathered loess was rapid and extensive. The Chinese loess record is entirely consistent with the glacial aridity model proposed earlier in this chapter, but, as with every broad generalization, it is the exceptions and the local variations which must ultimately be considered if we are to understand the full complexity of the Quaternary history of our deserts.

8 | Evidence from Terrestrial Flora and Fauna

They attacked the cedars – and while Gilgamesh felled the first of the trees of the forest Enkidu cleared their roots as far as the banks of the Euphrates.

Anon. (c. 2700 BC),
The Epic of Gilgamesh.
(Trans. N. K. Sanders, 1960).

Introduction

The fossilized remains of terrestrial flora and fauna are to be found in almost all global environments but, because of their abundance and the large number of sites available for preservation, the major focus of this line of investigation has been on humid temperate and tropical regions, which are poorly represented in the other lines of evidence for the reconstruction of Quaternary environments considered in this book. In order to survive under the relatively harsh conditions experienced on land, most terrestrial plants and animals possess some hard parts that preserve within sedimentary deposits. However, the usefulness of different organisms as palaeoenvironmental indicators varies as a result of a number of factors including the abundance and ease of identification of preserved parts, the degree of evolutionary change during the Quaternary, and the extent of their habitat ranges.

Since the pioneering work of Von Post in Scandinavia, early this century, the analysis of pollen and spores derived from higher land plants has become the most widely used technique for Quaternary ecological investigation. This is because these particles are well represented in most accumulating sediments, their wide dispersal provides a broad picture of surrounding vegetation, and their intricate and distinctive patterning allows identification to parent plants at a reasonable taxonomic level. From early studies broad regional vegetation-climate relationships were established from pollen studies in north-west Europe providing an important relative dating tool for archaeological and other events. The value of pollen analysis within this area declined in the 1950s with the development of numerical dating techniques, particularly radiocarbon dating (Appendix), but the ready application of this dating technique to pollen sequences opened up far more opportunities for it in palaeoenvironmental and palaeoecological investigation.

Other biotic materials generally serve to refine interpretations from pollen analysis or provide evidence from the few areas or sites where pollen is not preserved. Of significance are plant macrofossil remains, such as leaves and seeds, which can frequently be identified to a lower taxonomic level than pollen and provide more detail on the composition of particular communities, and animal remains, particularly beetles and vertebrates, that can respond more quickly than most plants to environmental change. Charcoal is valuable in providing direct evidence of fire while detailed environmental information can be provided by the nature and variation in thickness of tree rings preserved in the trunks of both living and dead trees.

This chapter is designed to reflect the relative importance of the different materials used for palaeoecological reconstruction and the techniques used in their analysis. The focus is primarily on the methodology and applications of pollen analysis. Other major floral and faunal remains are briefly examined in their role as providing additional and generally complementary information to that from pollen. The chapter concludes with an attempted overview of global vegetation during the Quaternary.

Pollen

Pollen analysis or palynology are general terms embracing the study of a variety of plant microfossils of which pollen grains and spores (particularly fern spores) are the most important. Pollen grains are produced in the anthers of male flowers of angiosperms (flowering plants) and gymnosperms (conifers) and their function is the transfer of genetic material to female flowers to effect fertilization and seed production, while spores are the dispersal units of ferns and lower plants that give rise directly to the next generation. Both these types of microfossil are small, about 5–100 micrometres, and their external coating is composed of a resistant substance known as sporopollenin. Many other types of microfossil can be analysed in association with pollen but, apart from charcoal, these provide evidence mainly of local site conditions rather than dry land environments.

There are a number of processes involved in the production of final pollen assemblages from the original vegetation which need to be taken into account in the reconstruction of this vegetation. These processes are identified in Figure 8.1 and outlined below.

Production

Plants are adapted to producing different amounts of pollen and,

PALYNOLOGICAL PROCESSES

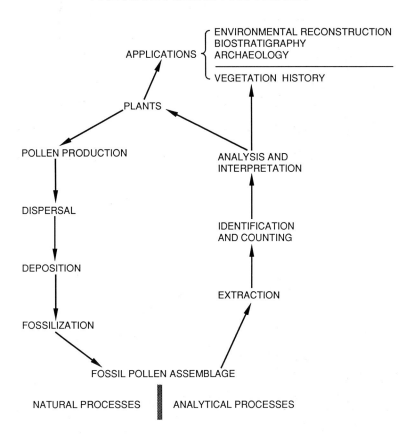

Fig. 8.1 Processes involved in the production of fossil pollen assemblages from parent plants and subsequent analysis and interpretation

in general terms, those producing the greatest quantities will be best represented in pollen samples. However, for any given species, production will vary according to a number of factors including climatic conditions and the plant's position within the vegetation. The majority of plants will flower most prolifically under optimal growing conditions but others may only flower or spore when stressed.

Dispersal

Only a proportion of pollen fulfils its intended function and much of it is available for transport to a suitable depositional site for accumulation in the fossil state. Dispersal is effected mainly by wind (anemophily), by nectar-feeding insects and birds (zoogamy) and water (hydrogamy). As animal pollination is generally very efficient, the majority of pollen grains reaching a deposition site are those dispersed by wind and also by water if there are streams running into it or there is easy inwash from surrounding slopes. There are many other variables involved in pollen transport.

A

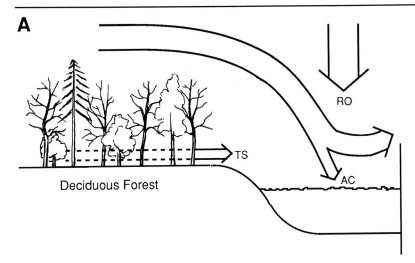

Deciduous Forest

B

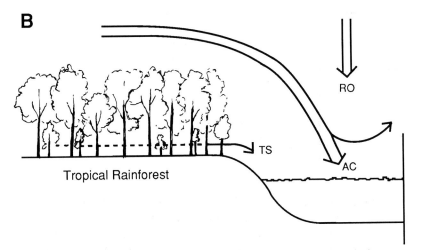

Tropical Rainforest

Fig. 8.2 A diagramatic representation of the relative importance of different modes of pollen transfer to a lake basin from
(A) mixed deciduous forest
(B) tropical rainforest
(based on results from Tauber 1967; and Kershaw and Hyland 1975)

These include attributes of the pollen itself such as size, shape and density, the position of plants in the vegetation, and climatic conditions. Representation of pollen will also be influenced by the proximity of the plants to a depositional basin and to the size of that basin.

A generalized model has been constructed which illustrates the ways in which different airborne pollen components can reach a depositional basin. This is shown in Figure 8.2 for small lakes within two different types of forest. The transported pollen is caught in specially designed traps floating on the surfaces of the lakes. There are three main modes of transport: by wind above the canopy, through the trunk space, and by rain after mixing in the atmosphere (Tauber 1967).

In deciduous forest, values for rainout (RO) and canopy components (AC) are relatively high because most species have wind-dispersed pollen, there are strong winds to transport the pollen and regular showers to wash the pollen from the atmosphere. The trunk-space (TS) component is also large because, although wind speeds and therefore carrying capacity are lower than outside the forest, the absence of foliage in early spring when many trees flower allows significant air flow. In addition, in autumn and winter when many trees have lost their leaves, there is a substantial reflotation of pollen that was trapped by the foliage earlier in the year.

In tropical rainforest, all components record much lower values because the majority of species are animal-pollinated and release little pollen into the atmosphere and wind speeds are lower. The trunk-space component is further reduced by the maintenance of a dense vegetation cover throughout the year while the rainout component is insignificant due to very regular showers that inhibit significant atmospheric mixing or to the existence of long dry seasons.

The three components tend to represent vegetation from different distances away from the site of deposition. The trunk-space component reflects vegetation growing closest to the site; the canopy component is derived mainly from canopy species within the general area; while the rainout component may include pollen from communities growing some distance away. The larger the lake and the further away from the lake edge that samples are taken for pollen analysis, the more important will be the pollen from the rainout-component relative particularly to the trunk-space pollen.

One important limitation to this model is that it fails to take account of waterborne pollen. Where there is significant stream inflow into a lake and particularly into marine environments, most pollen can be from this source and will emphasize stream-valley vegetation. In small lakes, a substantial component can be derived from inwash of pollen originally deposited on lakeside soil surfaces.

Awareness of the importance of the differential representation of vegetation at varying distances from a pollen site has led to the definition of distance components. These are local pollen, derived from communities growing at or close to the pollen site; extra-local pollen, most easily defined as pollen from plants growing round a depositional site and influenced by site hydrology; regional pollen, representing the vegetation characteristic of the area; and long-distance pollen, from vegetation types further afield (Janssen 1966).

The relative importance of these components at pollen sample sites in four similar-sized basins and catchments with varying site and vegetation characteristics is illustrated in Figure 8.3. In the two lake examples (A) and (B), pollen from regional vegetation is dominant. Extra-local pollen from lakeside communities has

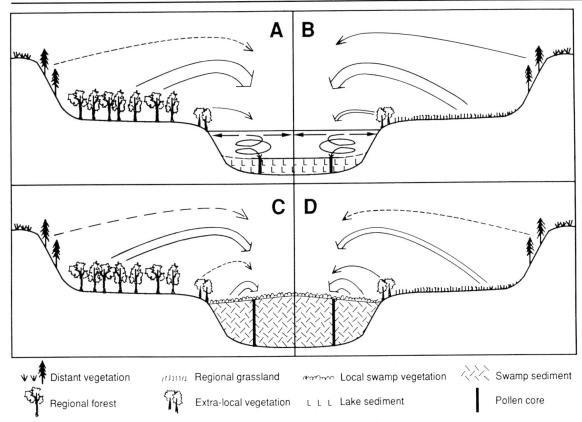

Distant vegetation   Regional grassland   Local swamp vegetation   Swamp sediment

Regional forest   Extra-local vegetation   L L L   Lake sediment   Pollen core

reasonable representation because of easy inwash of pollen into the lake. In the absence of plants growing in the basins, there is no local pollen component. There is a small component from distant forests and proportions for this and for extra-local pollen are larger than in case (B) because the low-growing regional grassland vegetation has lower production and dispersal capability than regional forest. There are similar relative proportions of extra-local, regional and long-distance components in the bog examples surrounded by grassland and forest, (C) and (D), but all are smaller than in the corresponding lake situation because of the abundance of locally derived pollen from plants growing on the bog surface.

Fig. 8.3 An indication of the relative representation of pollen from different components of two vegetation landscapes deposited in lake and swamp basins (see text for details)

Deposition

Figure 8.3 also serves to demonstrate different depositional processes within lakes and bogs. In bogs, deposited pollen tends to be caught effectively on the accumulating surface whereas in lakes there can be movement of pollen on the surface and within the lake due to wind and water-circulation patterns. These movements affect pollen types differentially producing different mixes in various parts of a lake. There may also be concentration in the centre of steep-sided lakes due to sediment focusing, while mixing of deposited pollen may be caused at times of lake overturn and by bottom-dwelling organisms.

Post-depositional changes

Once pollen has been deposited and incorporated into accumulating sediments, the very resistant outside wall or exine can be preserved almost indefinitely. However, if the depositional surface is subject to periodic drying, destruction due to oxidation can occur and if incorporation into the sediment body is slow, pollen grains can be exposed to microbial attack. Similarly, changes to pollen assemblages can occur if sediments are subsequently subjected to drying out or disturbance. Caution has to be exercised in reconstructing vegetation from samples that show signs of deterioration as pollen grains can be destroyed differentially and some are more readily identifiable in a corroded state than others.

Some post-depositional effects are shown in Figure 8.4, a section through a deposit in New Zealand. The section was originally prepared to reveal a selection of the large number of skeletons of large extinct flightless birds known as moas, preserved in the swamp. Although not shown here, these skeletons interrupt the pollen sequence and it is obvious that a great deal of sediment mixing would have occurred as the birds became trapped in the surface sediment. Similarly, the tree remains in the lower part of the sequence would have influenced sedimentation patterns. They demonstrate that the swamp was forested and this forest invasion may mean that the surface became sufficiently dry to allow some oxidation of the pollen. The topmost sediments have also been disturbed, most likely by tree-root penetration, and record distortion could be caused if younger sediments have filled the space originally occupied by these roots.

Site selection and sampling

There is probably no ideal site for most pollen studies as places of sediment accumulation are generally unrepresentative of the pollen catchment area. However, sites vary greatly in their suitability for various studies. For vegetation reconstruction, the choice is generally between lakes and mires (swamps and bogs). Deep lakes are frequently preferred because of constancy in the depositional environment and therefore sediment sequences are likely to be continuous. Swamps or bog sites on the other hand are more likely to have had a more varied history because of their sensitivity to changing hydrological conditions and because of successional changes resulting from the accumulation of sediments to maximum attainable heights for particular communities. Pyramid Valley Swamp (Figure 8.4) provides a good example of

2620 BP

2930 BP

3720 BP

4280 BP

Figure 8.4 Section through Pyramid Valley Swamp, New Zealand, showing variation in accumulated swamp sediments

some of the problems that can be encountered with swamp sites. The stratigraphy clearly shows changes in the nature of the site from an initial swamp (indicated by the accumulation of sedge peat) to a lake, characterized by fine-grained algal mud, and finally back to a sedge swamp. The presence of wood also indicates invasion of the swamp phases by trees that may be inseparable from those dominating the regional vegetation. As previously mentioned, the trees would also have disturbed previously accumulated sediment.

However, swamps do provide better pollen-receptive surfaces than lakes, and, where there is continuous accumulation, have the potential to provide a much more detailed vegetation record. In environments where the most suitable kinds of site are rare, it is necessary to examine a range of sites with different characteristics to allow reconstruction of a regional vegetation picture.

For some studies, neither deep lakes nor bogs may be appropriate. For example, in archaeological studies, caves that show signs of habitation may be preferred. Although pollen is not likely to be as well preserved as under constantly waterlogged conditions, and there may be difficulties in relating the pollen to regional vegetation, insights may be gained into the plant materials utilized by cave dwellers and into the impact of these people on the local vegetation.

Once a site has been selected, it is important to undertake a stratigraphic survey in order to find the most appropriate spot for sampling. This will generally be close to the site centre which is likely to contain the longest and most continuous sequence. An exposed sediment face provides the ideal basis for selection as local disturbances such as root penetration, animal skeletons and wood remains can be avoided, as indicated in Pyramid Valley Swamp. However, in basins that are waterlogged, it is necessary to rely on the examination of discrete cores and a variety of samplers has been designed to collect material from the range of lake and swamp sites encountered.

Extraction

Samples for pollen analysis extracted from the sediment sequences are subjected to various physical and chemical treatments in order to reduce the sediment matrix and make the pollen clearly visible for microscope examination. A range of treatments is possible because pollen is more resistant to destruction than most other components of the sediment. Commonly the following sequence of treatments is undertaken: addition of potassium hydroxide to break down the structure of the organic matter and remove humic acids; sieving to remove coarser plant remains and larger inorganic fragments; hydrogen fluoride treatment to dissolve fine siliceous matter and acetolysis (a mixture of acetic hydroxide and sulphuric acid) to digest cellulose and other polysaccharides and to darken the grains for easy recognition. In addition, treatment with Shultz solution (employing a mixture of nitric acid and potassium chlorate) may be necessary to remove woody material or lignin while the use of a heavy liquid such as zinc bromide to separate remaining organic from inorganic material may be employed if the organic component containing the pollen is relatively small. Between stages, samples are centrifuged to concentrate residual material and washed free of strong acids and alkalies in miscible liquids. In addition to pollen, charcoal is preserved in this preparation process.

Prepared samples are mounted in a suitable medium such as glycerine jelly or silicone oil on microscope slides ready for examination.

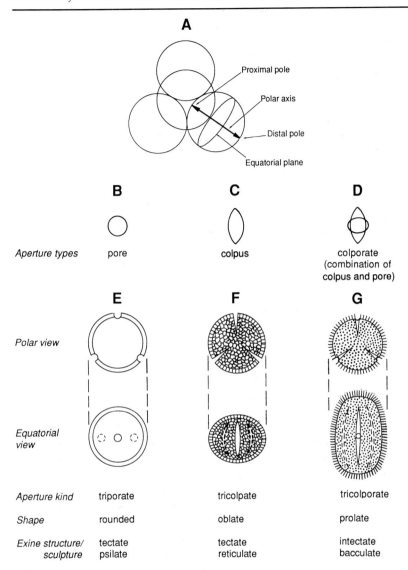

Fig. 8.5 Aspects of pollen morphology.
(A) The symmetry of pollen grains related to tetrad arrangement in the anther.
(B), (C), and (D) three types of aperture commonly found in pollen grains.
(E), (F), and (G) schematic representations of pollen grain types showing selected morphological features in polar and equatorial views

Identification and counting

Pollen grains and spores are identified from morphological features of the exine. Most pollen types have a symmetry that relates to their tetrad position during formation in the anther (see Figure 8.5(A)). Some of the major morphological features of pollen grains are illustrated in Figure 8.5(B)(C)(D).

Many other forms are illustrated in Figure 8.19. Those that are very different to the described forms include the gymnosperms *Pinus* and *Podocarpus* that have wings or bladders to assist in wind disperal and *Chenopodiaceae*, whose pores are evenly distributed over the whole grain surface. Identification to genus or family level and species level for selected taxa is routinely achieved.

Pollen counting is undertaken along viewed transects of the prepared slides under a microscope with high magnification until the desired number of grains has been recorded. Two major considerations in count size are the achievement of relatively constant percentage levels of major recorded taxa and inclusion of a substantial number of types present in the sample population. An illustration of how percentages for variously represented taxa in a sample level out with increased count size is shown in Figure 8.6(A). Below a total count of about 600 grains variability is great and if the count is too low, percentages will be unacceptably inaccurate. For poorly represented types such as *Lycopodium*, significant variability is maintained in counts above 600 grains and for most palynological studies necessary count levels would be unacceptably high to provide reliable estimates of percentages for these taxa. Statistical probability limits for calculated percentages of taxa with different count levels have been determined but unfortunately are seldom applied.

The manner in which the number of taxa recorded increases with count size is illustrated in Figure 8.6(B). There is a great deal of difference in pollen taxon diversity between samples from the different community types and it is probable that, in samples from lowland rainforest, many thousands of grains would have to be counted to ensure the inclusion of most taxa. It is suggested from these results that counts of about 550 grains are required to ensure the inclusion of a reasonable proportion of types represented with an absolute minimum of 150 grains.

Data presentation

Pollen counts from a site sequence are normally portrayed in the form of a pollen diagram where percentage values for recorded taxa are graphed in relation to stratigraphic depth. The diagram shown in Figure 8.7 illustrates many features common to more traditional types of pollen representation. Here, emphasis is on changes in the abundance of the dominant tree taxa as a basis for determination of gross vegetation changes. In line with this emphasis, the total pollen count for tree taxa from each sample forms the pollen sum on which all percentages are based. The counts for the understorey shrubs, *Corylus* and *Salix*, are omitted from the sum but are percentaged relative to the sum. This has resulted in percentages in excess of 200 per cent in some samples for the very abundant taxon, *Corylus*.

The diagram has been divided into zones based on changes in the representation of major taxa. This simplifies description of the diagram and aids subsequent interpretation. The selection of changes in major taxa are those that have been found to occur over

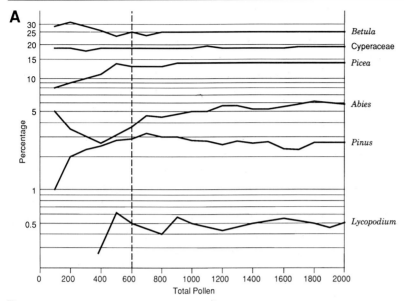

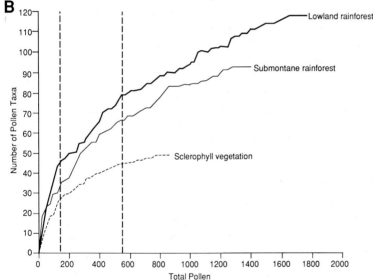

Fig. 8.6
(A) Changing percentages of selected taxa with increasing count size (from Birks and Birks 1981).
(B) Number of taxa counted as a function of sample size from selected pollen samples from north-east Queensland, Australia (from Kershaw, unpublished data)

a wider area. These include the beginning of a sustained rise in *Corylus* at the base of Zone V, a rise in *Ulmus* in Zone VIa, the achievement of greater levels of *Quercus* than *Ulmus* in Zone VIb, major increases in *Alnus* and *Tilia* at the base of Zone VIc, and a decline in *Ulmus* at the Zone VI/VII boundary. This provides a means of correlating between diagrams and consequently a relative time scale for dating associated events like changes in the nature of sediments, or materials such as archaeological artefacts, in the absence of any absolute dating control.

A more recently produced pollen diagram from the site (Figure

(a) **HOCKHAM MERE POLLEN DIAGRAMS**

Godwin and Tallantire (1951)

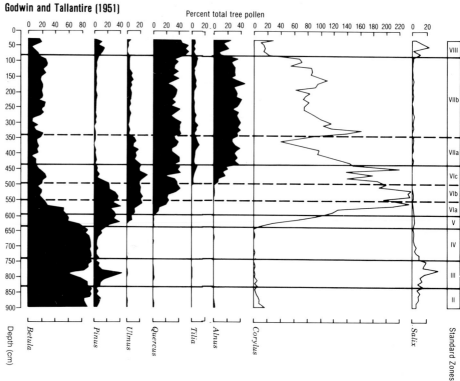

(b) **Bennett (1983)**

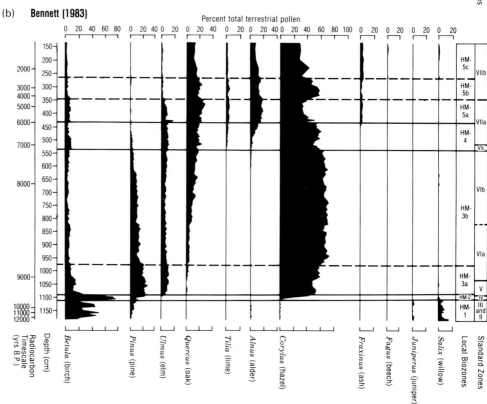

8.7) shows a number of differences. Some of these can be explained by selection of a core from another part of the site, demonstrating the significance of this variable, but the majority of differences are related to the application of more recently developed ideas on, and techniques of, pollen diagram construction. In the first place, the pollen sum is based on the total count of all taxa derived from dry land vegetation, rather than just the trees. This is an acknowledgment of the fact that the landscape has not always been forested and it also demonstrates increased interest in the ecology of all components of the system. Similarly, the use of local biozones, based on internal homogeneity in pollen composition of sections of the diagram independent of information from other sites, is a reflection of the importance of local site variation. Once individual diagrams have been interpreted independently, then comparisons with others can be made. The need for pollen as a stratigraphic tool has also been reduced with the development of radiocarbon dating. Finally, improved identification and much larger and more frequent counts have resulted in the recognition of a greater number of pollen types.

Interpretation

The first step in interpretation of most pollen diagrams is the reconstruction of the original vegetation. Once this has been accomplished, then explanations for the vegetation patterns and changes can be sought. Vegetation reconstruction is a difficult procedure considering the large number of processes involved in the production of the fossil pollen assemblage from the original plant cover. One way of accounting for many of the variables is to use modern pollen samples collected from known communities as a basis for interpretation. This not only provides a direct link between pollen and vegetation but, if modern and fossil samples are prepared and counted in a similar fashion, variation due to analytical processes is kept to a minimum.

An altitudinal sequence of modern pollen samples from the New Guinea Highlands (Figure 8.8) provides a good example of some of the values and also limitations of this approach. It has been used to assist interpretation of a number of pollen diagrams from the area. The vegetation types are clearly distinguished by significant representation of one or a combination of pollen types. The lower-altitude grasslands are dominated by grass pollen and can be separated from alpine grasslands on higher grass percentages and an absence of alpine indicators. Oak forest has the only high values of the 'oak' taxon *Lithocarpus/Castanopsis*, mixed

---

Fig. 8.7 Selected features of pollen diagrams from Hockham Mere, East Anglia, British Isles

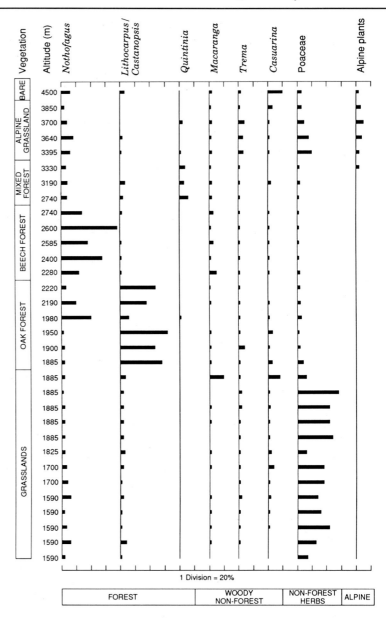

Fig. 8.8 Representation of major pollen taxa in relation to vegetation along an altitudinal surface sample transect in Papua New Guinea (modified from Flenley 1978)

forest is characterized by the only appreciable percentages of one component taxon, *Quintinia*, while the beech, *Nothofagus*, dominates beech forest. It is clear that most pollen is deposited close to its source but some wider dispersal does occur, particularly to higher altitudes. There is significant representation of *Nothofagus* in all higher-altitude samples than beech forest while this taxon, and the equally well dispersed *Casuarina*, are the most important components of, and are characteristic of, the highest altitude bare area which has no local pollen sources.

One factor complicating interpretation of climatically induced altitudinal movements recorded in pollen diagrams is the changes that have been brought about by people within the last few thousand years. It is considered that the lower grasslands are a result of deforestation and establishment of agricultural practices and that little evidence is preserved of the original forest cover. The effects of people are a common feature in pollen diagrams from many parts of the world and can prevent the application of modern pollen studies to the elucidation of natural causes of change, particularly climate. On the other hand, vegetation is a good indicator of the past impact of people, and in areas where vegetation alteration has not been so drastic, some assessment can be made of both natural and anthropogenically induced changes. In the Papua New Guinea diagram (Figure 8.8) the representation of the forest-disturbance taxa, *Macaranga*, *Trema*, and *Casuarina*, provides information on the degree of impact on the forests at various altitudes without destroying the basic altitudinal sequence.

Major refinements to interpretation have also been achieved by the use of absolute, rather than percentage, pollen data. With information in percentage form, a change in the proportion of any one taxon will influence the values of other taxa regardless of whether there have been concomitant changes in the distribution or abundance of parent plants. Where it is possible to calculate the number of grains deposited per unit volume of sediment, generally from sediments that have accumulated at a constant rate or where annual variations in sediment deposition can be detected, then measures of the real abundance of taxa can be derived. The usefulness of so-called pollen-influx data is well illustrated by a comparison of the percentage and absolute pollen diagrams from Roger's Lake (Figure 8.9).

The percentage diagram has much in common with those derived from Hockham Mere. Although from different continents, the basic composition of deciduous forests in these Northern Hemisphere mid-latitudes is similar, as is the pattern of forest development after the last glaciation. The late-glacial period, prior to about 10 ka ago, was dominated by herbs with significant representation by a small number of tree taxa. The postglacial period shows a gradual increase in forest taxa, due largely to increasing temperature levels, that eventually formed the present forests. The pollen-influx diagram demonstrates that some of the features of the percentage diagrams are apparent rather than real. There is no major change from herb- to tree-dominated vegetation from the late- to the postglacial but rather the trees increase in abundance while the herbs maintain their values. The cause of misinterpretation in the percentage diagram is due to very low influx values in the basal sediments. These demonstrate that the herbaceous vegetation was very open and that trees were probably

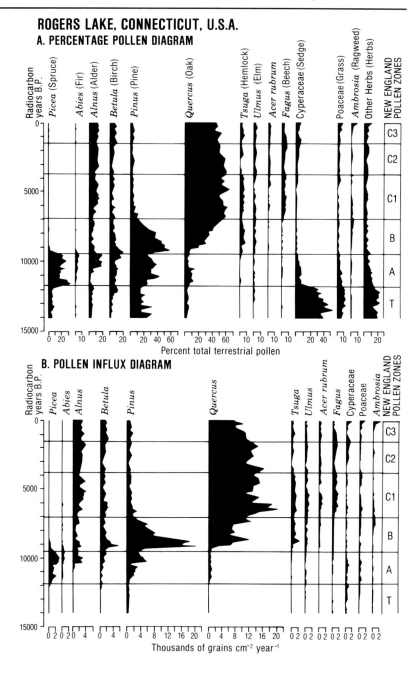

Fig. 8.9 A comparison of percentage and pollen-influx diagrams from Roger's Lake, Connecticut, USA (modified from Davis 1967)

absent from the area, their pollen being derived from distant sources. The pollen-influx data also provide a much more realistic picture of subsequent forest changes with spruce woodland clearly dominant about 12 ka ago, followed by mixed forest within which pine was very prominent before establishment of the present deciduous forest about 7 ka BP.

Fine-resolution palynology

Palynological reconstruction and interpretation has traditionally been at a relatively coarse temporal level with little attempt made to resolve variation on time scales less than hundreds or even thousands of years. However, recent developments in palynological methodology in association with increased precision in dating techniques and the application of sophisticated statistical methods are allowing more detailed investigations with resolutions at decadal or even yearly intervals from suitable sites (Green and Dolman 1988; Turner and Peglar 1988). The major thrusts of fine-resolution palynology, as it has been termed, have been on the documentation of human impact, often in combination with historical records, and on the examination of ecological processes that cannot be resolved adequately from standard pollen analytical methods or from the monitoring of extant vegetation.

An example that clearly illustrates the value of the method, even without statistical analysis, is provided by a study of the more recent seasonally laminated sediments from Crawford Lake, Ontario. The site is situated within mixed deciduous forests that have experienced disturbance from the agricultural activities of both Indian and European people. Contiguous samples taken at five- or ten-year intervals were prepared from 1000 to 1970 AD. Selected features of the percentage pollen diagram are shown in Figure 8.10. The Indian phase is clearly identified between about 1360 and 1660 AD by high grass levels, the presence of a few other weed species and by representation of pollen from cultivated corn. Variations in representation and abundance probably reflect the shifting nature of agriculture with new clearings being made about every 20 years as the nutrient status of soils in old fields became depleted. It is likely that agriculture around Crawford Lake itself was practised from about 1440 to 1460 AD where there is high representation of weed pollen combined with unusually thick laminae, the result of soil erosion from these nearby corn fields. Excavation and dating of an Indian village close to the lake confirms the local presence of agriculture at this time.

The European phase is marked by much more extensive and permanent agriculture. Earliest evidence, about 1820 AD, is consistent with the documented time for initial land acquisition in the area while the increase in weeds around 1840 AD corresponds with the beginning of occupation around the lake at this time. The peak in sorrel pollen at the same point as the increase in grass probably reflects the abundance of this weed in pioneer pastures while subsequent high levels of native ragweed in association with the presence of corn and introduced plantain can be related to the introduction of mechanised corn cultivation.

Tree-pollen values also provide evidence of the various agricultural activities. Initial Indian clearing is indicated by sustained

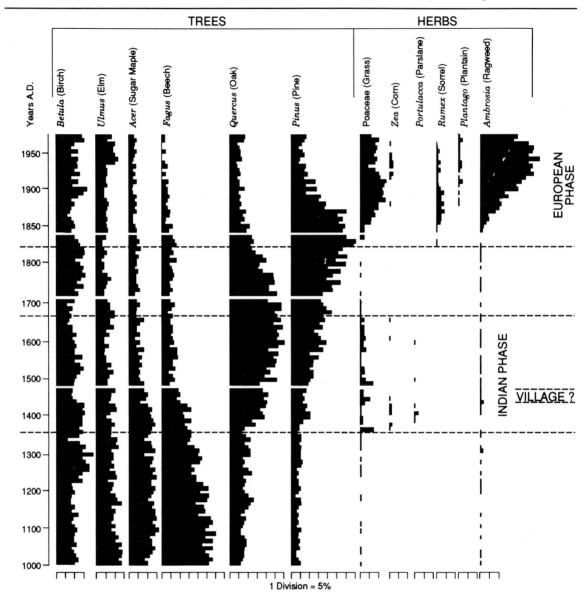

Fig. 8.10 Selected features of the fine-resolution pollen diagram from Crawford Lake, Ontario, Canada, illustrating agricultural indicators and changes in forest composition resulting from Indian and European occupation phases (modified from McAndrews and Boyko-Diakonow 1987)

decreases in the longer living 'climax' forest trees, sugar maple and beech. Regeneration of forest after the abandonment of fields is marked by increases in successional trees such as oak and pine with the latter subsequently being replaced by oak because of its longer life span, before both succumb to selective logging and clearing by European people. Beech suffers a further decline during this phase but some tree species maintain their values or even increase their representation with European farming.

Microscopic charcoal

Small, angular carbonized fragments, generally presumed to be the products of burning, have long been recognized in samples prepared for pollen analysis. However, it is only recently that charcoal counting has been included routinely in pollen analytical studies. This has come about partly as a result of an appreciation of the importance of fire as an important environmental variable in many of the world's vegetation types, and partly due to a concerted effort to understand the processes involved in relating sedimentary charcoal to individual fires and fire types (e.g., Clark 1983; Patterson *et al.* 1987). Many of the problems of charcoal analysis are similar to those of reconstructing past vegetation from pollen but some are unique to charcoal. These include:

(i) Difficulties in identifying charcoal to parent plants. Although large charcoal fragments can retain plant structure that allow identification in the same ways as other macroremains (see plant macrofossils, page 177) it is extremely difficult to identify the vegetation source of microscopic charcoal in many studies.

(ii) Identification of fire regimes. Charcoal, unlike pollen, is produced during relatively infrequent events and wide sampling intervals are likely to miss some fires and give a false impression of fire frequency. In addition, records are frequently blurred by substantial quantities of charcoal that wash into a basin from the surrounding soil surface after the fire events.

(iii) Post-depositional changes. Charcoal can break down into smaller fragments within the sedimentary matrix or during sample preparation, substantially altering the fire signal.

Despite these problems, charcoal has proved to be very important in helping to understand gross vegetation changes and in some cases has provided good insight into the dynamics of past and present vegetation communities, when employed in association with fine-resolution pollen analysis.

A palaeoecological study from the conifer-hardwood forests of North America demonstrates the application of charcoal analysis at different scales of resolution and indicates the likely role of fire in facilitating the climatically and anthropogenically induced changes to these forests noted earlier from Roger's Lake (Figure 8.9) and Crawford Lake (Figure 8.10) respectively.

The period from 10 to 6 ka BP in the pollen-influx diagram from Everitt Lake (Figure 8.11) illustrates the major changes in the vegetation that accompanied climatic amelioration after the last glacial period. The large charcoal peaks suggest fairly regular intense fires which probably resulted from stress imposed on the vegetation by the changing climate. These fires would have

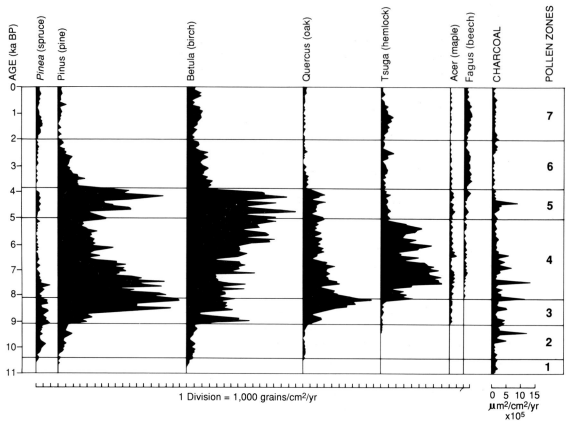

Fig. 8.11 Major features of the pollen-influx diagram from Everitt Lake, Nova Scotia, Canada (modified from Green 1981)

opened up the vegetation and allowed invasion by tree species migrating into the area, which were better adapted to the 'new' climatic conditions. After about 4500 years BP there is little evidence of large fires and it is considered by Green (1981) that the vegetation had reached a degree of equilibrium under relatively stable climatic conditions. The continued occurrence of small, less intense fires to the present day, suggests that the vegetation is adapted to and maintained by this kind of fire regime. Through the use of time-series analyses, Green was able to determine the detailed response of tree species to particular fire regimes. For example, from 4 to 2 ka small charcoal peaks suggest that fires recurred on average at 350-year intervals and that there was a progressive recovery of the vegetation after each fire with relatively early successional taxa such as *Picea* and *Pinus* exhibiting pollen peaks about 50 years, later ones such as *Quercus* about 150–200 years and latest successional plants like *Tsuge, Acer* and *Fagus* about 300–350 years after a fire. This knowledge of prehistoric fire regimes and vegetational successional responses is extremely important to future management of forests where fire control is now largely in the hands of people.

Plant macrofossils

The larger remains of plants, most commonly in the form of leaves, fruits and seeds, can be found in most sites suitable for pollen analysis. In fact, studies on macroremains preceded those on pollen. These tended to be concerned with past distributions of individual taxa rather than vegetation types and lacked quantification and stratigraphic control (Watts 1978). Those associated with early pollen studies focused on remains of local aquatic plants and contributed substantially to the documentation and understanding of bog and swamp development.

Increasingly, plant macrofossils are being examined with pollen in an attempt to refine records of terrestrial vegetation history. The preferred sites are either open lakes where dry land plant parts can be washed in or brought in by streams without impedance from swamp vegetation, or accumulations of animal refuse (middens) most commonly preserved in caves and rock shelters. One major value of macrofossils is that their presence can demonstrate conclusively that parent plants were growing in the vicinity of a site, an important consideration in reconstructing the vegetation of a small area or in the determination of migration rates where it is necessary to know exactly when a plant arrived at a particular site. This is often very difficult to determine from pollen that is widely dispersed. In certain instances, macrofossils can reveal the presence of species whose pollen is rarely recorded. An additional value of macrofossils is that many parts can be identified to a more refined taxonomic level than with pollen.

One of the most interesting and significant applications of macrofossil studies is in the elucidation of late Quaternary vegetation and environmental changes in arid areas, particularly the deserts of south-western North America. In this region, where continuous sedimentary sequences suitable for pollen analysis are often lacking, vegetation histories have been pieced together from the analysis of dated packrat middens. The middens, which are composed of plant material cemented to rock crevices by urine, can provide a remarkably accurate picture of a defined area of vegetation around a site because the packrats forage within a limited area, about 100 m from their dens, and collect plant material from a whole range of species. As middens are constructed over a limited period of time, each provides a time slice that, when radiocarbon dated, can be slotted into a sequence to provide a fairly continuous temporal record, sometimes extending beyond 40 ka BP.

Where pollen as well as middens are preserved, records have frequently been at odds, reflecting different catchment areas. This has allowed greater definition of the distribution of communities. In some instances though, such as that illustrated in Figure 8.12 where pollen and packrat midden macrofossils are from the same

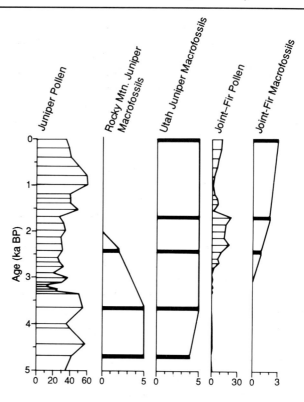

Fig. 8.12 Comparison between selected taxa from pollen spectra and macrofossil samples in packrat middens from Gatecliff Shelter, Nevada, USA (from Thompson and Krautz 1983, in Thomson 1988)

site, similar overall patterns are revealed. In this example the value of macrofossils is shown by the additional information on the representation of species contributing to the juniper component.

Tree rings

Prior to the development of fine-resolution sediment-based studies with acceptable dating control, the major method available for constructing terrestrial palaeoenvironments over the last few hundred or thousand years in some detail was through dendrochronology. In a narrow sense dendrochronology can be equated with the counting of annual rings, formed from a combination of early and late growing season cells in trees to provide an estimate of tree age. Important applications of the technique to Quaternary studies include the ageing of timber used in prehistoric structures by matching sequences of rings with living trees growing within the region today and calibrating the radiocarbon time scale (see Appendix). In a broader sense, dendrochronology includes dendroecology, which focuses on the dynamics of tree populations, and dendroclimatology, that is concerned with the reconstruction of past climates. These aspects are complementary in that a knowledge of past climatic conditions is necessary to a full understanding of population dynamics while population characteristics such as competition can affect the response of trees to climate. Climatological information is revealed largely by annual variation in the thickness of rings within trees that are responsive to some seasonal component of climate but can also be derived from an examination of wood density and the composition of carbon and oxygen isotopes within identified rings (see Bradley 1986).

Tree-ring samples from living trees are generally in the form of cores taken from the outside to the centre of the trees with an increment corer. These cores, smoothed with sandpaper to reveal the cellular structure, are examined under a microscope and the thickness of each ring is measured.

The development of a tree-ring chronology for a particular area is based on the cross-matching of the records from a number of trees within a species population. This is necessary to ensure that the chronology is complete, as individual trees may give incorrect ages due to missing rings, usually the response of the trees to extreme climatic conditions, or false rings when, due to variable conditions for growth, more than one ring may form in a year. Other problems involve unequal radial growth of trees that results in wedging or lobate growth. Some of these differences in growth pattern are illustrated in Figure 8.13.

Fig. 8.13 Tree-ring patterns of selected New Zealand conifers.
(A) Phyllocladus trichomanoides showing extremely clear rings with high year-to-year variability: ideal for ring counting, cross-dating and climatic reconstruction.
(B) Agathis australis exhibiting clear rings but uneven radial or lobate growth. In extreme cases this effectively prevents cross-dating.
(C) Podocarpus totara showing good ring definition but the wedging of individual rings which provides problems in cross-dating (from Dunwiddie 1979)

A

B

C

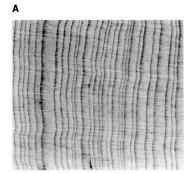
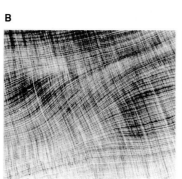
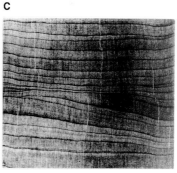

Most parts of the world that allow tree growth and have a seasonal component to the climate are suitable for tree-ring studies (see Figure 8.14). Within these areas the clearest and longest records have been obtained from conifers with the notable exception of *Quercus* in Europe and perhaps also *Nothofagus* in the temperate Southern Hemisphere rainforests. Most of these taxa have substantial pollen records and there is real potential to refine palaeoclimatic estimates from combined fine-resolution pollen and tree-ring studies. The exceptionally long records are from long-lived trees – over 4000 years in the case of the bristlecone pine – extended by cross-matching with the preserved

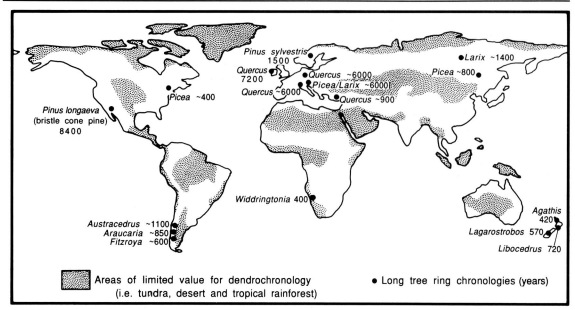

Areas of limited value for dendrochronology
(i.e. tundra, desert and tropical rainforest)

● Long tree ring chronologies (years)

Fig. 8.14 Location of long
tree-ring chronologies in areas
of the globe most suitable for
dendrochronology (modified
from Schweingruber 1988)

remains of dead trees. Preservation is facilitated in the bristlecone
pine by the dry, cool desert environment in addition to very
durable wood, and in the European trees by the anaerobic
conditions existing in swamp and alluvial sediments. The importance of bristlecone pine is primarily in calibrating the radiocarbon
time scale while the timberline species *Larix decidua* and *Picea
abies* are affording insights into annual weather conditions in the
European Alps. The *Quercus* sequences provide evidence of
changing landscapes and river flow patterns as well as dating
prehistoric dwellings and calibrating the radiocarbon record.

The reconstruction of past climates is potentially the most
valuable and the ultimate goal in the majority of tree-ring studies.
Within those areas most suited to dendrochronology, it is the
individuals close to the limits of the populations that are most
responsive to some climatic variable by exhibiting greatest annual
variability in tree-ring thickness. The critical variable in higher
latitudes and altitudes is generally summer temperature trends,
while mean annual precipitation is more important in lower
latitude semi-arid regions. Careful local site selection is also
necessary to minimize non-climatic effects such as water-table
fluctuations, competition, fire and disease. Ring widths also vary
with tree age and this effect is routinely corrected by fitting a
growth curve to each ring sequence and dividing each ring width
by the corresponding value of the curve (Fritts 1976). Once a
consistent chronology has been established from a number of
individual sequences, this is compared against meteorological
records and the relationships between tree-ring variability and a
particular climatic variable established. In ideal situations, the

tree-ring record can be calibrated from meteorological data and then used to provide quantitative climatic information for periods beyond recorded history, or for areas lacking meteorological data.

In many cases, it is difficult to extract a climatic signal from records but this can lead to some useful insights into the operation of other environmental variables. One of the more important of these is forest fires which, as previously mentioned, are often difficult to isolate from charcoal preserved in sedimentary sequences. The tree-ring record for the ponderosa pine in northern Arizona provides a good example of the effect on trees of the adoption of a fire-exclusion policy in the recent past (see Figure 8.15). Prior to about 100 years ago, fires occurred approximately

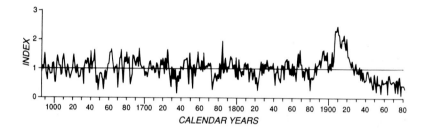

Fig. 8.15 Tree-ring chronology for ponderosa pine in northern Arizona, USA (after Kennedy-Sutherland 1983, in Schweingruber 1988)

every two to three years but the vegetation has not been burnt since. The trees responded to a period of high precipitation early this century by forming wide rings but, since this time, growth rates have slowed substantially. This is explained by a number of factors resulting from fire exclusion including competition for water from the many young individuals whose survival has been facilitated by the lack of fires and a shortage of nutrients which are locked up in the very slowly decomposing litter instead of being rapidly recycled by frequent fires. This example serves to reinforce the historic record and strongly suggests that the fire regime operating prior to 100 years ago had been in operation for at least the last few hundred years. However, it does not reveal direct evidence for fire. In some trees though, the presence of fire scars provides a more certain fire record. This is the case within the Australian treeline species *Eucalyptus pauciflora* (snowgum) where fire scars have been dated by ring counting (see Figure 8.16). Here, the frequency of fires is shown to have increased from the later part of the nineteenth century with the advent of grazing and then decreased in the later part of this century with the establishment of a fire-exclusion policy. The effect of European burning has changed the landscape in many areas from an open woodland of old trees to a denser scrub composed of young individuals.

A　　　　　　　　　　　　　　**B**

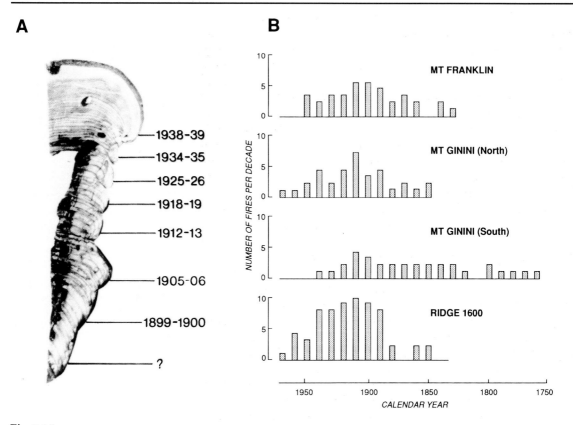

Fig. 8.16
(A) Section through a trunk of Eucalyptus pauciflora (snowgum) from the mountains of south-eastern Australia showing annual growth rings and fire scars that were formed in particular years (from Clark 1981).
(B) Fire frequencies determined from these fire scars, shown as number of fires per decade for a number of snowgum populations (from Adamson and Fox 1982). All information derived from the study of John Banks, Australian National University

Vertebrates

The preservation of bones and teeth in such sedimentary environments as semi-arid dunes, non-acidic lake margins and swamps, and particularly caves and rock shelters, has allowed the construction of fossil vertebrate assemblages from a wide variety of geographical areas. Although contributing significantly to a knowledge of Quaternary environments in general, the value of vertebrates for determination of detailed climatic and habitat change has been limited. This is due to a number of factors including relatively rapid rates of evolution, the nature or taphonomy of the assemblages and problems in identification of critical taxa and their ecologies (Baird 1991; Behrensmeyer and Kidwell 1985).

In comparison to plants and invertebrates, evolutionary rates including extinctions in many vertebrates appears to have been relatively rapid during the Quaternary. This has been of advantage in the construction of stratigraphies for terrestrial environments but has effectively limited the use of modern analogues for interpreting fossil assemblages and for environmental comparisons from fossil assemblages of different ages. The clearest example is provided by the large mammals that grew in size through the Quaternary until suffering massive extinction towards the end of the Pleistocene period. The actual cause of these megafaunal extinctions is hotly debated and discussion of it is most appropriately left until later (Chapter 9), when all relevant lines of evidence pertinent to this issue have been presented.

Despite the problems of evolutionary change in larger animals, one method of palaeoclimatic estimation has been devised that is based on morphological changes within large animals, particularly carnivores, that survived the extinction phase. This is based on the fact that body size for particular species tends to increase with decreased temperature in order to help conserve energy. The explanation is provided by Bergmann's Rule that as an animal increases in size its volume (responsible for heat production) grows more rapidly than its skin area (responsible for heat dissipation) (Klein 1986). The application of measurements on present-day populations to glacial assemblages based on teeth, the best preserved parts of the skeleton, has supported temperature lowering estimates derived from other lines of evidence.

Taphonomy, which in a broad sense includes the processes involved in the production of a fossil assemblage, has been of major concern to interpretation of assemblages of small vertebrates. The majority of sites studied are cave or rock shelters where the composition of assemblages depends very much on how the remains got there – whether the animals lived in the shelters or were brought in by animal predators or by people. Resolution of the origin of the faunas is not difficult with a knowledge of the ecology of the fossil vertebrates and of the eating habits of the predators which can be determined from the state of the bones. The important point is that there are many different taphonomic pathways and only similar ones can be compared usefully.

Identification of remains depends upon the degree of preservation and on the usefulness of skeletal characters for this purpose. The taxonomy of many small mammal species has been based on non-skeletal characteristics, such as colour, which are not evident in the fossil remains. It would appear that birds, although frequently overlooked in fossil assemblages, have tremendous potential for palaeoenvironmental reconstruction. The ecology of birds is generally well known, in comparison with small mammals which tend to be nocturnal or otherwise difficult to observe. In addition, it is frequently possible to identify birds to species level based on bone morphology, especially if the taphonomic agent is one which eats its prey whole so that the individual bones remain intact. The use of bird assemblages is proving very useful in south-eastern Australia where the previous existence of small subtropical rainforest patches, that are palynologically invisible due to the limited pollen production and dispersal capacity of parent plants, are being revealed by the remains of birds restricted to this habitat (Baird 1986).

Invertebrates

The majority of invertebrates used in palaeoenvironmental reconstruction live in aquatic environments (see Chapter 6 for details). A notable exception is the Coleoptera or beetles. This group contains many terrestrial members that possess a variety of features ideal for providing evidence of past climates (Coope 1970). They are taxonomically well-known, they exist in a whole variety of environments and preserve in abundance and diversity in a range of sedimentary deposits. In addition they appear to have been morphologically stable through much of the Quaternary. Their real importance though lies in their great mobility. Unlike plants which are generally long-lived or whose movements can be inhibited by unsuitable vegetation structure or soil conditions, and many animals that are 'tethered' to particular vegetation types, beetles can move rapidly and over long distances in response to changes in climate.

Their sensitivity to climate change in comparison with plants is well illustrated in a study from the middle part of the last glacial period in the British Isles (see Figure 8.17). Here, there is a short phase, around 43 ka ago, dominated by thermophilous beetles that are now confined to southern Euro-Asia, wedged between a cool, moist arctic climatic phase characterized by beetles now largely found in Arctic and subArctic regions and a cool continental phase containing east-central Asian fauna. It is estimated that temperatures must have achieved levels as high if not higher than those of today during the short period of amelioration yet there is no indication of this in pollen data from this period. The preferred explanation for the lack of response in pollen assemblages is that the phase was too short for trees to be able to migrate into the area from their glacial retreats, and that there is no easy means of separation, on palynological grounds, of true tundra and higher temperature treeless vegetation.

A global synthesis

So far in this chapter, information on Quaternary environments has been largely restricted to examples illustrating the different sources of evidence. Here an attempt is made to construct a broad regional picture particularly from studies in vegetation history.

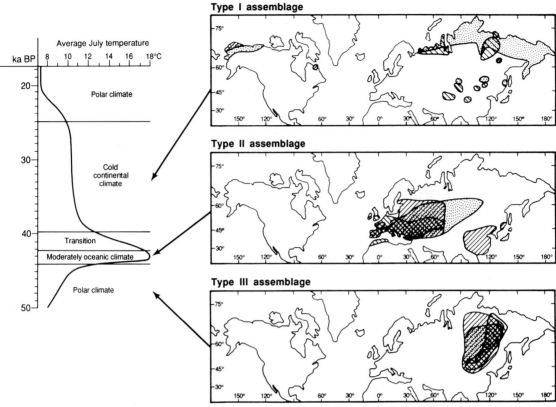

Type I assemblage

Type II assemblage

Type III assemblage

Fig. 8.17 Climate curve for the middle part of the last glacial period in the British Isles derived from present-day distributions of fossil beetle assemblages. The maps show the modern distributions of selected species characteristic of three major periods recognized (adapted from Coope 1975)

Present-day distributions

Present-day distributions of plants and animals are largely a product of Quaternary environmental changes. They also constitute the starting point for the reconstruction and interpretation of fossil assemblages. Some indication of the distribution of plants at least can be gained from an examination of major vegetation types of the world shown in Figure 8.18. Despite significant human modification of the landscape, the major vegetation types largely reflect present climatic variation. From the equator to the poles, where moisture is not limiting, there is a general gradient from tropical rainforest where temperatures are sufficiently high for year-round growth, through broad-leaf forest, where growth can be limited during the winter, and coniferous forest, where growth is confined to only a few months of the year, to tundra, where the growing season is too short to allow the survival of trees. This gradient is most complete in the land-dominated Northern Hemisphere. A similar gradient is evident in tropical montane areas though this becomes progressively truncated with increased latitude. Major moisture gradients are evident in the tropics and subtropics where evergreen rainforest gives way to semi-deciduous forest, shrubland and grass-dominated savanna with

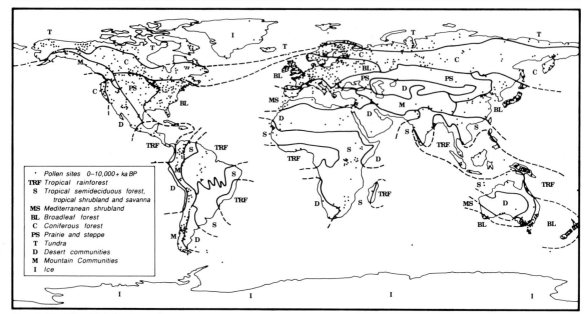

Fig. 8.18 The distribution of known palynological sites covering at least the Holocene in relation to major global vegetation types

lower and more seasonal rainfall and, in the Northern Hemisphere mid-latitudes where broad-leaf forest is replaced by herb-dominated prairie and steppe in continental interiors. Where moisture is most limiting, desert communities occur. The remaining vegetation type identified at this scale is Mediterranean shrubland that occurs in warm temperate latitudes which have a marked winter rainfall maximum.

Superimposed on this structural pattern of vegetation variation is the taxonomic composition of the flora that is a product of evolutionary history as well as present-day environmental conditions. Most evident are floristic differences between Northern and Southern Hemisphere mid- and high-latitude forests that most clearly reflect origins on the separate ancient landmasses of Laurasia and Gondwana respectively. Subsequent evolutionary development and floristic mixing as a consequence of continental movements and its climatic implications as well as the Quaternary climatic fluctuations have led to further global differentiation of floristic patterns, particularly with woody plants. Dispersal of herbs has been generally effective so that differences at the taxonomic level identifiable from fossil data are less pronounced from one region to another.

Modern pollen spectra

The recognition of past vegetation is largely based on pollen signatures derived from modern pollen samples taken within identified communities. Despite the diversity of plants within many communities, most pollen samples are composed predominantly of only a few taxa. Fortunately for vegetation and climatic reconstruction, these are generally the easily recognized and well-known canopy dominants that most clearly reflect macroclimatic variation. The major exceptions are the floristically very diverse tropical rainforests where the majority of trees are animal pollinated and little pollen disperses onto accumulating swamp and lake surfaces, and predominantly herbaceous vegetation where pollen spectra can contain a substantial component from surrounding forest vegetation. However, in these areas, the total pollen spectra generally allow identification of existing vegetation. A selection of the most commonly occurring pollen types and the vegetation types in which they occur is shown in Figure 8.19. These would probably allow the construction of broad vegetation histories from most parts of the world.

The fossil data base

Although not claimed to be complete by any means, the known

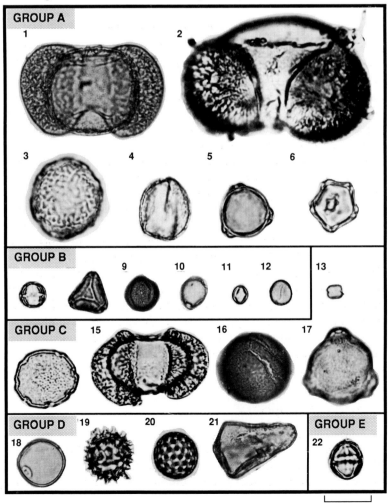

GROUP A
1 2
3 4 5 6

GROUP B
9 10 11 12 13

GROUP C 15 16 17

GROUP D 19 20 21 **GROUP E**
18 22

20 mm

Fig. 8.19 A selection of the major pollen types recorded in Quaternary records. The identified groups are characteristic of environments and floristic elements in different parts of the globe. Group A (1 = *Pinus*, 2 = *Picea*, 3 = *Ulmus*, 4 = *Quercus*, 5 = *Betula*, 6 = *Alnus*), predominantly Northern Hemisphere mid-latitude forests: Group B (7 = *Macaranga*, 8 = *Myrtaceae*, 9 = *Olea*, 10 = *Elaeocarpus*, 12 = *Moraceae*), predominantly Low latitude forests: Group C (13 = *Cunoniaceae*, 14 = *Podocarpus*, 15 = *Araucariaceae*, 16 = *Nothofagus*, 17 = *Casuarinaceae*), predominantly Southern Hemisphere mid-latitude forests: Group D (18 = *Poaceae*, 19 = *Cyperaceae*), predominantly dry or cool herbaceous vegetation: Group E (22 = *Rhizophoraceae*), mangroves.

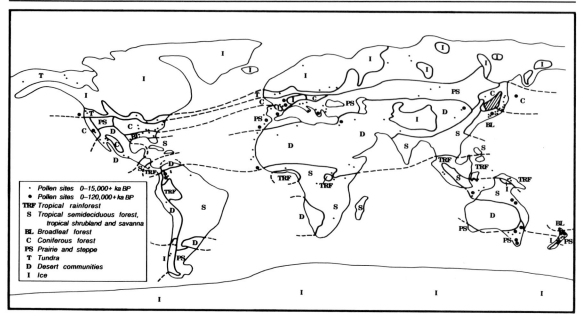

Fig. 8.20 The distribution of known palynological sites extending through the height of the last glacial period and those covering at least the last glacial–interglacial cycle in relation to inferred vegetation at the last glacial maximum. Vegetation distributions are largely from Adams *et al.* 1991

distribution of pollen records extending back from present to at least the early Holocene (see Figure 8.18) gives some impression of the extent of this kind of research within different environments. It is clear that there is relatively little information from many parts of the world. This is due to a number of factors including a lack of suitable sites for continuous sediment accumulation or pollen preservation in arid or seasonally dry environments, perceived difficulties in the application of pollen analysis to lowland tropical rainforests, and a lack of local interest or expertise. Sites are concentrated in traditional areas of research, particularly North America and Europe, where glacial erosion and deposition has resulted in the formation of numerous very suitable sites for investigation. The number and distribution of sites extending back to the last glacial period is very different (see Figure 8.20). Many fewer sites provide a record of this period because conditions were generally much drier in low and middle latitudes while the high latitudes of the Northern Hemisphere were covered in ice. The highest frequencies of sites are found in the middle latitudes of both hemispheres and in the montane tropics. Only a handful of records cover the range of conditions experienced through at least one glacial cycle (Figure 8.20). Of these, the longest records are from large subsiding lake basins and several of these, including those from the high plains of Bogota in South America, the Jordan–Dead Sea Rift Valley, Lake Biwa in Japan and Lake George in Australia provide reasonably continuous records through the whole of the Quaternary. Other substantial records, particularly from Europe, are not illustrated because they do not extend to the present day.

The late Tertiary/Quaternary transition

There is good evidence from many parts of the world of increasing replacement of forest by herbaceous vegetation through the latter part of the Tertiary period in response to a general trend towards drier and cooler conditions. This trend also led to the evolution of diverse assemblages of animals able to take advantage of expanding grazing land. This culminated in the megafauna of the Quaternary. The cooler conditions at higher latitudes, resulting from a steepening of the latitudinal temperature gradient (see Chapter 2) together with increased climatic variability accompanying the oscillatory movements of the developing ice sheets, caused a contraction of warmer elements towards the present tropical zone. Changes appear to have taken place at different rates and at different times depending on the attainment of critical threshold levels although the data for this period are rather patchy and the dating often uncertain.

The most reliable information for this period is derived from the more continuous pollen records which can be tied into the palaeomagnetic time scale. A number of these, particularly from Europe, indicate substantial changes around the proposed Pliocene/Pleistocene boundary about 2.4 Ma ago. Here a number of trees common in the Tertiary disappear with the establishment for the first time of tundra and steppe-like vegetation, indicating the first pronounced cold period. This is well illustrated by the Meinweg section from the Netherlands (see Figure 8.21). Around the same time, Lake George in Australia registers the last appearance of rainforest within the region, which appears to correspond with a change from a summer to winter rainfall regime (McEwen–

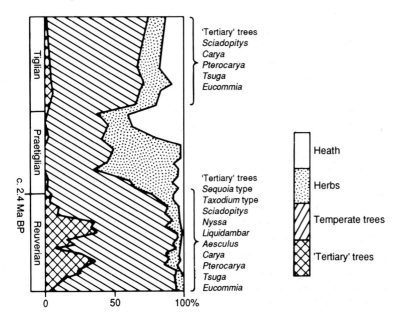

Fig. 8.21 Summary pollen diagram from Meinweg, showing the effect of the first major cool period around 2.4 million years ago on the impoverishment of the tree flora of Europe (modified from Zagwijn 1957)

Mason 1991). In the Colombian Andes, the record demonstrates a change to lower frequency and higher-amplitude climatic fluctuations. Some sustained changes in the nature of the vegetation communities also occur at this time and subsequently, but these may have been more associated with the introduction of Northern Hemisphere elements into the flora with the closure of the isthmus of Panama that occurred in the Pliocene, than with climatic change (Hooghiemstra 1989).

Within Europe, the progressive disappearance of warmth-loving taxa continued with each cold period until the late Pleistocene. Many of these taxa can still be found at equivalent latitudes in North America and Asia, suggesting that temperature oscillations alone were not responsible for their demise. Instead, the major reason is considered to be the existence of east–west mountain barriers which prevented their southward retreat and consequent survival in suitable environments during the glacials.

Glacial–interglacial cycles

There is evidence for cyclical fluctuations through the whole of the Quaternary period but apparent differences between records and generally poor dating control have prohibited detailed correlation prior to the mid-Pleistocene. Since this time the consistent low-frequency, high-amplitude nature of the cycles has allowed more realistic comparisons to be made between terrestrial sequences and general correlations with the established marine oxygen isotope stratigraphy. By this stage also the vegetation appears to have become adjusted to these cyclical oscillations, providing consistent climatic estimates from the pollen data. Summary diagrams for four terrestrial sequences are compared with the deep-sea core record in Figure 8.22.

It is interesting to note that there are major similarities between these records despite the fact that the pollen components are very different and they have been interpreted with respect to different climatic variables. Lake Biwa and Funza are considered to reflect predominantly changes in temperature, Hula basin moisture and Lake George a combination of moisture and temperature. From these and other records it is evident that forests generally expanded during interglacial periods as a result of higher temperatures at higher latitudes and altitudes and an associated increase in effective precipitation at lower latitudes. The Hula basin record is an exception and there are several parts of the world where driest conditions occurred during interglacials.

The Lake George core illustrates the changing role of fire within the recorded period. Through much of the sequence, charcoal peaks correspond with interglacials when it is considered that the vegetation, composed predominantly of sclerophyll forest or woodland, was more able to carry fire than during the cool dry

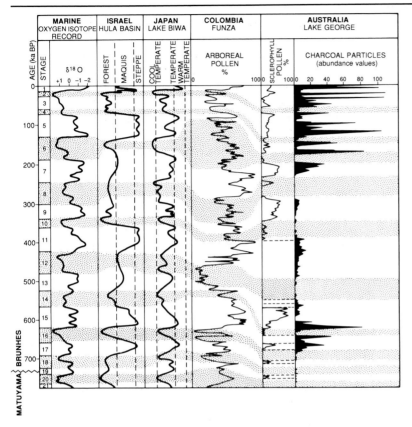

Fig. 8.22 Comparison of four Quaternary records – Hula basin, Israel; Lake Biwa, Japan (Fuji and Horowitz 1989); Funza, Colombian Andes (Hooghiemstra 1988); and Lake George, Australia (Singh and Geissler 1985) – in relation to the marine oxygen isotope record (Imbrie *et al.* 1984) through the last 800 000 years

glacials which supported only an open herbaceous vegetation. This burning pattern is used to infer the status of periods corresponding with deep-sea core Stages 11 and 12 where pollen is not preserved. However, the pattern changes from the period considered to represent Stage 5 (the last interglacial) to one of more intense and continuous burning within both glacial and interglacial periods. The major effect on the vegetation was to make it more fire prone with the replacement of relatively fire-sensitive *Casuarina* that had previously dominated the forests, with fire-promoting *Eucalyptus* and myrtaceous shrubs.

A more detailed picture of changing patterns of vegetation and burning within the last two glacial–interglacial cycles is provided by another record from Australia, Lynch's Crater (see Figure 8.23). This site is a volcanic crater swamp situated within the largest expanse of rainforest in north-eastern Queensland. Within the record, complex rainforest, defined on high levels of pollen from angiosperms, particularly *Cunoniaceae* and *Elaeocarpus*, dominates during the wetter interglacials. This alternates with the drier vegetation types Araucarian rainforest and sclerophyll woodland, identified by high percentages of the gymnosperms *Araucaria* and *Podocarpus*, and *Casuarina* and *Eucalyptus* respectively that become prominent during glacial periods.

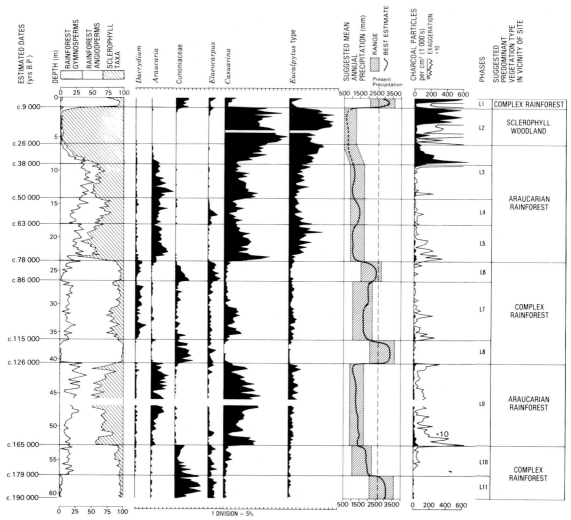

Fig. 8.23 Major features of the pollen diagram from Lynch's Crater, north-eastern Queensland, Australia (modified from Kershaw 1986)

Transitions between glacials and interglacials appear to be abrupt and, in the earlier part of the record they are accompanied by minor peaks in charcoal. It could be the case that the vegetation put under stress from changing climate is more prone to fire and that burning accelerates the change to a vegetation type more in balance with the new climatic conditions. This scenario is similar to that proposed for early Holocene vegetation changes around Everitt Lake, as indicated in an earlier section (page 176).

The general pattern of vegetation variation is altered between about 38 000 and 26 000 years ago, towards the end of the last glacial period, with the total replacement of Araucarian forest by sclerophyll vegetation. Although this corresponds with the beginning of the last glacial maximum, it is considered that climate alone would not have been the cause as no similar change occurred towards the end of the penultimate glacial period. In addition, the gradual nature of the transition stands in marked contrast to other, more obviously climate-induced changes in the record. The sharp increase in charcoal abundance at this time points to fire being the critical factor and this is supported by the change from fire-insensitive rainforest to fire-resistant sclerophyll woodland. This appears to have been a broad regional change as moist Araucarian rainforest now has a very restricted distribution despite the Holocene expansion of complex rainforest. One important component, the conifer *Dacrydium*, is no longer present at all on the Australian mainland and this is probably the most significant range reduction for a plant taxon within the mid to late Pleistocene anywhere in the world (Kershaw 1984).

At both Lake George and Lynch's Crater, increased burning during the late Pleistocene has been attributed to the activities of Aboriginal people. This is despite the antiquity of the Lake George evidence for burning, well beyond the earliest dated evidence for the arrival of people. However, there is some controversy over the dating of the Lake George record and the earliest date for Aboriginal presence is still being pushed further back in time. Despite these uncertainties, some explanation for major sustained changes in the vegetation of Australia, which have not been recognized elsewhere in the world, is required.

The development of the present vegetation pattern

The most intensively studied period of the Quaternary is that from the height of the last glacial period, about 18 000 years ago, to present. This is the period for which most sequences are available, dating is most exact, and which is of greatest relevance for understanding present-day patterns and processes. It is also the most complex in that the landscape has been substantially influenced by human activities. However, on a broad regional scale, climate is still considered to have been the major determinant on vegetation, even in the much-modified landscape of Europe (Huntley 1990).

At the height of the last glacial period when, on a global scale, both temperatures and precipitation reached their lowest levels, terrestrial ice-free landscapes were dominated by essentially treeless tundra and prairie-steppe at high and mid-latitudes and open savanna woodlands and grasslands at low latitudes. Forest vegetation had a very restricted distribution, and although small areas of the major forest types are shown in Figure 8.20, their extent in many places is inferred. Where evidence of forest types does exist, they bore only superficial resemblance to those existing today.

One major area of interest in Quaternary investigations is where and how the components of present-day forest types survived the last and previous glacial periods. With a dearth of fossil evidence,

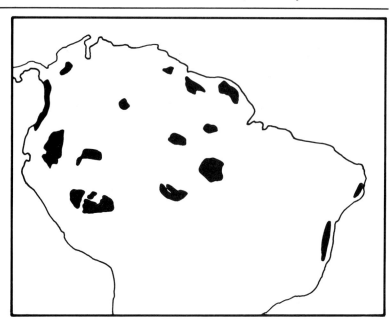

Fig. 8.24 Overlapping centres of endemism of butterflies, plants and birds in South American rainforests (from Whitmore and Prance 1987)

most discussion has centred on likely retreat areas or refugia, constructed from present-day landscape features, particularly the modern distributions of plants and animals. Debate has been most intense over the situation in the Amazon region of South America where the identification of possible refugia has been applied not only to tropical rainforest survival but also to the explanation of high species diversity and endemism within tropical rainforest systems. In the refuge theory as originally envisaged by Haffer (1969) rainforest survived in areas now possessing high diversity and endemism within the great expanse of Amazonian rainforest. Inferred refugia, determined from the overlapping centres of endemism for three rainforest taxonomic groups (butterflies, plants and birds), are shown in Figure 8.24. Many of these are in highland areas where it is thought that sufficiently moist conditions were maintained during the glacial periods to allow rainforest survival while savanna vegetation occupied the lowlands. Speciation would then be enhanced by geographical isolation during these glacial periods and also by subsequent hybridization during interglacials along contacts where forests from expanding refugia met (see Figure 8.25).

This theory may have some merit in providing mechanisms for speciation and ecotypic variation in taxa within the Quaternary period but recent fossil evidence suggests that it is probably an unlikely model for explaining patterns of diversity within rainforests (Bush and Colinvaux 1990). In the first place, palynological data indicate that mountain areas in Amazonia and elsewhere in the tropics suffered a depression of temperatures by some

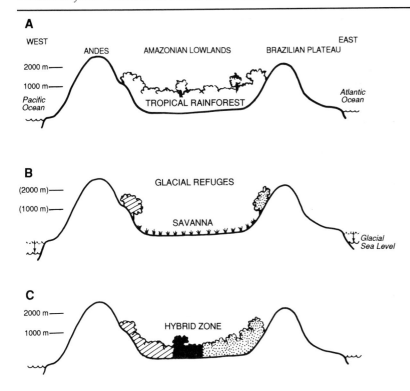

Fig. 8.25 Schematic diagram showing the major features of the refuge theory and modes of speciation.
(A) The pre-glacial situation: tropical rainforest covers the whole of the Amazonian lowlands to an altitude of approximately 1100 m above sea level.
(B) Glacial periods: the rainforest fragments and is restricted to isolated highland refugia where it experiences allopatric speciation.
(C) Interglacial periods: rainforest expands from refugia with the possibility of hybridization of related species occurring where refugial communities coalesce

4–6°C causing montane forests to descend to the altitudes considered to contain the lowland rainforest refugia. Consequently, although moisture levels are likely to have been suitable, tropical rainforest would have been restricted to the lowlands as a result of low temperatures. This suggests that either moisture levels were sufficient to support rainforest in some parts of the lowlands or that this vegetation survived in very localized moist areas such as river valleys and seepages. It would appear then that rainforest has expanded only recently into those areas that were regarded as refugia.

Areas of high species richness and endemism might be better explained by the combination of a whole range of factors including habitat diversity, frequent disturbances such as floods, streams and occasional fires and barriers formed by changing river patterns in addition to changing distribution patterns resulting from glacial–interglacial climatic oscillations. In any case, it needs to be remembered that complex rainforests were much more extensive during the more stable Tertiary period and regular disturbances or fluctuating climates may not have to be inferred to explain diversity in many rainforest components.

Bush and Colinvaux also attack the basic concept of refugia on the grounds that it implies the existence of discrete communities that expand and contract as a whole, an idea that is no longer tenable in the light of palynological and other evidence on the changing nature of plant communities. The most substantial evidence on long-term patterns of community change comes from analyses undertaken on pollen records since the last glacial, particularly those on the extensive data sets in North America and Europe. Here the development of forests can be traced as the ice sheets waned and the climate ameliorated.

Davis (1976) plotted the rates of migration of individual trees that make up the present composition of mixed deciduous forests of eastern North America from the dates of arrival at each of the pollen sites. From the selected examples shown in Figure 8.26, it is clear that each taxon has behaved individually, not as a member of an integrated community. Each appears to have had a different range during the glacial period and to have expanded at different rates since that time. In line with this, the composition of the vegetation has been changing continuously and there is little reason to believe that any kind of stability has been achieved with the present vegetation pattern.

The apparent lack of integrity of communities and differential migration rates have major implications for the reconstruction of climate from past vegetation. To what extent is the vegetation reflecting, or in balance with, prevailing climatic conditions at any point in time? It has already been suggested from the study of fossil insect assemblages that there may be a substantial lag in response of trees to climatic change. This could be caused by a number of factors including the limited dispersal capacity of seeds, competition from existing vegetation and the unsuitability of environmental attributes apart from climate. Several studies have identified this latter factor as important. In an attempt to explain the curious late Holocene expansion of hemlock westward within the last 5000 years (see Figure 8.26). Davis *et al.*

Fig. 8.26 Migration of selected mixed deciduous forest trees in eastern North America at 1000-year intervals from areas occupied during the last glacial period (from Davis 1976). (Dashed lines indicate the present distribution)

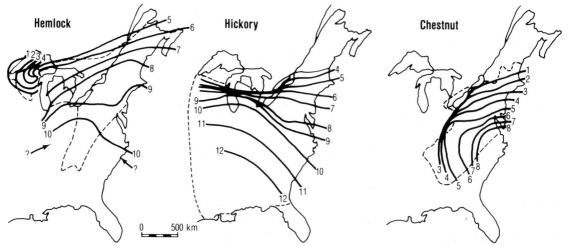

(1986) examined in detail a large number of sites within the Great Lakes region. It was concluded that this expansion was unlikely to have been the result of climate becoming suitable; rather it was most probably caused by problems experienced by hemlock in crossing Lake Michigan or the xeric prairie around its southern margin that delayed expansion for up to 1000 years. In another example, a lag in colonization of Britain by a species of birch after the last glacial period is attributed to the time required for the development of suitable soils on the glacial sediments (Pennington 1986). In a different environment, there appears to have been a lag of some 3000 years between the attainment of climatic conditions suitable for rainforest colonization and its arrival at a site on the Atherton Tableland of north-east Queensland, after the sclerophyll woodland phase noted in the Lynch's Crater pollen diagram (Figure 8.23). This delay was initially attributed to the slow intrinsic migration rate of rainforest from retreats occupied during the sclerophyll phase but fine-resolution pollen and charcoal studies have indicated that fire probably inhibited the expansion process (Walker and Chen 1987).

In the same study, the mechanism of invasion by newly arrived tree taxa into established forest was investigated. From pollen-influx data, it was determined that most populations increased exponentially indicating little inhibition by other taxa to their expansion. Similar results were obtained from an examination of the expansion of deciduous forest taxa at Hockham Mere (Bennett 1983b). These studies suggest that competition from existing forest components may not seriously inhibit the spread of new species once climatic conditions become suitable for their establishment.

Although the problem of migration rates has not been fully resolved, there is growing confidence that, on time scales of thousands of years, climate is the major determinant of plant distributions and that the complex and varying associations of plants through time is a reflection of continuously changing climate with many past climates lacking modern analogues (Webb 1986). On this assumption, a number of palynologists have been active in the construction of detailed, quantitative climatic estimates from pollen data using the transfer function approach similar to that developed for interpretation of assemblages of marine biota (see Chapter 5). The first step is the construction of isopoll maps (contoured maps based on equal percentages) for individual pollen taxa at the present day and the examination of the relationships between pollen abundances and climatic parameters. Those taxa showing climate-related variation are selected for the calculation of transfer functions which express the pollen-climate relationships. These transfer functions are then applied to the determination of climatic conditions from fossil pollen spectra and climate curves are constructed for individual pollen

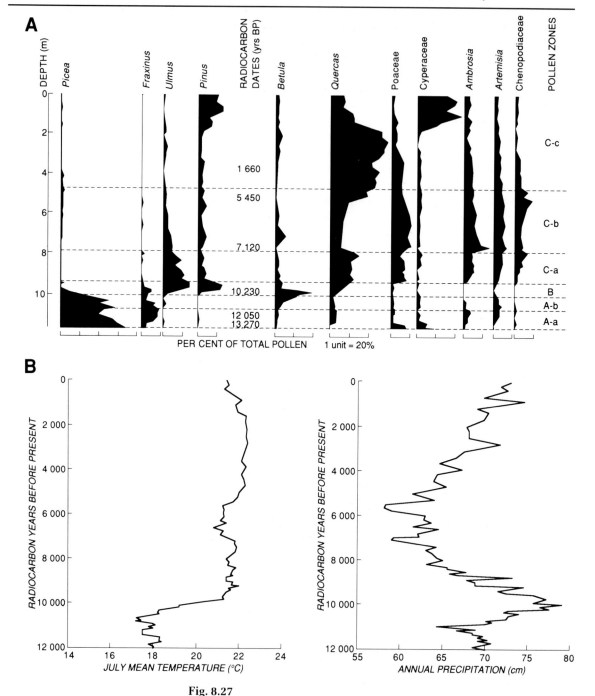

Fig. 8.27
(A) Pollen diagram from Kirchner Marsh, Minnesota, featuring those pollen types used in the reconstruction of climate from the site using transfer functions derived from the extensive set of modern pollen spectra from North America.
(B) Derived estimates of temperature and precipitation from the calibrated Kirchner Marsh pollen record (from Webb 1980)

diagrams. The results of this exercise on one pollen diagram using the North American data base are shown in Figure 8.27.

The application of isopoll mapping and transfer functions to the large data sets of North America and Europe is revealing substantial regional variation in late Quaternary climates that was previously either not suspected or difficult to substantiate from studies of restricted areas. A recent example from Europe, although expressing climatic conditions only in qualitative terms, illustrates the value of being able to extract information on seasonal climatic variation that was seldom possible with more traditional methods (Figure 8.28).

It might be expected that the development of similar data bases in other parts of the world will reveal similar insights into the regional nature of vegetation and climatic change. However, it must be remembered that these Northern Hemisphere regions are unique in that glaciation effectively removed vegetation from large areas and has excluded those plants incapable of rapid and large-scale migration. In many other parts of the world, it is becoming increasingly apparent that floras have remained relatively stationary with only changes in relative abundance occurring as a result of Quaternary climatic oscillations. This would have allowed the persistence of a greater range of response-types to environmental change including relict taxon populations. Consequently many of the concepts that have been developed from the study of vegetation in once glaciated regions may not be universally applicable and it may be difficult to construct broad-scale regional climatic changes from pollen data using the transfer function approach. If this is the case then the focus of research in these areas should continue to be placed on the intensive study of individual sites or small regions. If these sites are sensitively situated in relation to substantial climatic gradients, then they may still provide reliable climatic information and perhaps not be as affected by the migrational lag problem as once glaciated areas.

Further reading

Betancourt, J. L., Van Devender, T. R. and Martin, P. S. 1990: *Packrat Middens: the last 40 000 years of biotic change.* Tucson: The University of Arizona Press.

Birks, H. J. B. and Birks, H. H. 1981: *Quaternary palaeoecology.* London: Edward Arnold.

Bradley, R. S. 1985: *Quaternary paleoclimatology.* Mass: Allen and Unwin.

Faegri, K., Kaland, P. E. and Krzywinski, K. 1989: *Textbook of pollen analysis*, Fourth Edition. Chichester: Wiley.

Fritts, H. C. 1976: *Tree rings and climate.* London: Academic Press.

Huntley, B. and Webb, T. 1988: *Vegetation history.* Dordrecht: Kluwer Academic Publishers.

Summer Temperature	Winter Temperature	Annual Temperature

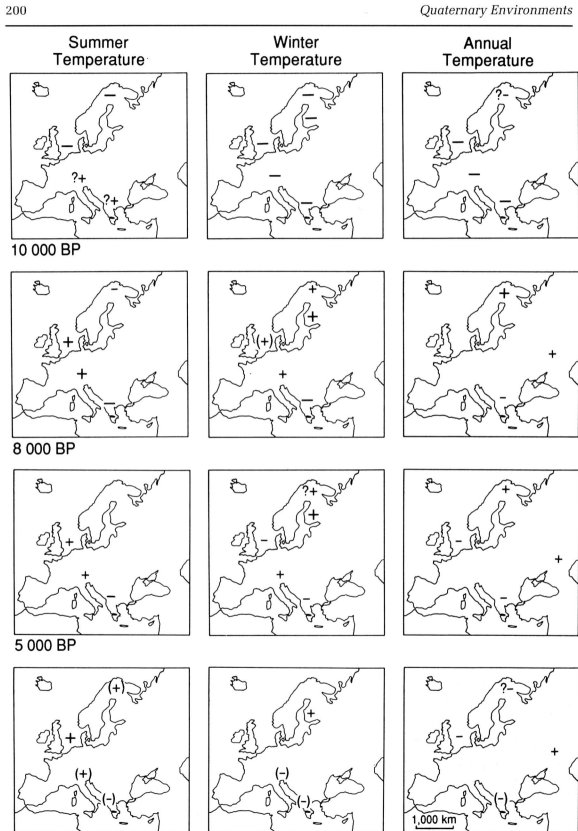

10 000 BP

8 000 BP

5 000 BP

2 000 BP

1,000 km

Moore, P. D., Webb, J. A. and Collinson, M. E. 1991: *Pollen analysis.* Second edition. Oxford: Blackwell.

Schweingruber, F. H. 1988: *Tree rings: basics and applications of dendrochronology.* Dordrecht: D. Reidel.

Walker, D. and Guppy, J. C. 1978: *Biology and Quaternary environments.* Canberra: Australian Academy of Science.

Fig. 8.28 Regional climatic changes during the Holocene of Europe derived from changing vegetation patterns deduced from pollen data. The size of the symbols indicates the relative magnitude of the differences from present-day values (from Huntley 1990)

9 Human Origins, Innovations and Migrations

Knowledge may have its purposes, but guessing is always more fun than knowing.

W. H. Auden (1907–1974),
Archaeology.

The Miocene hominoids of Africa and Eurasia

Throughout most of the Miocene epoch, for nearly 20 million years, a number of quadrupedal, vegetarian hominoids roamed the forests and woodlands of Africa and Eurasia (Laporte and Zihlman 1983; Hill and Ward 1988). They were cosmopolitan in their distribution (Figure 9.1) suggesting that there were very few major geographical barriers inhibiting their ability to range freely across this vast region at that time. Among the better known of these Miocene hominoids is the large, forest-dwelling genus *Gigantopithecus*, and the smaller genus *Ramapithecus*, which appears to have foraged successfully across the ecotone between the forest and grassland, at least during the later Miocene. The

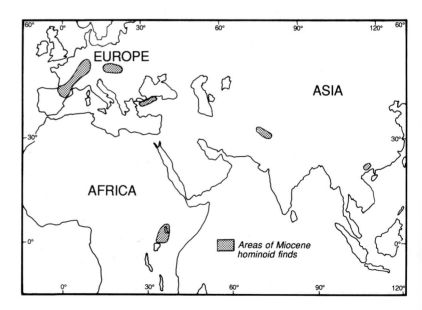

Fig. 9.1 Distribution of Miocene hominoid fossil discoveries in Africa, Asia and Europe (after Scarre *et al.* 1988)

Miocene hominoids disappeared from the fossil record about 7 Ma ago, and there is a tantalizing gap of several million years before the first appearance of the Pliocene hominids in Africa between 5 and 4 Ma ago.

The Pliocene hominids of Africa

Only in Africa have Pliocene hominids been found so far (White *et al.* 1981; Harris 1983; Johanson 1989). The best dated and most abundant Pliocene hominid fossils have come from the rift deposits of Ethiopia (the Afar Depression), Kenya (especially Koobi Fora near Lake Turkana) and Tanzania (most notably, Olduvai Gorge), but important early discoveries have also come from certain limestone caves in southern Africa (Figure 9.2) (White and Harris 1977; Walker *et al.* 1986; Johanson *et al.* 1987; Harris *et al.* 1988; Feibel *et al.* 1989).

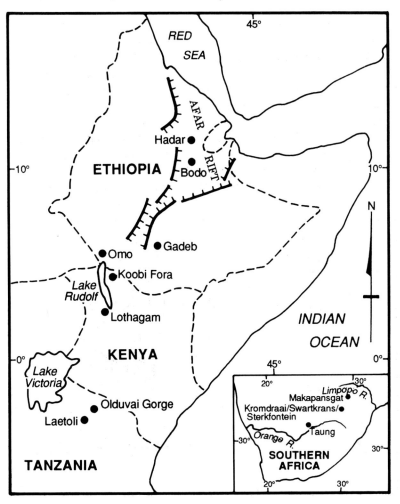

Fig. 9.2 Location of major hominid sites in Africa. Except for Gadeb (an Early Stone Age site in the Ethiopian uplands with possible early use of fire by *Homo erectus*), all the sites shown have yielded Australopithecine fossils ranging in age from 4 to 1 Ma

Recent advances in molecular biology are helping to clarify the vexed question of when the pongid and hominid lineages began to diverge (Tuttle 1988). (The primate 'superfamily' *Hominidea* consists of all modern and ancestral *Pongidae* and *Hominidae*.) Similarities in immunological response and in plasma protein structures point to divergence between the pongid and hominid families somewhere between 6 and 3 Ma ago (Sarich and Wilson 1967). The oldest securely dated *Australopithecus afarensis* fossils, recovered in late 1981 from the Middle Awash Valley of the southern Afar Rift in Ethiopia, come from 11 m beneath a primary airfall tuff dated by $^{40}Ar/^{39}Ar$ and zircon fission track analyses to 3.8–4.0 Ma ago (Clark *et al.* 1984; Hall *et al.* 1984). Although small-brained, with a cranial capacity roughly similar to that of a modern chimpanzee, these early Pliocene *A. afarensis* hominids were indisputably bipedal, a conclusion based on detailed anatomical observations (White and Suwa 1987) and strikingly confirmed by the remarkable 3.5 Ma-old fossil footprints of *A. afarensis* in carbonatite ash at the prehistoric site of Laetoli in Tanzania (Leakey and Hay 1979).

In connection with these hominid discoveries from Africa, it is interesting to call to mind some of Charles Darwin's comments about early human origins. In *The Descent of Man*, Darwin (1871) wrote as follows: 'In each great region of the world the living mammals are closely related to the extinct species of the same region. It is therefore probable that Africa was formerly inhabited by extinct apes closely allied to the gorilla and chimpanzee: and as these two species are now man's nearest allies, it is somewhat more probable that our early progenitors lived on the African continent than elsewhere'. He went on to say that 'those regions which are the most likely to afford remains connecting man with some extinct, ape-like creature, have not as yet been searched by geologists' (*op. cit.*, 520–1).

A century later, Darwin's prediction has been amply vindicated, first with Raymond Dart's discovery of *Australopithecus africanus* in South Africa in 1924, then with Robert Broom's discovery of *Australopithecus robustus* in 1938, from the South African cave site of Kromdraai, and more recently with the spectacular finds by Mary and Louis Leakey from Olduvai Gorge, which span nearly two million years of human evolution (Gowlett 1984a; Weaver 1985; Fagan 1989).

Very considerable effort has been devoted to reconstructing the pattern of global tectonic and climatic events which may have been associated, directly or indirectly, with the emergence of the Australopithecine hominids in Africa (for comprehensive reviews see Axelrod and Raven 1978; Brain 1981; Behrensmeyer 1982; Laporte and Zihlman 1983; Van Zinderen Bakker and Mercer 1986; Adamson and Williams 1987; Singh 1988; Prentice and Denton 1988). If one is seeking a single major cause, it is tempting to blame the Messinian salinity crisis of 6–5 Ma (Ryan 1973; Hsü *et al.* 1977; Hsü 1983) for the genetic isolation of Africa from Eurasia which enabled the Australopithecine hominids to evolve in isolation from the rest of the world. Be that as it may, further speculation would be unprofitable until the gap in the hominoid fossil record between 7 and 5 Ma has been finally closed, and until the time of divergence between pongid and hominid is more precisely known.

Homo habilis: the first stone toolmaker

Like the present-day great apes, the early Pliocene Australopithecine hominids were opportunistic users of twigs, sticks and stones. Once used, these temporary tools were forthwith discarded, never to be used again. There is no evidence that any of these creatures ever conceived of the idea of deliberately modifying stone fragments in order to make stone tools until the very late Pliocene, roughly 2.5 Ma ago. At this time in the Gona Valley near Hadar in the southern Afar Rift of Ethiopia, stone tools made to a recognizable and replicated pattern made their first appearance in the archaeological record (Harris 1980, 1983). These tools were extremely simple and highly effective (Roche 1980; Gowlett 1984b). Several sharp flakes were dislodged from a large pebble by striking a single hard blow with a suitable hammerstone. Armed with a handful of such flakes, a scavenging hominid could cut through the tough hide of some large and recently dead herbivore, detach sizeable portions of meat and flee unscathed with this protein-rich food while the African carnivores were still dozing in the tropical heat of the day (Clark 1976a). The pebble tool or 'Oldowan tradition' persisted with minimal modification for a further million years, until 1.5 Ma ago, prompting us to ask why these early, small-brained, bipedal hominids first became toolmakers 2.5 Ma ago, rather than earlier or later.

Late Pliocene environmental changes

As we have seen in earlier chapters, a number of major environmental changes took place towards 2.5 Ma ago (for detailed discussions see Brain 1981; Behrensmeyer 1982; Adamson and Williams 1987; Vrba 1988). The most notable change, with far-flung repercussions for global climate, was the sudden accumulation of ice in North America and Europe. This rapid build-up of ice in the Northern Hemisphere is very evident in the oxygen isotope record of deep-sea cores, and is reasonably accurately dated to 2.5–2.4 Ma ago (Shackleton and Opdyke 1977; Shackleton *et al.* 1984).

There were several major side effects of this rapid accumulation of ice at high and middle latitudes in the Northern Hemisphere. At Pliocene Lake Gadeb in the south-eastern uplands of Ethiopia abundant pollen grains preserved in the lacustrine diatomites indicate that the montane climate in this equatorial region became significantly cooler and drier towards 2.5–2.35 Ma ago (Bonnefille 1983). In China, loess began to accumulate 2.4 Ma ago, when the late Pliocene climate became cold and dry (Heller and Liu 1982). In southern Europe, pollen evidence shows that the seasonally fluctuating Mediterranean type of climate began about 2.3 Ma ago (Suc 1984).

Prehistoric meat-eating and butchery sites

It is tempting to speculate that late Pliocene intertropical desiccation associated with global cooling and ice-cap growth at high altitudes, may have displaced the African vegetation belts and so initiated competition for food among the robust and gracile Australopithecine hominids (Assefa *et al.* 1982). In any event, the possessors of sharp stone flakes would have been able to supplement their mostly vegetarian diet with meat cut from animals that they themselves may not have killed. Several days of decay are needed before the hides of the larger African herbivores can be tackled effectively by carnivores (Assefa *et al.* 1982), time enough for opportunistic tool-using hominids to scavenge some midday meat.

Although Binford (1981, 1983) has severely criticized some of the evidence relating to possible prehistoric big-game hunting and butchery sites, there is no doubt that as time went on, meat became an increasingly important part of the diet of such ancestral humans as *Homo habilis* and *Homo erectus*. Before a prehistoric butchery site can be accepted as such, four sets of criteria need to be met. The first and most obvious prerequisite is a single carcass in undisturbed or primary context (Clark and Haynes 1970). A second requirement is a recognizable spatial pattern of stone tools, also in primary contact and in some form of functional association with the bones of the butchered animal (Clark and Kurashina 1979). Ideally, the bones should have visible cutmarks close to their extremities and roughly perpendicular to the long axis of the disarticulated bones (Jones 1980; Potts and Shipman 1981; Bunn 1981). Pseudo-cutmarks can be produced by trampling animals and by various other natural processes (Behrensmeyer *et al.* 1986) so that a fourth and final prerequisite is micro-wear on the cutting edge of discarded flakes known from experimental work to be solely caused by cutting through hide, ligaments and flesh (Keeley 1980; Keeley and Toth 1981).

Homo erectus: fire and the Acheulian tradition

Homo habilis is widely believed to be the first stone toolmaker, and the originator of the Oldowan (or pebble tool) tradition (Figure 9.3). For a million years, from roughly 2.5 to 1.5 Ma BP, there was virtually no change in the stone tools, but a dramatic change took place at 1.5 Ma BP, associated with the first appearance of *Homo erectus*, a creature with a larger and more complex brain than its predecessor, *Homo habilis* (Figure 9.3). The new stone toolmaking technique consisted in detaching large flakes from a big lump of rock; the flakes were then fashioned into symmetrical handaxes and cleavers 15–30 cm long by removing smaller flakes from both sides ('bifacial flaking') all around the

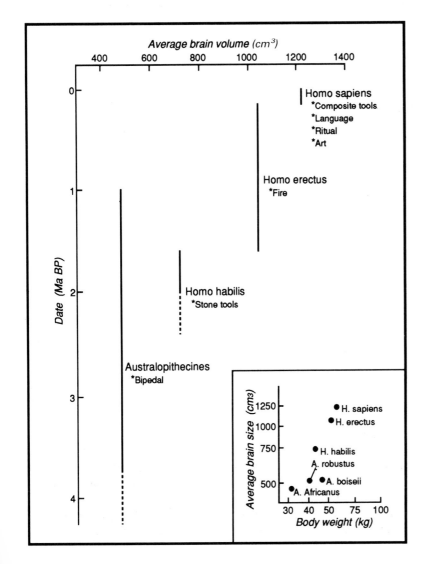

Fig. 9.3 Changes in hominid physical and cultural development from Early Pliocene to Late Pleistocene (after Fagan 1989). The inset shows that the *Homo* line has a much higher ratio of cranial capacity to body weight than any of the Australopithecines (after Pilbeam and Gould 1974)

GEOLOGICAL
TIME SCALE (ka BP)

TOOL TECHNOLOGY

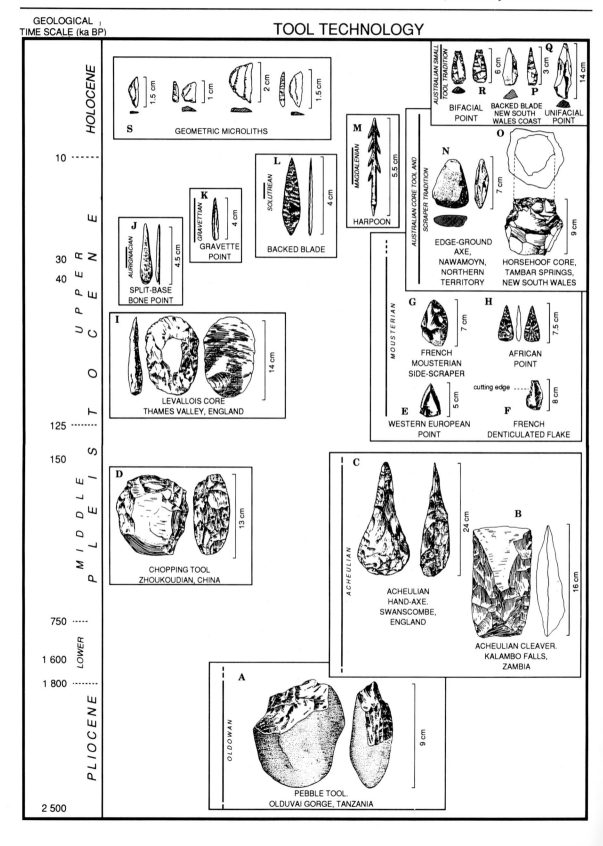

HOLOCENE

S GEOMETRIC MICROLITHS
1.5 cm   1 cm   2 cm   1.5 cm

AUSTRALIAN SMALL TOOL TRADITION

R BIFACIAL POINT   **P** BACKED BLADE NEW SOUTH WALES COAST   **Q** UNIFACIAL POINT
6 cm   3 cm   14 cm

M *MAGDALENIAN* HARPOON   5.5 cm

10 ------

L *SOLUTREAN* BACKED BLADE   4 cm

AUSTRALIAN CORE TOOL AND SCRAPER TRADITION

N EDGE-GROUND AXE, NAWAMOYN, NORTHERN TERRITORY   7 cm

O HORSEHOOF CORE, TAMBAR SPRINGS, NEW SOUTH WALES   9 cm

K *GRAVETTIAN* GRAVETTE POINT   4 cm

J *AURIGNACIAN* SPLIT-BASE BONE POINT   4.5 cm

30
40

MOUSTERIAN

G FRENCH MOUSTERIAN SIDE-SCRAPER   7 cm

H AFRICAN POINT   7.5 cm

I LEVALLOIS CORE THAMES VALLEY, ENGLAND   14 cm

E WESTERN EUROPEAN POINT   5 cm

F FRENCH DENTICULATED FLAKE   cutting edge   8 cm

125 --------

150 ------

D CHOPPING TOOL ZHOUKOUDIAN, CHINA   13 cm

ACHEULIAN

C ACHEULIAN HAND-AXE. SWANSCOMBE, ENGLAND   24 cm

B ACHEULIAN CLEAVER. KALAMBO FALLS, ZAMBIA   16 cm

750 ------
1 600
1 800 --------

LOWER

A *OLDOWAN* PEBBLE TOOL. OLDUVAI GORGE, TANZANIA   9 cm

2 500

U P P E R   P L E I S T O C E N E   **M I D D L E   P L E I S T O C E N E**   **P L I O C E N E**

periphery to give a sharp, serrated edge (Figure 9.4). These bifacially worked cleavers and handaxes are characteristic of the 'Acheulian tradition', which persisted, with progressive refinements, until about 150 ka ago, together with an improved version of the pebble-tool tradition, termed the 'Developed Oldowan'. (Together, the Oldowan and Acheulian comprise the 'early Stone Age' or 'the lower Palaeolithic'.) A second major technological discovery attributable to *Homo erectus* is that of fire (Barbetti *et al.* 1980; Gowlett *et al.* 1981; Clark and Harris 1985; Brain and Sillen 1988). The use of fire not only had important implications for diet, ease of mastication, ease of digestion and the curing and storage of smoked meat (Clark and Harris 1985) but also provided greater security, warmth and the possibility of greater social interaction by all members of the group after sunset, including an increasing reliance on verbal communication. It is noteworthy that with the discovery of fire, small bands of *Homo erectus* moved out of the tropical savanna lowlands to occupy high-altitude grasslands such as the Gadeb plains in the east-central highlands of Ethiopia (Clark 1987) as well as moving out of Africa to occupy new sites at higher latitudes in Europe and Asia, roughly a million years ago (Figure 9.5).

Migration of *Homo erectus* from Africa to Eurasia

One of the best known *Homo erectus* sites in Asia is the 'Peking Man' site of Choukoutien 50 km south-west of Beijing in north-eastern China. Occupation of the limestone cave began towards 460 ka ago and continued intermittently until roughly 230 ka BP, indicating a movement into cooler latitudes by at least 0.5 Ma ago (Liu 1983; Wu and Lin 1983). Two older *Homo erectus* sites near Lantian in central China have recently been redated palaeomagnetically (see Appendix) to 0.65 Ma BP for the Chenjiawo mandible and to 1.15 Ma BP for the Gongwangling cranium (An and Ho 1989). The *Homo erectus* sites from Java are still poorly dated but on the basis of cranial morphology and brain size they appear to be somewhat younger than the three oldest *Homo erectus* sites in China.

In Syria, Acheulian bifaces are associated with an extinct middle Pleistocene fauna of *Elephas*, *Equus* and *Hippo*, yielding an approximate age range from c. 0.7 Ma to 125 ka BP. Upper Acheulian artefacts in the Golan Heights region of Syria/Israel are

Fig. 9.4 The origin and development of Palaeolithic tool technology from 2.5 Ma to 10 ka. Neolithic polished stone tools are not shown here (after Mulvaney 1975; Clark 1977; White and O'Connell 1982; Gowlett 1984a; Fagan 1984)

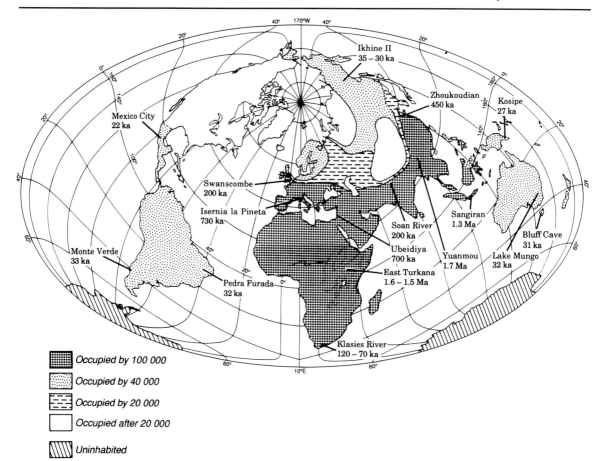

Occupied by 100 000

Occupied by 40 000

Occupied by 20 000

Occupied after 20 000

Uninhabited

Fig. 9.5 Successive stages in the prehistoric settlement of the world, showing the very late entry into Australia and the Americas (after Fagan 1989; Barraclough 1982; White and O'Connell 1982; Davis 1986; Scarre *et al.* 1988; Dillehay and Collins 1988; Cosgrove 1989)

bracketed by lavas dated to 470 and 230 ka BP. In the Jordan Valley of Israel, the 'Ubeidiya Formation contains chopper tools and flakes comparable to those in Upper Bed II at Olduvai, suggesting a possible age of 1.4 Ma and a bracketing age range of 1.5 to 0.7 Ma BP (Tchernov 1987). In France, hearths in the Escale cave in the Massif Central date back to the start of the Middle Pleistocene (0.7 Ma BP), and in Hungary and Romania, flaked pebbles, flake tools and hearths occur in Middle Pleistocene formations dated between c. 0.7 Ma and 125 ka.

On the basis of the foregoing evidence it seems fair to conclude that *Homo erectus* was certainly present in the warmer parts of western and central Europe, beyond the maximum limits of the ice, by at least 0.7 Ma, and somewhat earlier in the Middle East and China. The cranial capacity of *Homo erectus* showed a steady increase during this time from around 775 cc at 1.6 Ma to 1300 cc by 0.2 Ma ago, but there was remarkably little change in the complexity of the Acheulian and Developed Oldowan tool kits during this long interval of time (Figure 9.3). To quote Clark (1976a, 47): "*Homo erectus*, though he exploited a wide range of resources, did so only at a very low level of efficiency and with minimal ability to specialize". The paradox here, as Clark has noted, is that with the emergence of *Homo erectus* there was a drastic change in man's ability to adapt to a wide range of environments, but only a relatively small change in stone-tool technology.

Despite the lack of any major change in stone-tool technology between about 1.5 and 0.2 Ma, the multi-purpose Acheulian and Developed Oldowan tool kits enabled the early Stone Age peoples to acquire enough plant and animal foods to survive and proliferate. As noted earlier, an additional and very important item in the material repertoire of *Homo erectus* was fire, the use of which dates to 1.0–1.5 Ma at Swartkrans cave in South Africa (Brain and Sillen 1988), to perhaps 1.5 Ma at Chesowanja in Kenya (Gowlett *et al.* 1981) and possibly also at Gadeb in the Ethiopian uplands (Barbetti *et al.* 1980), and to c. 0.7 Ma at Escale cave in France and 0.5 Ma at Choukoutien cave in China (Gowlett 1984a). With the warmth and protection it afforded, fire enabled small bands of *Homo erectus* hunters to venture into hitherto unoccupied parts of Europe and Asia where the long cold winters required more effective shelters than were needed in the African savanna.

From *Homo erectus* to *Homo sapiens*

Apart from a gradual increase in cranial capacity between 1.5 Ma and roughly 0.3 Ma there is very little evidence of any progressive morphological changes in the skeletal anatomy of *Homo erectus* until about 0.2 and 0.3 Ma ago when populations showing a combination of *Homo erectus* and *Homo sapiens* traits began to appear in Africa. At such sites as Laetoli in Tanzania, Omo in southern Ethiopia/northern Kenya, and Broken Hill in Zambia, the skulls have a relatively large cranial capacity of around 1200 cc but retain the large brow ridges, low sloping frontals and thick bones characteristic of *Homo erectus*. All of these sites are c. 150–100 ka old. By 90–100 ka, anatomically modern humans were present at Border Cave in South Africa and at Es Skhul in Israel, indicating that the transition from *Homo erectus* to *Homo sapiens* must have taken place between about 200 and 100 ka ago (Stringer *et al.* 1989).

From Early Stone Age to Middle Stone Age

By 250–130 ka two distinct toolmaking traditions were widespread in Africa and Eurasia. The Acheulian tradition involved the use of large, standardized bifacial tools – typified by the cleavers and handaxes illustrated in Figure 9.4. An assemblage of very variable choppers, scrapers and stone knives was characteristic of the Developed Oldowan tradition. Equipped with one or both of these Middle Pleistocene tool kits, the *Homo erectus* populations were able to occupy every type of habitat apart from deserts and lowland and montane evergreen forests. The Early Stone Age (or Lower Palaeolithic) was characterized by a very slow rate of cultural change, a limited range of foraging activities and an unspecialized use of natural resources. Flakes were used to obtain meat, wooden sticks to dig up roots, tubers and rhizomes, and hammer stones to crack nuts (Clark 1975, 1976a).

The Middle Stone Age (or Middle Palaeolithic) is characterized by a revolutionary new technique which involved prior flaking and preparation of the parent corestone to allow greater control over the form of the finished flake. These Levallois cores and flakes are diagnostic of the Middle Palaeolithic industries of Europe and Asia as well as of the Middle Stone Age assemblages throughout Africa (Clark 1988). The co-occurrence at many cave sites of small bifacial implements (Mousterian points) with fragmented bones of game animals suggests that the stone points were being used as spear tips – a good indication that composite tools of wood and stone were now being widely used.

It would be misleading to claim a simple equivalence between the inception of the Middle Stone Age and the demise of *Homo erectus* but it is worth noting that this second major innovation in stone toolmaking (the first being the large bifacially worked flake tools of the Acheulian) coincides with the emergence of recognizably modern humans – *Homo sapiens* – some 150–100 ka ago. The increase in brain complexity at this time strongly suggests a greater development of language skills, with an associated increase in the degree of social cohesion necessary in group hunting and sharing (Trinkhaus and Howells 1979). Whether the ability to articulate words was the same in all *Homo sapiens* populations is still a matter of debate (Trinkhaus 1986) but the recent discovery of a 60 ka Middle Palaeolithic hyoid bone at Kebara cave on Mount Carmel in Israel shows that humans were certainly morphologically capable of fully modern speech at that time (Arensburg *et al.* 1989).

Middle Stone Age ritual and art

The middle Stone Age was also a time of more intensive settlement in Africa and Eurasia (Figure 9.5) and a time when a number of distinctively human cultural traits first became apparent (Figure 9.6). The excavations at Shanidar in the Zagros hills of northern Iraq strongly suggest that these Neanderthal hunters were not only caring for their sick and elderly but were also burying their dead over 60 ka ago (Trinkhaus 1983), although

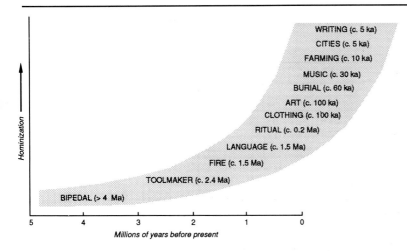

Fig. 9.6 The development of human culture during the late Pliocene and Quaternary showing increasing 'hominization' through time (after Tobias 1979; Williams 1985)

Gargett (1989) has recently called into question the validity of some of the evidence for deliberate Middle Palaeolithic disposal of the dead. The problem is slightly compounded by the recent thermoluminescence dating of Neanderthal remains from Saint-Césaire in France to 36.3 ± 2.7 ka, which makes them much younger than some of the upper Palaeolithic (Aurignacian) sites in Spain which are believed to have been occupied by modern humans (Mercier *et al.* 1991; Stringer and Grün 1991) so that a reappraisal of Neanderthal sites now seems in order (Diamond 1989).

Recent uranium-thorium dating of the cave travertine at the Mousterian site of Tata in Hungary shows that the cave was occupied towards the end of the last interglacial some 100 ka ago. Among the artefacts was a polished ivory plaque 11 cm long carved from a mammoth molar and rubbed with red ochre (Schwarcz and Skoflek 1982). The *Homo sapiens neanderthalensis* sculptor responsible for this plaque had a cranial capacity of 1400–1500 cc, which is fully equivalent to that of present-day humans.

Ritual and art did not originate in the Middle Stone Age, as the curious case of the defleshed Bodo skull indicates. Recovered in 1976 from Bodo catchment in the Middle Awash Valley of Ethiopia, this robust cranium belongs to a late Middle Pleistocene hominid transitional in appearance between *Homo erectus* and *Homo sapiens* (Conroy *et al.* 1978). When cleaned of its secondary calcium carbonate coating and examined in Addis Ababa by Dr Tim White in 1981, the skull revealed cut marks on the facial bones and inside the eye sockets, indicating deliberate defleshing of the face and removal of the scalp, presumably for ritual purposes (White 1986). Since the skull dates to roughly 0.3 Ma, the ritual defleshing must be of similar antiquity.

Late Stone Age diversity and migrations

As time progressed, increasing use was made of materials other than stone or wood, including bone, antler, mammoth ivory and shell (Straus 1985; Mellars 1989). Many of these materials were fashioned into pendants or carved into animal and human figures, particularly during Late Stone Age or Upper Palaeolithic times, when small, beautifully made stone engraving and cutting tools began to proliferate from about 40–35 ka onwards in Europe and western Siberia (Gowlett 1984a; Fagan 1989).

It was towards the close of the Middle Palaeolithic and the start of the Upper Palaeolithic in Europe and Asia that the last great prehistoric migrations took place (Figure 9.5). Northern Australia was first occupied some 50–60 ka ago (Robert *et al.* 1990). By 35–40 ka the first Australians had reached as far south as Swan River in Western Australia (Pearce and Barbetti 1981) and Lake Mungo in semi-arid New South Wales (Barbetti and Allen 1972; Bowler *et al.* 1972), implying a series of prior sea journeys between mainland Asia and Northern Australia by a competent maritime people (Hallam 1977; White and O'Connell 1978; Jones 1979; Flood 1989).

Eastern Siberia was not occupied until about 30–35 ka (Klein 1975), putting a maximum age of c. 30 ka for the first entry into Alaska via the Bering land bridge during times of glacially lowered sea level (see Chapter 4, Figure 4.6(B)), always assuming that the first Americans were tundra dwellers who preferred to follow the herds on foot rather than to voyage by boat. Since they were primarily hunters who depended upon the great herds of mammoths for their food, fuel and raw materials for shelter, it seems highly probable that they did walk across the Bering land bridge which would then have been a grassy plain nearly 1500 km wide from north to south.

Very few sites older than about 20–25 ka have yet been found in Central and South America although there is growing evidence that some prehistoric sites in Brazil may be as old as 35–40 ka (Guidon and Delibrias 1985, 1986; Bednarik 1989). In North America the majority of prehistoric sites are younger than 12–14 ka, indicating a major influx of Palaeo-Indians after 12 ka only, once the MacKenzie Valley had become free of ice (Martin 1973).

Pleistocene faunal extinctions

Ever since the well-publicized excavations of Boucher de Perthes (1847, 1857, 1864) in the Somme Valley of northern France in the 1840s and 1850s, geologists on both sides of the English Channel were alerted to the coexistence in Europe of Palaeolithic stone tools and a now extinct fauna, including woolly rhinoceros, sabre-toothed tigers and mammoths (Evans 1860; Prestwich 1860; Lyell 1873). At about the same time, the fossil bones of large extinct marsupials were being recovered from now dry lake and swamp deposits in south-eastern Australia (Owen 1870). The initial reaction of natural scientists in both Europe and Australia was to invoke climatic change as the sole cause of these extinctions. However, growing recognition of the efficiency of Late Stone Age hunting and trapping skills, and the apparent synchroneity between the demise of mammoths in North America and the first major influx of Palaeo-Indian mammoth hunters has led many researchers to abandon a climatic explanation in favour of human predation and what Paul Martin so vividly refers to as 'Pleistocene overkill' (see Martin and Wright 1967; and its monumental successor, Martin and Klein 1984).

Controversy about the causes of the Pleistocene faunal extinctions continues unabated, and although some of the debate is as polarized today as it was last century, a greater measure of consensus now appears to be emerging. We will comment briefly on the four major hypotheses proposed so far, for each may be valid in certain places at particular times.

The climatic change hypothesis has the merit that the geological, biological and isotopic evidence of repeated fluctuations in Quaternary temperature and precipitation is often of high quality and accurately dated. A recent variation on this theme is Mörner's (1978) suggestion that marine regressions will lead to a regional fall in groundwater levels as sea level drops, resulting in the drying out of springs and lakes, causing widespread faunal extinctions. In Australia, Horton (1980, 1984) has drawn attention to the coincidence between formerly wooded areas and megafaunal fossil finds, arguing that late Pleistocene aridity destroyed the woodland habitat of the large browsing marsupials, tethering them to dwindling waterholes, around which they soon consumed all the available forage. The site of Lancefield Swamp in Victoria strongly appears to support this hypothesis (Gillespie *et al.* 1978). A weakness of this hypothesis is that it does not convincingly account for why the fauna survived repeated previous Quaternary droughts, but succumbed to the latest major drought, by which time humans were also present in Australia.

The fact that the Pleistocene fauna of Australia and North America appeared to have survived unscathed the climatic vicissitudes of the Middle and early Late-Pleistocene only to disappear shortly after the arrival of humans in those two continents gave rise to the hypothesis of 'Man the destroyer' espoused by Merrilees (1968) in Australia and by Martin (1967; and Martin and Klein 1984) in North America. Although on the face of it a logical and plausible explanation, the 'overkill' hypothesis has several flaws. First, as many Australian workers have been at pains to emphasize, there is a curious absence of kill sites, and second, there was a very long period of co-existence between humans and megafauna well in excess of 20 000 years so that Martin's (1984) 'Blitzkrieg' model of overkill is certainly not applicable to Australia (Gillespie *et al.* 1978; McIntyre and Hope 1978; Sanson *et al.* 1980; Gorecki *et al.* 1984).

Although mindful of Flint's dictum that 'absence of evidence is not evidence of absence', many researchers accepted the apparent lack of butchery sites and the long period of coexistence between megafauna and prehistoric hunters as convincing evidence that direct human predation was not a sufficient and necessary cause of late Pleistocene faunal extinctions. However, they were not convinced that climatic change was an adequate explanation either, for the reasons outlined earlier, and so sought a third explanation: human modification of the original vegetation through the use of fire (Jones 1968, 1969). There is certainly good evidence from northern Queensland that some species of tropical trees did become extinct in the late Pleistocene at a time when high charcoal concentrations indicate more intense and perhaps more frequent fires (Kershaw 1978, 1984). However, as Horton (1982) has pointed out, fires may promote new plant growth and may enhance the grazing potential, so that wallabies and kangaroos may benefit rather than suffer from the modification to their habitat caused by fire.

Haynes (1991) has reviewed the evidence for an interval of widespread drought in the western United States between 11.3 and 10.9 ka. This drought coincided with the time of the cold and dry Younger Dryas event in northern Europe as well as the very time when the clovis Palaeo-Indians were hunting and butchering mammoths and bisons concentrated around dwindling water holes. The extinction of the mammoth (*Mammuthus columbi*) at this time may therefore reflect both climatic desiccation and human predation (Haynes 1991).

Unconvinced by the efficiency of either climatic change or hunting or burning as the major agents of Pleistocene faunal extinctions, an increasing number of natural scientists are now invoking a more complex and multi-causal explanation. Long-term changes in plate movements cause changes in atmospheric and ocean circulation, in the distribution of land and sea, and in marine regressions and transgressions (see Chapters 2 and 4). The record of Cainozoic plant and animal extinctions reflects the interplay of physical and biological factors (Butzer 1982). The result is a repatterning of food supplies (which may favour grazers at the expense of browsers, as in Australia); changing competition between different groups of plants and animals; and cyclic swings from complex ecosystems with a diverse fauna and flora to simpler ecosystems dominated by fewer species. Against this background of progressive long-term changes in fauna and flora, it is easy to envisage that rapid Quaternary climatic fluctuations, human predation, and fire, acting in concert, will selectively destroy those species already made vulnerable by the longer-term changes in climate and habitat caused by Cainozoic plate movements. In short, the arrival of humans in Australia towards 50 ka was but one of the factors responsible for the demise of the megafauna, but may have been the final and decisive factor accelerating the processes of extinction. J. H. Calaby's (1976) verdict is worth citing in this context: "... the weight of evidence favours climatic changes as the ultimate major cause of extinction and the most that Pleistocene men may have done was to hasten the extinction of the remaining already doomed species that were still around when he arrived."

Isotopic evidence of palaeodiet

One of the major reasons for the continuing debate over prehistoric faunal extinctions is the lack of precise quantitative information about what these creatures ate. Were they browsers or grazers or mixed feeders? It is difficult to argue convincingly that many of the now extinct large Pleistocene marsupials of Australia were dominantly browsers when this hypothesis is simply asserted and never adequately tested. Fortunately, we now have the means to test whether the giant kangaroos like Procoptodon ate grass or browsed on shrubs and trees by analysing the $^{13}C/^{12}C$ ratios in unmineralized samples of bone.

Plants can fix atmospheric carbon by photosynthesis in one of three possible ways. All trees, most shrubs, and grasses growing in shaded forests or temperate climates follow the Calvin or C_3 pathway of photosynthesis (Van der Merwe 1982). Grasses adapted to growing in strong sunlight, including most tropical grasses, follow the Hatch-Slack or C_4 pathway of photosynthesis and most succulent plants follow the third pathway, which involves crassulacean acid metabolism (CAM pathway of photosynthesis). All three photosynthetic systems fractionate the carbon isotope ratio of atmospheric CO_2 in quite different ways (Vogel *et al.* 1978; Vogel 1978; Van der Merwe 1982).

The outcome of this differential fractionation of the carbon isotopes during photosynthetic fixation of carbon is that C_4 plants have $\delta^{13}C$ values of -9 to $-16‰$ (average $-12.5‰$), C_3 plants have δ^{13} values of -20 to $-35‰$ (average $-16.5‰$), and CAM plants have mean $\delta^{13}C$ values of roughly $-16.5‰$ (Van der Merwe 1982).

Further fractionation ensues when the plants are eaten by animals, including humans. Bone collagen is enriched by roughly $5‰$ relative to the mean ratio of the plant food eaten. The $\delta^{13}C$ value of the bone collagen of browsing animals (such as kudu in South Africa) is about $-21.5‰$, since they only eat leaves of trees and shrubs, i.e., C_3 plants (Van der Merwe 1982). Grazing animals, which eat only C_4 grasses, will have $\delta^{13}C$ values of -8 to $-10‰$, and mixed feeders, such as sable antelope, values of -13 to $-15‰$. Isotopic measurements on bone or teeth therefore provide a means of assessing the proportions of C_3 and C_4 plants eaten by prehistoric animals and people. Ambrose and De Niro (1986) have recently used both carbon and nitrogen isotope ratios in bone collagen to distinguish between human diets in Africa, including groups eating marine foods, cereal grains, and pastoralists. In the latter case, they could clearly distinguish between camel pastoralists and capri-bovine pastoralists.

An additional factor needs to be considered when using plant carbon isotopes to infer animal and human diet. The recent research by Martinelli *et al.* (1991) has revealed that in certain environments such as the floodplain forests of the Amazon, much of the biogenic CO_2 may be recycled before it is mixed completely into the atmosphere. The intensity of this recycling is greater in the eastern than in the western Amazon basin, and increases systematically inland, demonstrating that carbon isotope gradients can vary across the same ecosystem, both between different species and within the same species.

One of the most intriguing aspects of human diet is the dramatic way in which it changed at the end of the Pleistocene. The inception of plant and animal domestication towards 10 ka denotes the end of the Palaeolithic and the start of the Neolithic tradition in every inhabited continent except Australia. With the Neolithic came a dramatic change in food production which ushered in some of the most revolutionary changes ever experienced by human societies during the long saga of prehistoric cultural evolution.

Neolithic plant and animal domestication

For over 2 million years the genus *Homo* lived by gathering plant foods and hunting wild animals. *Homo sapiens* remained a hunter-gatherer for 100 ka, until about the start of the Holocene, 10 000 years ago. Within the next 5000 years or so, virtually the entire human population had become predominantly farmers or herders (see Figure 9.7), although wild animals continued to be hunted and wild plants to be gathered long after the emergence of agriculture (Legge and Rowly-Conway 1987). What prompted the change?

Cohen (1977) has argued that agriculture does not provide a more secure, more palatable, more nutritious or more varied diet than that of most hunter-gatherers, but farming does ensure that more calories of food can be obtained from a given area of land in a given time than by simple collecting of wild foods. He went on to argue that growing pressure upon food resources toward the

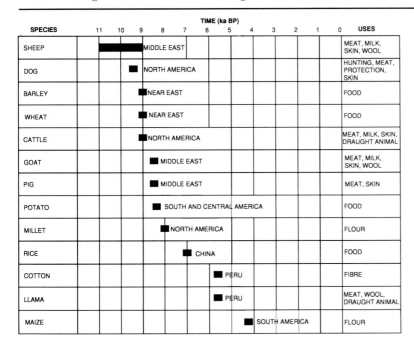

SPECIES	TIME (ka BP)												USES
	11	10	9	8	7	6	5	4	3	2	1	0	
SHEEP		MIDDLE EAST											MEAT, MILK, SKIN, WOOL
DOG		NORTH AMERICA											HUNTING, MEAT, PROTECTION, SKIN
BARLEY			NEAR EAST										FOOD
WHEAT			NEAR EAST										FOOD
CATTLE			NORTH AMERICA										MEAT, MILK, SKIN, DRAUGHT ANIMAL
GOAT				MIDDLE EAST									MEAT, MILK, SKIN, WOOL
PIG				MIDDLE EAST									MEAT, SKIN
POTATO				SOUTH AND CENTRAL AMERICA									FOOD
MILLET				NORTH AMERICA									FLOUR
RICE					CHINA								FOOD
COTTON						PERU							FIBRE
LLAMA						PERU							MEAT, WOOL, DRAUGHT ANIMAL
MAIZE							SOUTH AMERICA						FLOUR

Fig. 9.7 Origins of domesticated plants and animals (after Bray and Trump 1970; Barraclough 1982; Gowlett 1984a; Scarre *et al.* 1988; Fagan 1989)

end of the Pleistocene, caused by a continually growing world population, was the primary stimulus which caused former hunters and gatherers to become pastoralists and farmers.

Cohen's views have been challenged by a number of archaeologists, some of whom have argued that Neolithic methods of food production were a cause rather than a consequence of human population growth (Hassan 1980). It is worth noting that world population 10 ka ago is estimated at around 5 million, increasing by a factor of 20 to some 100 million some 5 ka ago (May 1978). By 300 years ago, world population amounted to around 500 million, and reached 1000 million (1 billion) by 1850, 2 billion by 1930, over 4 billion by 1978 and over 5 billion by 1990. Such dramatic increase was only possible with the change in methods of food production initiated during the Neolithic.

It is highly unlikely that any one explanation will adequately account for the origins of agriculture. Here, as elsewhere, we should be sceptical of unicausal explanations, and reject over-facile correlations between different events as proofs of a causal explanation. Whether the onset of farming followed a wet or dry climate is not, in itself, proof that the change in climate initiated agriculture in that locality. After all, many previous similar climatic fluctuations did not. Far too many attempts to explain agricultural origins succumb to the *post hoc ergo propter hoc* fallacy by claiming that because B followed A, A must have caused B.

The ultimate causes of early farming and herding are probably a variable combination of social, economic, technological and environmental factors.

The actual process of domestication is illuminating. Stemler (1980) has emphasized that the two major prerequisites for cereal domestication were the harvesting of whole grain-heads of mutant plants with a tough rachis and the later sowing of grains from such mutant plants. With tough-stemmed tropical panicoid grasses like sorghum and millet, an efficient harvesting tool was essential (Stemler 1980). Clark (1971, 1980) has pointed out that early farming in the Nile Valley was preceded by a long interval of pre-adaptation to agriculture. Certainly, efficient sickles were already widely used for harvesting wild grasses in the Nile Valley and the Near East during the Late Upper Palaeolithic at which time grindstones also became abundant (Close 1989). Where environmental stress was great, as at Dhar Tichitt in Mauritania, the transition from collecting wild grasses to harvesting domesticated cereals was remarkably rapid, in this instance a few centuries only (Munson 1976; Stemler 1980; Williams 1985, Figure 6).

In Africa and many parts of the Middle East and Asia, archaeologically visible evidence of early farming and herding coincides with a change from a cold, dry and windy terminal Pleistocene climate to a warm and wet early Holocene climate (Adamson *et al.* 1980; Wendorf and Schild 1980; Williams 1984). Careful scrutiny of the onset of plant and animal domestication in North Africa shows that the beginning of agriculture cannot be directly linked to climatic change, since it was strongly time-transgressive (William 1985). Furthermore, plant and animal domestication often started at quite different times at the same site, and the onset of agriculture was often quite different at localities in relatively close proximity to one another (Williams 1985, Figure 5).

Regardless of whether the inception of early farming and herding was associated with demographic pressure upon resources (Cohen 1977), or to a technical capacity to harvest, store and sow appropriate cereal grains (Stemler 1980), or to a combination of climatic, ecological and technological factors (Clark 1976b, 1980, 1984b), the net effects were generally the same.

As noted earlier, relative to Upper Palaeolithic hunting and gathering. Neolithic agriculture would have provided more food calories per unit area of land per unit time (Cohen 1977). In addition, the combination of milk and a porridge of cooked cereal grains was ideal food for young children, and easier for them to digest than meat, especially if the latter was inadequately cooked (Stemler 1980). Cultivation, to be effective, requires a more sedentary lifestyle than herding or hunting, hence allowing a far higher population density than non-agricultural societies would find viable. The consequences of the 'Neolithic Revolution' were

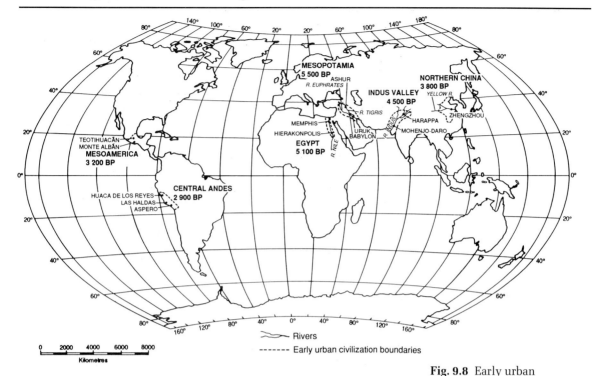

Scale:
0 2000 4000 6000 8000
Kilometres

Rivers
------- Early urban civilization boundaries

Fig. 9.8 Early urban civilizations of the world (after Barraclough 1982; Scarre *et al.* 1988; Fagan 1989; Higham 1989)

an increase in sedentary living and in population, leading to the emergence of urban civilizations during the mid to late Holocene in widely separated parts of the world, including South America, the Nile Valley, Mesopotamia, the Indus Valley, and China (Figure 9.8).

"From seasonal lakeshore dwellings to settled villages is a reflection of a profound change in social structure baldly summarized by the adjectives Mesolithic and Neolithic. Evolutionary changes in stone-tool technology brought about revolutionary changes in food production. From Enkidu to Gilgamesh, from individualistic to hierarchical, from ephemeral camp to enduring city, the Neolithic foundations for our present urban civilization had now been laid" (Williams 1985, 183).

Further reading

Clark, J. D. and Brandt, S. A. (eds) 1984: *From Hunters to Farmers: The Causes and Consequences of Food Production in Africa.* Berkeley: University of California Press. (A scholarly, wide-ranging and thought-provoking set of essays.)

Fagan, B. M. 1985: *In the Beginning: An Introduction to Archaeology.* Fifth edition. Boston: Little, Brown and Company. (A clear and comprehensive explanation of archaeological techniques.)

Fagan, B. M. 1989: *People of the Earth: An Introduction to World*

Prehistory. Sixth edition. Boston: Scott, Foresman and Company. (Perhaps the best single introduction to world prehistory at undergraduate level.)

Flood, J. 1989: *Archaeology of the Dreamtime. The Story of Prehistoric Australia and its People*. Revised edition. Sydney: Collins. (A good clear, up-to-date account of Australian prehistory.)

Gowlett, J. 1984: *Ascent to Civilization: The Archaeology of Early Man*. London: Collins. (A beautifully illustrated and clear account of world prehistory by an active practitioner.)

Scarre, C. (ed.) 1988: *Past Worlds: The Times Atlas of Archaeology*. London: Times Books Limited. (A superbly illustrated and comprehensive work of reference.)

Sherratt, A. (ed.) 1980: *The Cambridge Encyclopaedia of Archaeology*. Cambridge: Cambridge University Press. (Good, scholarly, wide-ranging essays, effectively illustrated.)

Williams, M. A. J. and Faure, H. (eds) 1980: *The Sahara and the Nile. Quaternary Environments and Prehistoric Occupation in Northern Africa*. Rotterdam: Balkema. (A well-indexed work of reference with articles on the Sahara, the Nile and North African prehistory by leading scholars in the field.)

10 Atmospheric Circulation during the Quaternary

Blow, winds, and crack your cheeks!
rage! blow!
You cataracts and hurricanoes, spout
Till you have drench'd our steeples,
drown'd the cocks!

William Shakespeare (1564–1616),
King Lear.

Introduction

Climatic change is the hallmark of the Quaternary. Relative to the previous 60 million or so years of Tertiary geological history, the climatic fluctuations of the Quaternary were unprecedented in terms of the speed and amplitude of global temperature oscillations. The mean annual temperature difference between last glacial maximum (18 ± 3 ka) and the early Holocene 'climatic optimum' or 'hypsithermal' (9 ± 3 ka) amounted to about 10°C in many temperate land areas of the world (see Chapter 3), which is roughly equivalent to the inferred long-term drop in land and sea temperatures between early Eocene and early Pliocene in these same regions (Figure 2.3, Chapter 2). The surface of the Southern Ocean cooled by about 10–15°C over a time-span of roughly 40 million years (Shackleton and Kennett 1975). Temperature drops of similar magnitude were evident in eastern Australia over the much shorter time-span of about 100 000 years which separated the last interglacial (125 ± 5 ka) from the last glacial maximum in this region. The Vostok ice core from Antarctica (Figure 3.10, Chapter 3) displays a similar pattern of rapid fluctuations in temperature over the past 160 ka, as well as dramatic contrasts in the concentration of atmospheric carbon dioxide and methane, with much lower amounts of both gases being present during times of lower global temperature (Barnola *et al.* 1987, 1991; Jouzel *et al.* 1989; Chappellaz *et al.* 1990).

In their attempts to understand the nature and causes of Quaternary climatic fluctuations, research workers have adopted two different approaches, which have too often been poorly integrated with one another despite being essentially complementary. The palaeoclimatic modellers use a variety of global circulation models (GCMs) to simulate likely patterns of past climatic change. Some of these models incorporate quite realistic past geographies; some are coupled ocean-atmosphere models; but few are capable of precise spatial resolution and none can yet cope adequately with cloud dynamics. The palaeogeographers and palaeoecologists, on the other hand, within the time resolution of the dating methods used (see Appendix) and the spatial resolution of the evidence, whether geological, biological, archaeological or geochemical, can usually contrive to reconstruct past changes in local, regional or global marine and terrestrial environments reasonably well, albeit with varying degrees of quantitative accuracy. Earlier chapters in this book have already discussed the scope and limitations of the evidence used to reconstruct former environments. A criticism frequently levelled at the producers of palaeoclimatic reconstructions based on such proxy data is that they proceed from the primary data (for example, pollen evidence), directly to an inference about climate, when it might be more appropriate to ignore climate and simply deduce what one can about that aspect of the environment (in this case, vegetation) with which the evidence is most directly concerned. Such criticism is often justified, and just as often forgotten.

A notable exception to many of the above strictures are the attempts by the CLIMAP project members to reconstruct seasonal changes in global geography at the last glacial maximum, taken to be 18 ± 3 ka (CLIMAP 1976, 1981). Much of this work was based on the great improvements in our understanding of past ocean behaviour made possible by advances in marine biostratigraphy and in oxygen isotope analyses of suitable deep-sea cores (Hays, Lozano *et al.* 1976a; McIntyre, Kipp *et al.* 1976; see also Chapter 5). The CLIMAP palaeogeographical reconstructions, especially of winter and summer sea-surface temperatures, provided useful constraints for GCM models (Gates 1976), stimulating fresh modelling experiments, some of them devoted to testing the role of past variations in the obliquity and precession of the earth in controlling or influencing Quaternary climatic changes (Hays *et al.* 1976; Imbrie and Imbrie 1979; Kerr 1986; Kutzbach and Guetter 1986; Ruddiman *et al.* 1986; COHMAP 1988). There have been some very fruitful by-products from the CLIMAP and COHMAP projects and related efforts to test the models against the observational data, and vice versa. One very useful outcome has been to expose possible discrepancies between palaeotemperature estimates from land and sea (Rind and Peteet 1985), thereby helping to improve future models. However, the ultimate test of any global circulation model is how well it can simulate present-day climatic patterns, so we begin with a brief account of our present climate.

Present-day global atmospheric and oceanic circulation patterns

Global atmospheric circulation is a function of the amount of solar energy received at the surface of the earth. Insolation is at a maximum in intertropical latitudes, but the incidence of incoming short-wave solar radiation varies seasonally as a result of the tilt in the earth's axis, and the annual apparent migration of the overhead sun between the two tropics. Variations in the distribution of land, sea and ice (see Chapter 1), as well as in cloud cover, will also modify the heat budget of the earth, so that in some latitudes or regions there is a net heat loss from the surface of the earth by outgoing long-wave radiation. Feedbacks are shown in Figure 1.6, and average global climate in Figures 1.3 and 1.4 of Chapter 1. Our concern here is with seasonal variation. Figure 10.1 shows the mean temperature distribution over land and sea during January and July, reflecting the impact of the northern winter/southern summer and northern summer/southern winter respectively. The near-surface changes in wind patterns associated with these seasonal temperature (and pressure) patterns are illustrated in Figure 10.2. The rotation of the earth and the variation in speed of surface rotation with latitude (fast at the equator, slow at the poles) modify the simple Hadley circulation model outlined in Chapter 7 and are responsible for the oblique orientation of westerlies and Trade Winds. Figure 10.3 is a more detailed portrayal of sea-surface temperatures, shown here for February and August to allow for the slower heating and cooling of the oceans relative to the land. Between them, Figures 1.3, 1.4, 10.1, 10.2 and 10.3 give a highly generalized picture of the geographical and seasonal variations in temperature and precipitation, and of the global distribution of presently wet, dry and cold climatic zones in relation to atmospheric and oceanic circulation. Earlier chapters have described changes in the late Quaternary distribution of land, sea and ice, showing how the areas then occupied by deserts and by humid regions show a significant contrast with the present. We now enlarge on this contrast, focusing initially upon the CLIMAP reconstructions of sea-surface temperatures.

Sea-surface temperatures during the last glacial maximum

Figure 10.4 is a simplified version of the CLIMAP (1976) reconstruction of sea-surface temperatures during the northern summer (August) 18 000 years ago, but replotted on the map projection used throughout much of this book. It was prepared by converting numerical estimates of three planktonic groups (foraminifera,

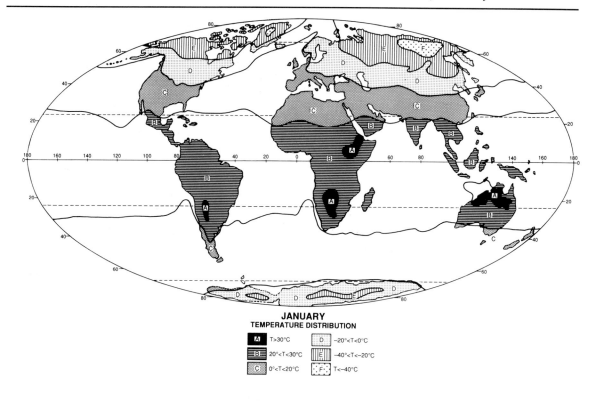

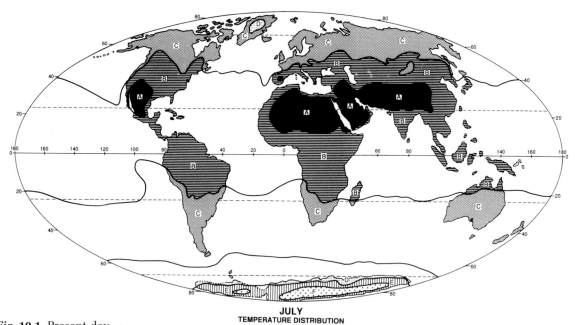

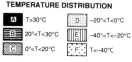

Fig. 10.1 Present-day temperature distribution in January and July (after Tanke and Gulik 1989)

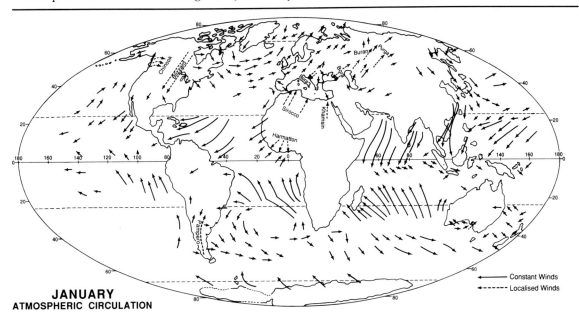

JANUARY
ATMOSPHERIC CIRCULATION

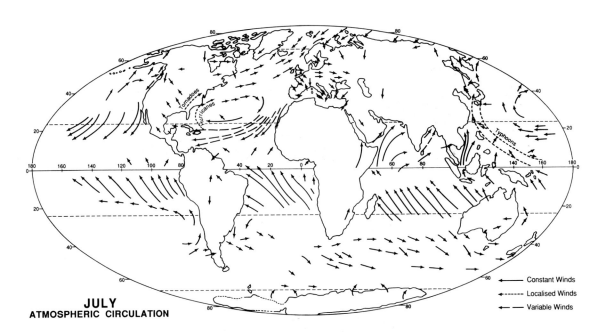

JULY
ATMOSPHERIC CIRCULATION

Fig. 10.2 Present-day atmospheric circulation in January and July (after Tanke and Gulik 1989)

(a)

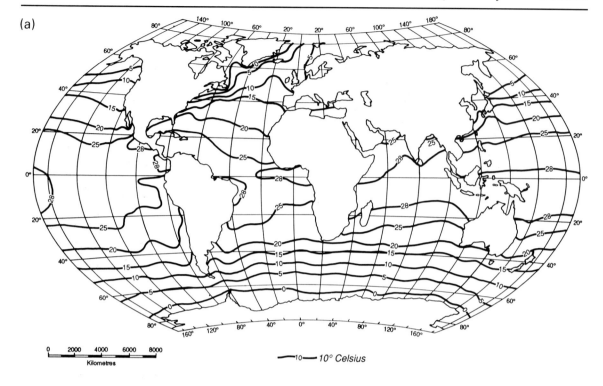

0   2000   4000   6000   8000
Kilometres

———10——— *10° Celsius*

(b)

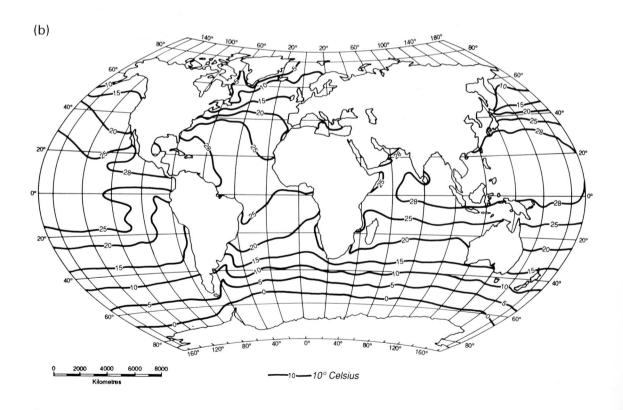

0   2000   4000   6000   8000
Kilometres

———10——— *10° Celsius*

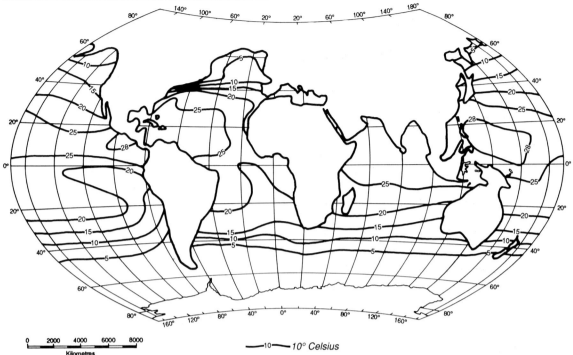

—10— *10° Celsius*

Fig. 10.4 Reconstructed August sea-surface temperatures during last glacial maximum (18 ka) (after CLIMAP 1976)

radiolaria and coccoliths), the fossil remains of which were preserved in deep-sea sediments laid down towards 18 ka, into estimates of sea-surface temperatures at last glacial maximum using the transfer function statistical technique described in Chapter 5. Only in the Pacific were all three planktonic groups used, and there were occasional major discrepancies ($\geq 1.6°C$) between temperatures in the eastern equatorial Pacific estimated from the zooplankton and phytoplankton groups, in which case the coccolith values were used. These and other sources of error are discussed in detail by the CLIMAP team (CLIMAP 1976).

The next step is to compare the differences between measured present-day sea-surface temperature values and those deduced for the last glacial maximum. Figure 10.5 does this for February and August, using the CLIMAP (1981) data as evaluated by Rind and Peteet (1985), who found that the 18 ka temperatures near Hawaii and in the Pacific subtropical gyre were up to 2°C warmer then than now. They concluded that the CLIMAP (1981) estimates of last glacial maximum sea-surface temperatures at low and subtropical latitudes were inconsistent with estimates of 18 ka land temperatures in those latitudes and were probably up to 2°C too

Fig. 10.3 Present-day sea-surface temperatures in
(a) February
(b) August (after Bartholomew 1980, Plate 3; Gorshkov 1978, 129 and 135; Japan Meteorological Agency 1975, 7 and 9)

(a)

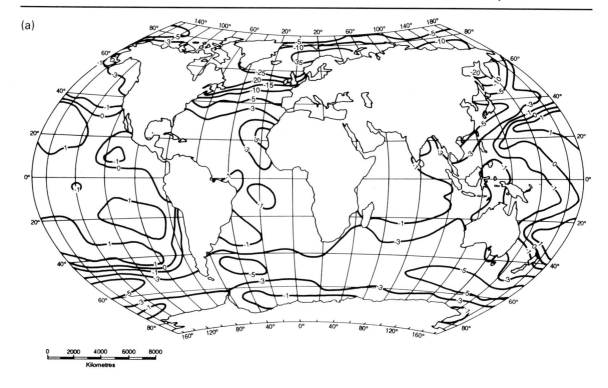

(b)

temperatures in those latitudes and were probably up to 2°C too warm. A problem with this conclusion is that the comparison between 18 ka sea-surface temperatures and 18 ka land temperatures estimated from tropical upland pollen spectra and snowline depression on tropical mountains (Hawaii, Colombia, East Africa and New Guinea) may not be appropriate. Indeed, a more relevant comparison is with lowland pollen taxa blown offshore and preserved in 18 ka marine sediments, since the glacial lowering of temperature was apparently less (−2 to −4°C) in certain tropical lowlands than in the adjacent high tropical mountains (−5 to −15°C) and much more in line with the adjacent marine record (Bonnefille *et al.* 1990; Van Campo *et al.* 1990). To reconstruct former changes in land and sea temperatures is a prerequisite towards reconstructing the associated changes in global atmospheric circulation patterns, for only thus can we really come to grips with possible causes of Quaternary climate change.

Atmospheric circulation patterns during the late Quaternary

Attempts to model Quaternary global atmospheric circulation patterns are necessarily hypothetical, and are usually limited to the last 20 000 years or so by the lack of well-dated marine and terrestrial evidence. One approach much favoured by palaeoclimatologists is to compare the last glacial maximum (18 ± 3 ka) with the early Holocene (9 ± 2 ka), which is often considered the equivalent of an interglacial.

An example of this approach is the attempt by Nicholson and Flohn (1980) to identify changes in the position of the intertropical convergence zone (ITCZ) over Africa during the northern summer and winter at three contrasted times (18 ka, 10–8 ka, 6.5–4.5 ka; Figure 10.6). The merit of such an attempt, however much one may dispute the detail, is that it is inherently testable and refutable. The presence or absence of warm or cold oceanic water offshore at particular times and of aridity on land can be tested against the evidence of ocean cores (Sarnthein *et al.* 1982; Pokras and Mix 1985; Diester-Haass *et al.* 1988). Pollen spectra from marine and terrestrial sediments and lake-level histories can be used to check estimates of the efficacy and areal extent of monsoon rains and the ITCZ (Butzer *et al.* 1972; Street and Grove 1979; Hastenrath and Kutzbach 1983; Bonnefille *et al.* 1990; Gasse *et al.* 1990; Lézine 1991). Several workers have also noted that the

---

Fig. 10.5 Sea-surface temperature differences between present-day and last glacial maximum (18 ka)
(a) February
(b) August (after Rind and Peteet 1985, Figure 3)

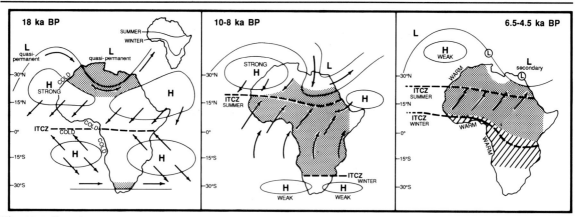

Fig. 10.6 Putative atmospheric circulation scheme for Africa during the very late Pleistocene (20–12 ka), the early Holocene (10–8 ka) and the mid-Holocene (6.5–4.5 ka) (after Nicholson and Flohn 1980, Figures 8, 9 and 10)

late Quaternary African monsoons were highly sensitive to changes in the earth's orbital geometry, and this is reflected in fluctuations in Nile discharge over the past 464 000 years (Rossignol-Strick 1983) as well as in fluctuations in lake levels throughout the tropics during the past 18 000 years (Kutzbach and Street-Perrott 1985).

The strong contrast between the mainly warm and wet early Holocene climate of Africa and the mainly cold, dry and windy terminal Pleistocene climate (Williams 1985; and Figures 6.7 and 10.6) has its counterpart in peninsular India (Duplessy 1982; Williams and Clarke 1984) as well as in Australia (Figure 10.7). As a result of lower sea level during last glacial maximum, conservatively estimated at −135 m although some estimates opt for a lowering of −175 m for northern Australia, the land area of Australia was increased by a fifth and what is now the Arafura Sea and the Gulf of Carpentaria was a land bridge over 1200 km wide linking the much enlarged mainland of Australia to a slightly bigger Papua New Guinea. Both summer and winter rainfall were well below present levels in the tropical north-east (Kershaw 1978) and the temperate south-east (Bowler 1978), and many of the now vegetated dune systems were then active as shown in Figure 10.7(A). Pollen evidence from Papua New Guinea, north-eastern and south-eastern Australia, supplemented by geomorphic evidence from elsewhere in Australia, and by geochemical, microfossil, trace element and isotopic analyses of certain volcanic lakes in Victoria (see Chapter 6 and Figure 6.6), all point to warmer and wetter early Holocene climates in much of Australia. Times of maximum precipitation may not have been synchronous throughout Australia, for the peak in tropical summer rainfall towards the middle of the Holocene seems to coincide with a phase of reduced winter rainfall in the Victorian maar lakes of the far south-east (De Deckker *et al.* 1988).

Figure 10.7 illustrates another aspect of Quaternary environmental change alluded to in Chapter 4. The late Pleistocene land

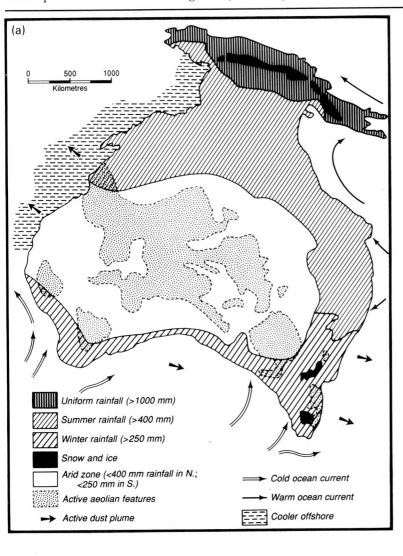

(a)

Uniform rainfall (>1000 mm)

Summer rainfall (>400 mm)

Winter rainfall (>250 mm)

Snow and ice

Arid zone (<400 mm rainfall in N.; <250 mm in S.)

Active aeolian features

Active dust plume

Cold ocean current

Warm ocean current

Cooler offshore

Fig. 10.7 Morphoclimatic map of Australia–Papua New Guinea, during
(a) the last glacial maximum (18 ka)
(b) the early Holocene (9 ka)
(after Williams 1984). See p. 234.

bridge connecting Australia with Papua New Guinea diverted the warm south equatorial current from its path between these two large islands, thereby depriving northern Australia of a major source of moist maritime air, further accentuating late Pleistocene aridity in that region. Subaerial exposure of the Great Barrier Reef had a similar effect, and it was not until the very early Holocene that the postglacial rise in sea level finally submerged both reef and land bridge, causing a sudden influx of moist air to what are now the coastal lowlands of tropical northern Australia. The early Holocene increase in summer rainfall is reflected in the sudden expansion of rainforest in north-eastern Australia into localities previously covered in eucalypt forest or woodland (Kershaw 1978).

A final example will suffice to show the value of using a

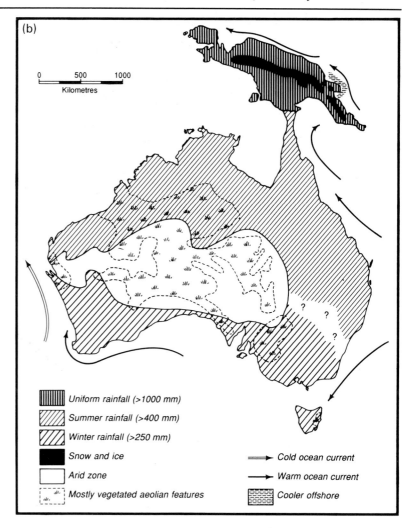

(b)

Uniform rainfall (>1000 mm)

Summer rainfall (>400 mm)

Winter rainfall (>250 mm)

Snow and ice

Arid zone

Mostly vegetated aeolian features

Cold ocean current

Warm ocean current

Cooler offshore

combination of field evidence and theoretical models to reconstruct former Quaternary environments in different parts of the world. The modelled reconstruction of likely wind directions over North America, the North Atlantic and Europe during the last glacial maximum winter (18 ka, January) by Kutzbach and Wright (1985; and Figure 10.9) is based on prior knowledge of the probable extent and thickness of ice caps and sea-ice in this region at this time (see Chapter 3). Unlike the 18 ka climate reconstruction for Africa (Nicholson and Flohn 1980; and Figure 10.6), this modelled reconstruction shows both high- and low-altitude winds, and identifies the probable winter and summer temperature and moisture status of a number of localities relative to the present-day. It is interesting to note that even during the mostly colder and drier last glacial maximum, certain localities are inferred to have been seasonally warmer then than now. Predic-

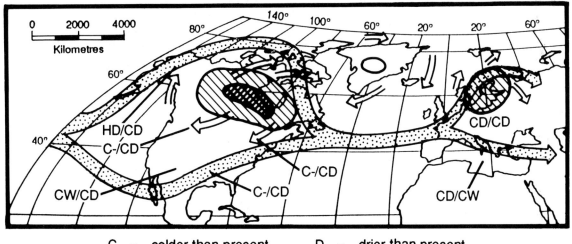

C = colder than present   D = drier than present

H = hotter than present   W = wetter than present

tions of this sort are testable, and can be a useful guide for further field research. The 9.5 ka wind directions shown in Figure 10.9 differ from those shown on Figure 10.8 in being based on early Holocene dune alignments around the margins of a much-shrunken Laurentide ice sheet. Nevertheless, the anticyclonic wind directions associated with the ice cap-induced high pressure system are still very much in evidence, enhancing the credibility of the 18 ka reconstruction of surface wind directions.

Fig. 10.8 Modelled reconstruction of high-altitude winds (continuous dotted arrows) and low-altitude winds (short arrows) for January during the last glacial maximum (18 ka) over North America, the North Atlantic and Europe. Also shown are the January (before backslash) and July (after backslash) temperature and precipitation anomalies (after Kutzbach and Wright 1985, Figure 11)

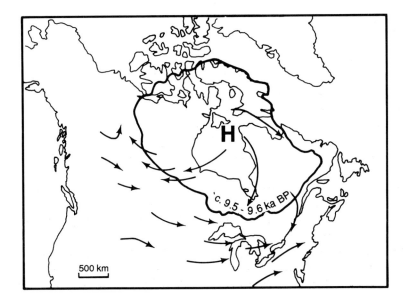

Fig. 10.9 Map of the early Holocene (9.5 ka) Laurentide ice sheet with anticylonic wind directions as inferred from alignments of early Holocene dunes (after Kutzbach and Wright 1985, Figure 15)

Global palaeohydrology and links between oceanic and atmospheric circulation

Mean global precipitation on land and sea amounts to 857 mm/a, equivalent to roughly 400×10^3 km³/a over the oceans and 100×10^3 km³/a over the land. Although 80 per cent of all precipitation falls over the oceans, evaporation from the oceans yields 440×10^3 km³/a of water equivalent, or about 86 per cent of the land precipitation. In other words, the oceans supply an extra 6 per cent of the land precipitation, amounting to 36×10^3 km³/a, which is returned to the oceans as a similar volume of annual runoff (Bloom 1978, 95). Of the total precipitation over land, roughly one-third is from maritime sources and two-thirds from continental evapotranspiration from plants, lakes, swamps and rivers. During interglacial (and present) times, some 97.6 per cent of the world's water supply is stored in the oceans, and about 2 per cent in ice caps. During glacial maxima, about 10 per cent is stored in ice sheets and 89.5 per cent in the oceans (Flint 1971, 83–4; and Chapter 3). The present ice sheets contain 26×10^6 km³ of ice, equivalent to a sea-level change of 65 m for a world ocean surface area of 362×10^6 km². During the last glacial maximum the total ice volume was about 77×10^6 km³, equivalent to 197 m of sea-level change (Flint 1971, 84). The sea level fell about 135 m below present level at 18 ka (see Chapter 4) which is remarkably close to Flint's (1971) estimate of around 132 m $(197 - 65$ m).

The global water budget summarized above has oscillated between interglacial and glacial modes at orbitally controlled intervals throughout the Quaternary. Within these two extremes there were countless smaller fluctuations, many of them linked to changes in oceanic circulation patterns which in turn influenced the atmospheric circulation over land and sea (Broecker and Denton 1989, 1990). Although some of the attempts to relate sudden climatic fluctuations to changes in the North Atlantic deep-water flux during the late Quaternary perforce remain speculative (Broecker *et al.* 1985; Street-Perrott and Perrott 1990; Charles and Fairbanks 1991), they are based on an impressive and growing body of well-dated palaeo-oceanographic evidence (Prell 1984; Boyle 1990; Hasselmann 1991).

Several of these attempts to interpret possible late Quaternary interactions between ocean, atmosphere and land deserve more detailed comment. In Chapter 6 we discussed the historic influence of El Niño–Southern Oscillation events upon droughts and floods in Africa, Australia, India and China as a classic example of how ocean-atmosphere interactions can control extreme climatic events on land at scales of 10^{-1} to 10^2 years. Here we focus on late Quaternary climatic changes at scales of 10^2 to 10^4 years.

Paradoxically, the best records of late Quaternary climates in peninsular India are those furnished by deep-sea cores from the northern Indian Ocean and Arabian Sea (Prell *et al.* 1980; Duplessy 1982; Prell 1984). Duplessy (1982) used differences in the oxygen isotopic composition of planktonic foraminifera (see Chapter 5) from the northern Indian Ocean to reconstruct the probable Holocene and late Pleistocene climates on land. He concluded that the south-west summer monsoon was weaker at 18 ka than it is today, for the upwelling along the southern coast of Arabia had disappeared at that time, implying much weaker south-westerly winds, and so much reduced summer rainfall. He also deduced that the Ganges and Brahmaputra contributed much less water to the ocean at 18 ka, since the salinity gradient in the Bay of Bengal was very much steeper than today. The inference of glacial aridity in India drawn from the ocean cores is entirely consistent with the aeolian and pollen evidence from north-west India (Goudie *et al.* 1973; Singh *et al.* 1974) and with the alluvial and fossil evidence from the Son and Belan rivers in north-central India which suggests that they were more seasonal and had more sparsely vegetated catchments towards 18 ka than during the Holocene (Williams and Royce 1982; Williams and Clarke 1984). Prell (1984) has used the presence or absence of cold upwelling along the southern coast of Arabia, as deduced from deep-sea cores, to reconstruct a longer history of monsoons in this region. Phases of weakened summer monsoons appear to coincide with the 21 ka orbital precession cycle (Prell 1984).

With improvements in the resolution with which late Quaternary marine cores can now be dated, many workers now accept that the most recent deglaciation (c. 16–7 ka) consisted of several distinct climatic oscillations, with rapid ice melting at 14–12 ka and 10–7 ka separated by an interval of no or very little melting (Duplessy *et al.* 1981; Berger *et al.* 1985; Fairbanks 1989; Jansen and Vena 1990). Particular interest has been focused on the Younger Dryas event (see Chapter 3), long recognized by Scandinavian palynologists as a return to a cold, dry, near-glacial climate in north-west Europe for a few centuries between 11 and 10 ka. The Younger Dryas cold phase is also evident in North Atlantic deep-sea cores, in the melting history of the Laurentide ice sheet, in Greenland ice cores, in African and Tibetan lake histories, in the alluvial history of the White Nile, and even in the isotopic composition of freshwater mollusca from southern Africa (Leventer *et al.* 1982; Dansgaard *et al.* 1989; Broecker *et al.* 1989).

An ingenious attempt to explain the abrupt climatic oscillations which characterize the last deglaciation is that of Broecker and his colleagues, who argue that the global oceanic thermohaline circulation is controlled by fluctuations in Atlantic salinity (Broecker *et al.* 1985). How far such Atlantic salinity changes reflect changes in glacial meltwater influx is still a moot point, but there is no longer any argument about the importance of Quater-

nary ocean-atmosphere interactions in controlling climatic
fluctuations on land, whether modulated by Milankovitch forcing
or not.

Environmental Changes: Past, Present, Future

The evidence suggests that agriculture has been practised for 7000 years, and the consequent reclamation and cultivation of land has completely destroyed the natural vegetation in many areas.

Zhao Ji, Zheng Guangmei, Wang Huadong, Xu Jialin,
The Natural History of China, 1990.

Introduction

We have now reviewed the characteristics of the major global environments that existed during the Quaternary. The inescapable conclusion is that, during these 1.8 Ma, change was almost ubiquitous. Environments regionally and locally reflected diverse influences arising from progressive cooling and ice build-up or warming and deglaciation; increasing dryness or more widespread pluvial conditions; falling or rising sea level; changes in windiness, continentality, summer solar radiation, and dustiness of the atmosphere. Terrestrial environments displayed varying responses to these changing stimuli: snowlines fluctuated through a height range of more than 1 km; vegetation communities adapted repeatedly to altered conditions; lakes dried out or overflowed; rivers changed in character from stable meandering systems to aggrading braided ones and vice versa; dunes were periodically active or stable, and soils were affected by deflation, glacial scour, erosion by meltwater floods, and ground-ice effects. In the oceans, temperatures, carbonate balance, circulation, oxygenation, nutrient availability and biological productivity were all subject to significant variation. The diverse activities of people must now be included in the list of potential triggers of environmental change. We are responsible for a diversity of activities affecting the environment, and in this chapter we will only be able to examine a selection of these. These examples should, however, serve to make the point that we are now capable of influencing the future course of environmental changes, but must employ our developing knowledge of the mechanisms lying behind the global environment and its behaviour in order to foresee and manage our potential impacts on the environment. As we shall see, wrought carelessly and in ignorance, changes to environmental processes caused by people may seriously upset mechanisms which previously regulated aspects of the environment.

Our present environment, in the global sense, is governed by the kinds of conditions which have existed in earlier brief, warm interglacial periods: the sea level is high, the atmosphere moist and warm (and hence relatively enriched in the natural greenhouse gases), and land and sea ice are restricted largely to the high northern and southern latitudes. Under the continuing influence of the Milankovitch solar forcing, we may expect that in due course these conditions will give way gradually to increasing cold and the renewed development of glacial conditions. Indeed, much of the evidence reviewed earlier offers the possibility that the enic of the present interglacial interval may have been reached about 4 ka BP, and that we are already experiencing descent into the next glacial stage.

As we have seen, the Quaternary is also distinguished by the development of humankind and the diversification of our culture. Exceptionally rapid growth in the human population and its cultural richness has occurred during the last glacial and the Holocene (Chapter 9). These developments have certainly been fostered by the climatic amelioration of the late glacial and Holocene, and the concomitant development of organized Neolithic agricultural food production which it has permitted. We now exist as a technologically powerful, numerous, and resource-consuming group of organisms such as has not been present in any earlier interglacial. Growth of the human population until the present displays an ever-increasing acceleration that is more than exponential. The present net rate of global population increase is nearly 100×10^6 per annum (i.e., about 2 million additional people per week!). Simultaneously, as technological and cultural development continues, our demands for energy and resources are also accelerating (Gilland 1988; Simmons 1989).

What are the implications of this special circumstance? An obvious one arises from our generation of greenhouse gases, and the inadvertent climate change that this may induce, which we consider briefly below. The height of an interglacial is the least desirable time to force climatic warming: the global climate is already at or near the peak of its natural temperature variation (at least on the Quaternary time scale), and the added heat arising from the anthropogenic greenhouse effect is indeed generating a gigantic global experiment. Fortunately, our awareness of our relationship to the environment, and our dependence upon it, have also evolved to a level not seen before. We have much to learn about the global environment, but already possess significant knowledge which can be brought to bear on many issues now facing us as a species. The perspective of the Quaternary is a vital one here, and permits us to understand the history of our environment and to see many of the ways in which it has responded to previous stimuli. In addition, many of our landscapes, and the soils which support us, our animals and our crops, are legacies of events occurring during the Quaternary, and our growing understanding of these events will be important in aiding our stewardship of the earth's resources for the future.

As described in earlier chapters, the 8–10 ka period of warming that marked the end of the last glacial brought global temperatures back to levels that had not been experienced since the preceding interglacial, 100 ka before. This was thus a rapid temperature swing to which biota had to adjust. Plant and animal communities in glacial or interglacial times survive ensuing climatic changes only in areas which provide a habitat where they are able to reproduce. Temperate forests of the interglacials survive glacial conditions only at low altitudes and latitudes; cold-climate communities survive the interglacials only at high altitudes and latitudes. Such areas where organisms may continue to exist despite shifts in their habitat are termed 'refugia'. There are organisms which have survived the Tertiary climatic cooling and desiccation in such refugia, and certainly remnants of glacial communities also exist today through this mechanism. During glacial conditions, the oak, pistachio and olive woodlands of the Mediterranean contracted to refugia which probably included Israel and Jordan (Roberts 1989) and possibly other areas of the south-eastern Mediterranean. During postglacial times, tree species recolonized in a westerly direction, replacing the glacial herb-steppe communities. Low-latitude rainforests in Africa, South America and Australia were similarly much more restricted in their extent during the dry full-glacial conditions, retreating to as yet unknown refugia (Nicholson 1989). Refugia from glacial cold and altered food availability must also have been exploited by many animal groups. The rates at which forest recolonization from refugia takes place have been estimated in several areas, and range up to several kilometres per annum; for some species, such as North American beech, the rates are only a tenth of this (Roberts 1988). Cold-tolerant species exist today at higher altitudes in interglacial refugia, much as they must have done at 125 ka BP. This poses particular problems for conservation, owing to the limited extent of some of these communities and their fragmentation among multiple isolated sites. Mapping and identification of the flora and fauna are necessary precursors to informed management of these remaining communities. However, the realization that continual change is what really characterizes the environment must guide attitudes which suggest that we must attempt to lock up and preserve (e.g., in national parks) environments in their present, transitory configurations. Now that many ecosystems are isolated by settlements, croplands, or plantation forests, opportunities for migration are greatly restricted. Future environmental changes, perhaps anthropogenically caused, will thus pose a new hazard for biota, and human intervention (perhaps through the creation of migration corridors, or the artificial relocation of organisms to more suitable environments) may be required to preserve communities and genetic diversity. We must recognize too that the composition of many ecosystems has been deliberately modified already through human intervention. The introduction of foreign organisms for cultivation, or for the biological control of pest organisms deliberately or accidentally transported, has been widespread during the last two centuries especially.

Agriculture and accelerated soil erosion

Human use of the soil for crop growing and for the production of animal fodder has intensified greatly during the past few thousand years, and more dramatically during the last century. Over vast areas, the natural vegetation has been removed for the sowing of crops; the repeated cultivation of the soil has left it exposed to stripping by both wind and water, and in recent times, the extensive mechanization of agriculture has led to the progressive modification of the internal fabric of the soil, its 'structure', by the breakdown of natural aggregates of mineral particles and of pore spaces by the weight of machinery. Catastrophic damage to agricultural land, such as witnessed in the American 'dust bowl' conditions of the 1930s, is not uncommon. However, even well-managed continued cropping also strips nutrients from the soil, which must be replaced by fertilizers. Similarly, a wide range of chemicals is added to crops and soils to control pests. Consequently, runoff, or continued water and wind erosion, carry these chemical materials away with water and soil particles, so that widespread off-farm impacts and water pollution are additional consequences of modern large-scale agriculture.

The accelerated soil erosion that is associated with unprotected soils after cropping, or which occurs among row crops (including major crops such as corn and soybeans), has averaged more than 18 t/ha/a in the USA, according to the 1982 assessment of the National Resources Inventory (NRI), and 44 per cent of the 170 million hectares under cultivation there has lost soil at rates above those regarded as tolerable for sustained use (Lee 1984; Soule, Carré and Jackson 1990). This survey also reported a mean loss of sediment of 1.7 t/ha/a for ungrazed forest land – a striking 10 times lower than the mean for cropland. Brown (1981) estimated that worldwide, in excess of 21×10^9 t of soil is lost to erosion annually, at rates in many areas above those that can be sustained if cropping is to continue. Similar conclusions are drawn from the compilation of international soil loss data of

Brown (1984), who also estimates that the global agricultural soil reserve is being lost at a rate of about 0.7 per cent per annum (i.e., 7 per cent per decade). Soil conservation measures, including the reduced use of soil tillage between harvest and re-sowing, are fortunately being ever more widely adopted. We must accept, though, that in order to produce sufficient food for the burgeoning global population, the intensity of pastoral and agricultural land use will need to increase, so that soil degradation problems will undoubtedly remain a hallmark of modern food production.

Drought, overgrazing and desertification

Human use of the landscape has, for millennia, wrought changes in that landscape beyond mere soil erosion, which may be identified under the general heading of 'land degradation' (Blaikie and Brookfield 1987). The Mediterranean region provides abundant examples of deforestation, soil erosion, and loss of ecosystem productivity resulting from human occupation. The forests of this region have long been exploited for fuel and building materials, and cleared for agriculture. Destruction of the vegetation has been exacerbated by the numerous flocks of grazing animals. In many areas it is only possible to speculate about the pre-disturbance condition of the vegetation.

In the extensive dry regions across North Africa and into the Middle East, the remains of once-productive agricultural land remind us of these changes. The general deterioration of the soil and vegetation cover caused by human occupation, and associated with falling crop productivity which has caused several infamous famines, has been termed 'desertification'. It is a phenomenon which is now widespread in and around the margins of the world's drylands (Figure 11.1). However, it is necessary to recall our Quaternary perspective: in parallel with the effects of human occupation of the landscape are the ongoing climatic shifts associated with external causes such as the Milankovitch mechanism. It is thus necessary to isolate the separate effects of these agencies before the true human impact on the landscape can be judged.

Let us consider as an example the famine-prone Sahel region, which runs east–west across Africa, including parts of Senegal, Mauritania, Mali, Niger, Chad, Burkina Faso, Nigeria, Sudan and Ethiopia. This region derives its name from an Arabic word meaning shore or coast, and is indeed the southern margin of the hyperarid Sahara which dominates Algeria, Libya and Egypt to the north. Across the Sahel, the annual average rainfall declines northward at a mean rate of 1 mm/km; annual totals in the south are around 600 mm, and in the north, around 100 mm (Le Houérou 1989). People have had an impact on the environments

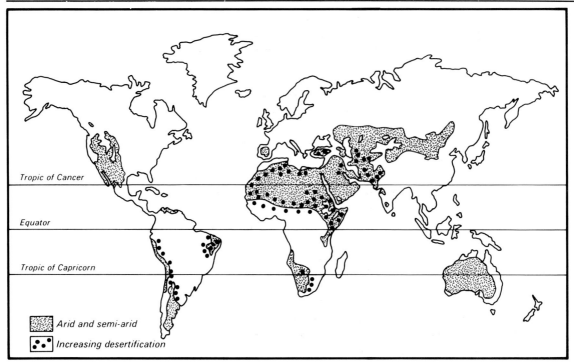

Tropic of Cancer

Equator

Tropic of Capricorn

Arid and semi-arid

Increasing desertification

Fig. 11.1 Regions of the world where there has been a marked increase in desertification between 1977 and 1984 (after Williams 1986, based mainly on data tabulated by Mabbutt 1985)

of the Sahel for millennia, but this is now increasing because of high population growth rates which have developed especially in the last few decades. Cropping is progressively replacing low-density rangeland grazing, and fallow periods are being reduced to permit more frequent harvests (Le Houérou 1989). In addition, land in the drier northern regions (where the coefficient of variability of the annual total rainfall – inversely proportional to its reliability – is higher) is being pressed into use. For example, in Niger cultivation 40 years ago was restricted to areas south of 15°N (annual rainfall about 400 mm or more) but has now spread north to 16°20′N, where the annual rainfall is less than 250 mm; high incidence of crop failure has resulted, together with the loss of the previously useful grassland (Le Houérou 1989). Sharply increasing stock numbers when combined with drought have left the soil surface unprotected against wind, rain splash, rill and sheet erosion. The soil surface becomes compacted and impervious to water; soil moisture levels fall and large areas become bare in the typical sequence of events involved in desertification. Through events of this kind, the southern margin of the Sahara has moved 80–100 km southward in the Sudan between 1958 and 1975; similar trends are displayed in Mali, Chad, and elsewhere.

As argued above, it is necessary to examine the possible role of external climate change as a fundamental cause of the desert expansion. Petit-Maire (1990a,b) has done this in the context of what is known of the Quaternary environments of Africa. At 130 ka BP, during interglacial conditions, lakes existed throughout the Sahara; these dried up progressively as the Wisconsin–Weichsel glacial developed, and at 18 ka BP the arid Sahara extended as far south as 14°N, where derelict dunefields have been mapped (Petit-Maire 1990a). Forests retreated to the upland refugia referred to earlier. Human activity in the landscape cannot have been responsible for this environmental change – population numbers were too low, and in any case the same sequence of climatic events is essentially global in its occurrence (see Chapter 8). During the late glacial and early Holocene, rainfall across the Sahel increased, lakes refilled, the hyperarid Sahara retreated northward, and conditions improved until about 9 ka BP. However, the interval of benevolent climate was short-lived: by about 6 ka BP, drought was repeatedly affecting northern Mali, and the southern margin of the Saharan environment has today extended to about 17°N (Petit-Maire 1990a). It seems unlikely that the Neolithic peoples of this area, who were not cattle herders, and only populated the area after about 7 ka BP (Petit-Maire 1990b), can be responsible for the late Holocene environmental deterioration (see Chapter 8). Rather, if indeed the peak of the present interglacial has passed, a trend towards aridification is what must be expected, as conditions slowly cool and become more like those of the last glacial. However, it must be the case that the overgrazing, soil deterioration, and resultant loss of grasses and woody species act to reinforce the externally driven tendency. It has been hypothesized that the higher albedo of the desertified landscapes acts through altered fluxes of heat and moisture into the atmosphere, as well as through reduced surface roughness. Atmospheric circulation models predict declining rainfall (especially over the Sahel) and southward displacement of wetter areas if surface albedo is increased (Nicholson 1989), but empirical verification of these ideas is still required. A considerably more significant role may be played by fluctuations in sea-surface temperatures, and links with global-scale phenomena, such as the El Niño–Southern Oscillation phenomenon (ENSO), may in due course provide us with some ability to forecast periods of drought in areas like the Sahel (Rasmusson 1987). Thus, a working conclusion at present is that there are probably external factors which are promoting drought in areas like the Sahel, but that the conditions are almost certainly made worse by the anthropogenic desertification.

Globally, the significance of the albedo effect and related changes involved in desertification continue to be investigated. Certainly, desertification has affected large areas, including the Sahara and its margins, the Rajasthan desert in India, and smaller famous cases such as the deforestation of Lebanon. Many of these cases are discussed by Sagan *et al.* (1979), who conclude that globally, in excess of 9×10^6 km^2 has been affected; they estimated the rate of growth of desertified land as about 0.1 per cent of the land surface of the globe per decade. The albedo of such surfaces increases from perhaps 0.16 to 0.35 (Sagan *et al.* 1979); the effect of this is to lower the atmospheric temperature, and, on the basis of Quaternary analogues, this in turn must be expected to result in lower rainfall. A similar role for dust deflated from the overgrazed surfaces has been hypothesized: that it may block incoming solar radiation, hence cooling the surface and resulting in descending air and reduced opportunity for precipitation (Hansen and Lacis 1990). Sagan *et al.* (1979) speculate that the long history of anthropogenic desertification, which together with salinization and deforestation has now affected about 15 per cent of the global land area, may indeed have contributed to global cooling since the relative warmth of the earlier Holocene. They estimate that the anthropogenic component of this cooling amounts to 1°C.

Irrigation and salinization

To permit cropping in dry regions, people have long employed irrigation. In the Negev desert of Israel, Nabatean peoples 2 ka ago directed runoff water onto their fields by major earthworks and stone embankments which 'harvested' water from extensive areas in the surrounding hills. They were able to grow grain crops and fruits successfully, and some of their techniques are being reintroduced at the present day, together with newer techniques to permit the growth of trees in what is termed the 'savannization' of the desert (Evenari *et al.* 1971; Israel Land Development Authority 1990). Larger-scale irrigation, employing water diverted from rivers and dams, is however associated with the problem of 'salinization'. The irrigation water brings with it dissolved materials which accumulate in the soil to increase the salt content there. Rising water tables produced by the irrigation, combined with strong evaporation at the surface, progressively result in salts being set down in the upper layers of the soils where they produce conditions hostile to continued plant growth. More than 50 per cent of the world's irrigated land is salinized to some degree (Rice and Vandermeer 1990). Techniques to cope with the salt accumulation, including the use of evaporation ponds where the salts are harvested, become essential where the salinization is significantly reducing crop yields. Severe salinization, which converts productive land to increasingly bare salinity flats, produces a similar albedo increase to that of desertification already described, and undoubtedly also contributes to global environmental change.

Irrigation also represents a disturbance to the hydrology of the landscape surface, in which water is diverted from one area to another, and lost to evaporation and transpiration. Consequences are often felt in the ecosystems which suffer water loss. However, in areas where irrigation water, or water for stock or domestic use, is derived from groundwater, additional problems may arise. In many areas, such as the Great Artesian Basin of Australia, the groundwater that is harvested from bores has been moving slowly through the aquifers for tens or hundreds of ka. Much of it must have entered through groundwater recharge areas during wet intervals of the Quaternary. The water, thus, in reality is a store of 'fossil' groundwater which may not be replaced in the contemporary climatic environment at a rate which can sustain the rate of extraction. If this is so, then extensive use of groundwater from such aquifers will progressively deplete the groundwater store, and will be unsustainable. In Australia, where much of the pastoral country relies heavily on groundwater, there are 400 000 bores in use (Habermehl 1985); in some aquifers, extraction is in approximate equilibrium with recharge, but in others, falling yields imply that extraction is proceeding too rapidly. In the Great Artesian Basin, dating of the water by isotopic methods has shown that much of it dates from recharge occurring since the middle Quaternary (Bentley *et al.* 1986).

Human effects on the atmosphere

The human impact on the composition of the global atmosphere may be the most far-reaching of our environmental disturbances. Many of the activities already mentioned result in the release of infra-red absorbing greenhouse gases, which reduce heat loss from the earth-atmosphere system (Table 11.1; Gribbin 1986; Pearman 1988; Schneider 1989; Leggett 1990). The metabolism of cattle, for example, releases methane, a very effective greenhouse gas.

Table 11.1 Summary of key greenhouse gases affected by human activities (from Houghton *et al.* 1990)

	Carbon dioxide	Methane	CFC-11	CFC-12	Nitrous oxide
Atmospheric concentration	ppmv	ppmv	pptv	pptv	ppbv
Pre-industrial (1750–1800)	280	0.8	0	0	288
Present day (1990)	353	1.72	280	484	310
Current rate of change per year	1.8 (0.5%)	0.015 (0.9%)	9.5 (4%)	17 (4%)	0.8 (0.25%)
Atmospheric lifetime (years)	(50–200)	10	65	130	150

ppmv = parts per million by volume
ppbv = parts per billion (thousand million) by volume
pptv = parts per trillion (million million) by volume

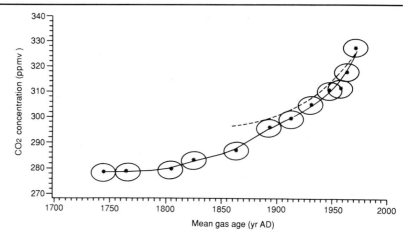

Fig. 11.2 Measured mean CO_2 concentration plotted against estimated age, based on analysis of air occluded in west Antarctic ice (after Neftel *et al.* 1985)

Modern plastics production and refrigeration plants release the chlorofluorocarbon gases (CFCs), which are potent greenhouse gases whose atmospheric abundance has roughly doubled since observations began in 1977 (Smil 1990). Deforestation and the combustion of fossil fuels all release CO_2, a less potent but nonetheless important greenhouse gas, as well as nitrous oxide (Cofer *et al.* 1981). Ice-core records (see Chapter Three; Figure 11.2) together with contemporary observation programmes (Figures 11.3, 11.4, 11.5) clearly indicate that human activities are increasing the concentration of CO_2 and other gases in the atmosphere, and a tendency towards global warming is the almost certain consequence. This tendency is reinforced by the increasing absolute humidity of warmer air, since water vapour is also a greenhouse gas (Del Genio *et al.* 1991; Rind *et al.* 1991).

Release of CO_2 has resulted from forest clearance and the resulting oxidation of litter and soil organic matter. Faure *et al.* (1990) have estimated that 33–47 per cent of the Holocene phytomass, equal to 275–490 GT of carbon (1 GT equals 10^{15}g), has been passed into or through the atmosphere by this mechanism. This leaves a present phytomass carbon store of about 580 GT of carbon, compared to the estimated glacial phytomass store of 300 GT and the interglacial value of 900 GT (Faure 1990). There have also been changes in the soil carbon store and in peat deposits, assessed by Adams *et al.* (1990). In addition to phytomass loss, the combustion of the fossil fuels coal, oil and gas has released about 20×10^9 t of CO_2. Only about 40 per cent of the released CO_2 has evidently remained in the atmosphere, increasing the level in the atmosphere from the pre-Industrial Revolution value of 280 ppmv (parts per million by volume) to the present value of 350 ppmv. The remaining CO_2 has evidently been absorbed by the oceans or other parts of the biosphere in ways that are incompletely understood (Smil 1990).

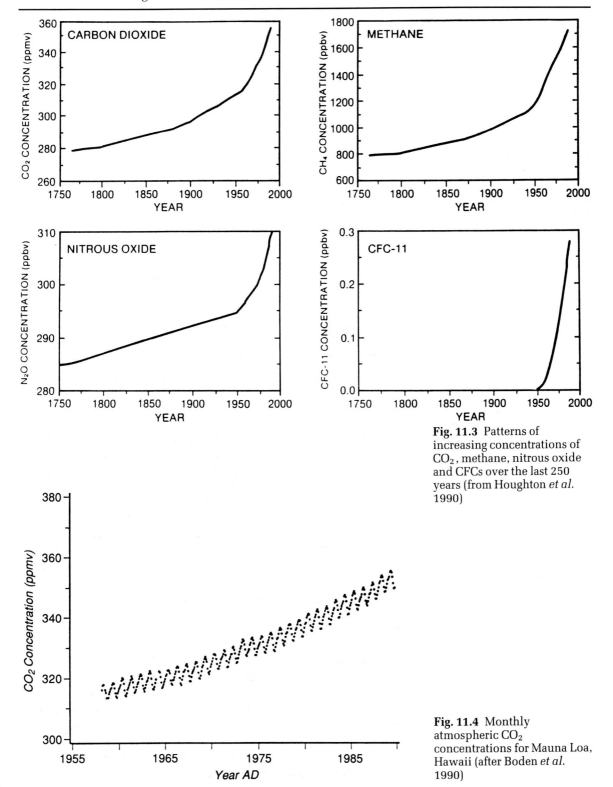

Fig. 11.3 Patterns of increasing concentrations of CO_2, methane, nitrous oxide and CFCs over the last 250 years (from Houghton *et al.* 1990)

Fig. 11.4 Monthly atmospheric CO_2 concentrations for Mauna Loa, Hawaii (after Boden *et al.* 1990)

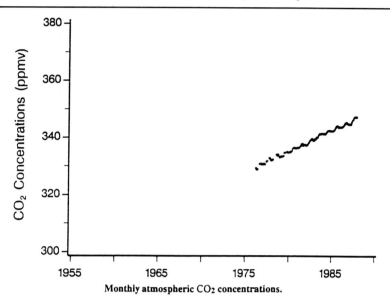

Fig. 11.5 Monthly atmospheric CO_2 concentrations for Cape Grim, Tasmania (after Boden *et al.* 1990)

The effects of the greenhouse gases on the global climate are estimated from numerical models of the global atmosphere. These are still developing and do not incorporate all of the processes which must be modelled, nor do they have the spatial resolution to forecast climatic outcomes at continental or regional scales reliably. In addition, the prediction of future climatic outcomes depends upon the adoption of an appropriate scenario describing the future generation of the gases involved. This is complex because it involves predictions of the growth of the global population, together with its demands for energy and resources. One scenario commonly employed is the 'business as usual' scenario which involves extrapolation of contemporary trends. Other scenarios involve the foreshadowed adoption of controls on both population and the generation of greenhouse gases; the rate of adoption of such controls varies from scenario to scenario. There is no way at present to know which (if any) scenario will approximate the final outcome.

The general conclusion from the 'business as usual' scenario is that atmospheric CO_2 levels may double in the next 50 years or so. While the climate models yield quite variable results for the climatic consequences of this, a general conclusion is that the mean atmospheric temperature will rise by $3.5 \pm 1.5°C$ (Figure 11.6). In consequence of this change, many other effects follow. There will be retreat of valley glaciers, contributing water to the oceans. The sea level will tend to rise in response, but largely because of the thermal expansion of the surface water (Wigley and Raper 1987; Mikolajewicz *et al.* 1990). However, it has been suggested that the warmer oceans will yield higher rainfalls in

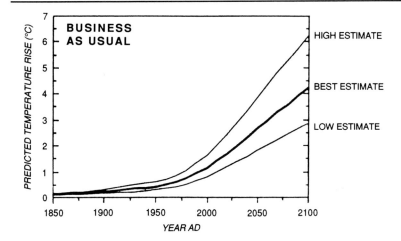

Fig. 11.6 Simulation of increases in global mean temperature from 1850 to 1990 produced by increased concentration of greenhouse gases and predictions of the continuing rise for 1990–2100 based upon 'business as usual' emissions (after Houghton *et al.* 1990)

many areas, possibly including the Antarctic ice caps, so that water may also be removed from the oceans. On the basis of the effects presently understood, the predicted overall sea-level response is for a rise of 8–29 cm, most probably 18 cm, by 2030, and of 21–71 cm, most probably 71 cm, by 2070 (Warrick and Oerlemans 1990). Sea level is presently rising, judged against stable costs, by 1–2 mm per annum (Warrick and Oerlemans 1990), and global temperatures are confirmed to be rising (Jones *et al.* 1986; see Figure 11.7). Increased coastal erosion, and greater storm-generated marine inundation of low-lying areas are among a multitude of potentially costly and damaging consequences of the predicted sea-level rise.

Ecosystem changes will also follow any greenhouse warming, and are likely to involve a mixture of undesirable and beneficial effects. Agriculture may benefit from increased rainfall in some areas (Figure 11.8). In locations which do not experience significantly increased rainfall, soil moisture levels will fall, and the suitability of land for crop production will change. Altered rainfall seasonality or storm intensity will have implications for surface erosional processes, particularly in agricultural areas and rangelands. Plant productivity increases in an atmosphere enriched in CO_2, and the effect increases in warmer conditions, so that there may be beneficial consequences for crop production: yields are estimated to increase in some areas by 50–60 per cent (Smil 1990; Adams *et al.* 1990). Also, frost damage may be reduced. However, microbial respiration also increases in warmer conditions, so that there may be a feedback mechanism here to release additional CO_2 (Melillo *et al.* 1990). This may be reinforced by the death of temperate forests and their replacement by grassland as a result of moisture stress. A particular concern with the prospective greenhouse climatic warming is that, unlike the

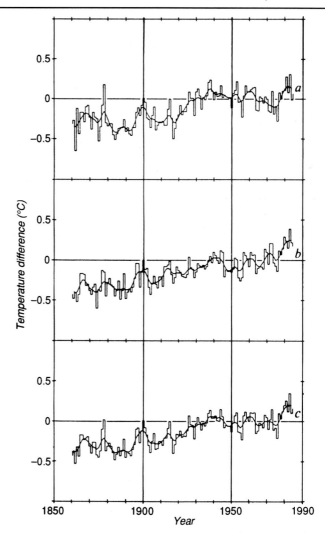

Fig. 11.7 Annual temperature variation since 1861, based on sea-surface temperature data. Smooth curves are 10-year Gaussian filtered values. (a) Northern Hemisphere; (b) Southern Hemisphere; (c) global (after Jones *et al.* 1986)

warming which led from the last glacial phase into the Holocene, it will be very rapid. The anticipated rates of up to 0.3°C per decade (Melillo *et al.* 1990) may be too fast for an accommodating migration of ecosystems to occur, especially in view of the fact that potential migration pathways may not be available because of urban or agricultural development.

Many people have considered that the demand for food and for improved standards of living will ensure that fossil fuel consumption and deforestation will continue well into the future. Whether or not this proves, in the end, to be so, we would be foolhardy as a species to ignore the risk involved. A search for increasingly efficient forms of transportation and power generation, and plans to control global population, will be beneficial in extending the lifetime of available global resources whether or not such steps are

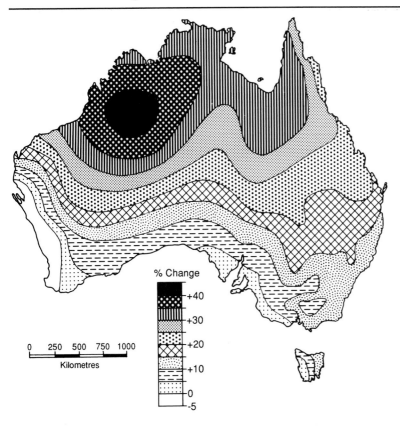

% Change

+40

+30

+20

+10

0

-5

0 250 500 750 1000
Kilometres

Fig. 11.8 Pattern of percentage change in net primary productivity relative to the present for a scenario of approximately doubled atmospheric CO_2 concentration (after Pittock and Nix 1986)

strictly required for the protection of the global climate. However, such action as we take may prove to be too little, too late. Therefore consideration is being given to ways in which the global environment might be deliberately modified to counteract undesirable changes – a somewhat anthropocentric view of the environment (Fyfe 1990). What kinds of actions could we take?

In order to control the CO_2 content of the atmosphere, Faure *et al.* (1990) have envisaged a number of strategies. By diverting water to dry regions and cultivating forests, it would be possible to store 10 GT of carbon in each 10^6 km^2 of forest. Alternatively, artificial peatlands could be created in continental depressions. The stimulation of corals and algae through the supply of extra nutrients to the water could result in the storage of additional carbonates in the marine environment, or vegetation could be submerged and buried offshore to simulate the fossil storage of carbon: 4 km^3 of vegetation buried would store about 1 GT of carbon. Finally, CO_2 could be stored in sea water by the deliberate pumping of deep cold water to the surface or by the break-up of sea ice to produce sinking cold water.

Continuing sea-level rise will pose enormous problems for coastal cities and in the other areas already mentioned. Deliberate manipulation of global sea level has been suggested by Newman and Fairbridge (1986), who argue that the construction of reservoirs during the past few decades has stored water equivalent to a 0.75 mm/a sea-level rise. Much larger systems of water storage could continue to be employed in this way: Newman and Fairbridge contemplate the pumping of sea water into major continental depressions such as the Imperial Valley of California, the Dead Sea rift between Israel and Jordan and the Qattara depression in Egypt. Additionally, the level of the Aral–Caspian sea, presently 28 m below sea level, could be lifted. Raising this by 10 m could store sufficient water to stabilize sea level for about a decade. There are many difficulties with these ideas, including the environmental cost of inundating the areas mentioned, many of which are scenically or culturally important. Further, evaporation from the proposed sites could lead to the deposition of salt, and continued delivery of sea water could produce long-term changes in ocean chemistry, with possible subsequent climatic consequences. Finally, the enormous additional mass loaded onto the crust may induce brittle failure there and result in increased earthquake activity, as has already been associated with some large water supply reservoirs.

A final potential impact of human activity upon the composition of the atmosphere must be mentioned. This concerns the effects of warfare in which significant numbers of nuclear warheads are detonated. It has been suggested that these detonations could result in enormous fires in urban areas, as well as the release of silt and clay particles from the soils at detonation sites. Smoke and dust particles may be injected into the atmosphere in sufficient quantities to reflect solar radiation to the extent that temperatures at the surface fall dramatically, perhaps to below freezing point over large areas. This is the 'nuclear winter' scenario (Harwell 1984; Pittock 1987). The effects of such particles in the atmosphere depend on their size, quantity, and location. If injected into the stratosphere, smoke particles might endure for years, whereas in the troposphere they are washed out more rapidly. Atmospheric models run on the basis of likely circumstances suggest that dramatic cooling could last from a period of weeks up to many months. Freezing conditions would result in a scarcity of water, and the germination of crops could be prevented, with famine following. The extinction of many species might also result.

These predictions are fortunately untested. Some investigators remain sceptical of the predictions that have been made, some indeed forecasting dramatic warming rather than cooling as the aftermath of nuclear war (e.g., Singer 1989). The consequences of this could be just as severe and either outcome must be strenuously avoided.

Conclusion

We have seen that people are now so numerous and have such widespread effects on the global environment that we are partly responsible for its behaviour. We have altered the composition of the atmosphere, changed the water balance and albedo of large areas of the land surface, and significantly perturbed the cycling of carbon through the ecosystems of the planet. The environmental changes which we are continuing to generate have analogues in the environmental changes of the Quaternary, but only in certain respects. Our knowledge of the events of the Quaternary is insufficient to identify and understand the multitude of interconnections and feedback linkages which were involved in the environmental instability which was its hallmark. However, we do possess a reasonable description of the general nature of environmental changes and of the rates at which they occurred. This reveals that the anthropogenically induced environmental changes which are presently under way are fundamentally different, at least in terms of their rate of occurrence, to those of the Quaternary. Greater stress will consequently be placed on organisms required to adapt to changing habitats. Our knowledge of the events of the Quaternary period also permits us to judge the magnitude of contemporary environmental changes against those of the past, and by analogy with the nature of the past environments, to gauge something of their probable consequences. There are, then, clear and compelling reasons to continue to refine our understanding of the Quaternary and its legacy, and to apply this knowledge to a growing understanding of our own place within the global environment, and of our responsibility for its management.

Appendix: Quaternary Dating Methods

Since geological time is not salami, slicing it up has no particular virtue. But if it is to be sliced there is no need to botch the job, and chronometric dating provides the guidelines.

Claudio Vita-Finzi,
Recent Earth History, 1973.

Introduction

A wide range of Quaternary dating methods is available, each restricted by its material applications, the events being dated and the type of result given. The latter provides a means of classifying dating methods. Numerical age estimates are those which alone give a numerical indication of the age based on rates of change. Relative age determination allows samples to be ordered with respect to age but requires calibration against samples of known age before numerical ages can be estimated. Correlation of different sequences is possible using age-equivalent marker horizons such as tephra or pollen, which have been dated by another method. Table A.1 compares a number of Quaternary dating methods. Recent reviews include Roberts (1989); Roth and Poty (1989); and Aitken (1990).

Radiocarbon dating

The principles of the radiocarbon dating technique were first published by Libby *et al.* (1949). Since then the technique has advanced rapidly into the most widely applied and accepted dating method for the latter part of the Quaternary.

Radiocarbon (carbon-14 or ^{14}C) is an unstable radioactive isotope of carbon which forms in the atmosphere as a result of the interaction of the secondary neutrons of cosmic radiation with ^{14}N nuclei:

$$^{14}N \xrightarrow[\text{emission of beta-particles (decay)}]{\text{cosmic ray bombardment}} {}^{14}C$$

Once formed, ^{14}C oxidizes to CO_2 in the atmosphere and enters the oceans and groundwater via exchange, plants via photosynth-

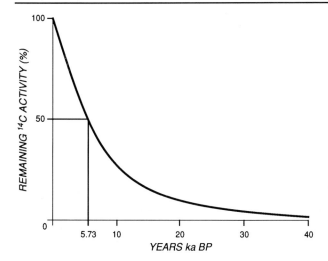

Fig. A.1 Radiocarbon decay curve

esis, and animals via the food chain. It is assumed that the ratio of $^{12}C:^{13}C:^{14}C$ has remained relatively constant in the atmosphere through time at 98.9 per cent ^{12}C, 1.1 per cent ^{13}C, and 1×10^{-10} per cent ^{14}C (Linick *et al.* 1989). This atmospheric carbon isotope ratio is assumed to remain constant in the biosphere and hydrosphere through continuous exchange and rapid mixing. Because ^{14}C is isotopically unstable, it decays to the stable element ^{14}N by emitting beta-particles (negatively charged electrons) at a fixed but exponentially declining decay rate, as illustrated in Figure A.1.

The decay rate is conventionally expressed as a half-life, the time required for the decay of half of the original ^{14}C content, and was originally calculated at 5568 ± 30 years by Libby *et al.* (1949). Polach and Golson (1968) more recently estimated the half-life as 5730 ± 40 years, a difference of 3 per cent from the accepted half-life. During the life of the organic matter the loss by decay is matched by ^{14}C uptake. At death, however, carbon replenishment stops and the ^{14}C content declines at the decay rate; the radiocarbon 'clock' has started. By comparing the remaining radioactivity (or ^{14}C content) of the sample with that of similar species living today (a contemporary reference standard), an indication of the elapsed time since the last ^{14}C uptake can be determined. The ^{14}C date obtained from the interior of the tree will be older than that obtained from the exterior. Any material containing carbon can be dated using the radiocarbon technique, including wood, charcoal, soil, peat, rock varnish, seed, bone, carbonate, shell, cloth and groundwater.

Gas proportional and liquid scintillation techniques have been used to measure the remaining radioactivity of a sample, based on the detection of the negative electrons of ^{14}C decay (radiometry), by voltage pulses and photo fluxes, respectively (Polach 1987). The decay rate thus determined can be converted to an estimate of

Table A.1 Comparison of Quaternary dating techniques (see text for references except where otherwise stated)

Technique	Parameter measured	Age range	Matrix	Application Event	Result
Radiocarbon	^{14}C activity or ^{14}C atoms (AMS)	100 a BP–50 ka BP	organic materials e.g., wood charcoal soil bone shells groundwater		relative ages and numerical age estimate
Thermoluminescence (TL)	TL emissions	100 a BP–1 Ma BP	quartz and feldspar mineral grains e.g., loess and sand dunes fluvial deposits pottery and ceramics tephra carbonates/ precipitates	thermal agent (volcanics, pottery firing) chemical precipitation sunbleaching (dune building, fluvial history)	numerical age estimate
Electronic Spin Resonance (ESR)	ESR of trapped electrons	2 ka BP–10 Ma BP	e.g., shells corals bones and teeth organics some precipitates	as above	numerical age estimate
Uranium-Series	concentrations or isotopic ratios of Uranium and its daughters	See Tables A.2 and A.3 for age range of each technique	wide variety of materials containing uranium or thorium (see Tables A.2 and A.3 for details)		numerical age estimate
Potassium-Argon (K-Ar)	$^{40}Ar^*$ accumulation ^{40}K content ($^{40}K/^{40}Ar^*$) $^{40}Ar/^{39}Ar$	100 ka BP–no upper age limit	materials containing potassium – volcanic, metamorphic and igneous rocks (see Figure A.9 for datable materials and their ranges)	rock/mineral formation of alteration (e.g., weathering) recrystallization faulting thermal history	numerical age estimate
Fission track	fission track density	100 ka BP – no upper age limit	materials containing ^{238}U e.g., zircon glass (i.e., obsidian) basalts volcanic pumice	volcanic events landscape evolution sea-floor spreading archaeological materials	numerical age estimate

Method	Measurement	Time range	Material	Application	Age type
Obsidian-hydration	thickness of the hydrated layer	1 ka BP–millions of years BP	obsidian	glacial moraine deposition archaeological materials	numerical age estimate
Historical archival archaeological	records	Present–tens of thousands of years BP		various / various – alluvial history	numerical age estimate / numerical age estimate
Dendrochronology	growth rings	Present–1 ka BP	trees		numerical age estimate
Varves and rhythmites	glacial melt laminae	Present–10 ka BP	lake sediments	glacial histories	numerical age estimate
Lichenometry		10 a BP–thousands of years BP	lichen	glacial histories	numerical age estimate
Palaeomagnetism – (a) polarity reversals (b) polarity excursions (c) secular variation	direction of remnant magnetic field	(a) 50 ka BP–1 Ma BP (b) 10 ka BP–50 ka BP (c) 2.5 ka BP–3 ka BP	magnetic minerals e.g., volcanics (including tephra) hearths kilns windblown and waterlain sediment tephra	sea-floor spreading sedimentation cooling of volcanic or archaeological samples	correlation (numerical age estimates possible using mastercurves)
Amino Acid Racemization (AAR)	D/L ratio	100 a BP–2 Ma BP	fossil materials containing protein breakdown products e.g., bones, teeth shells forams calcareous sediments and peats	palaeoclimates and death of the organism	relative age (conversion of numerical age is possible by correlation with materials of known age)
Tephrochronology		see K-Ar, ^{14}C or other approximate dating methods	tephra layers	volcanism	correlation
Weathering Pedogenesis	sed. character	1–10 ka BP	soils stratified units glacial deposits	soil formation glacial/interglacial history climate (palaeo-)	relative age and correlation (limited)

the radiocarbon age using the decay curve (Figure A.1). These conventional techniques have an upper age limit of 50 ka BP. Older samples decay too slowly for meaningful measurement or differentiation from background signals. To extend the measurement period it is possible to enrich the carbon isotopes, either by thermal diffusion or with lasers, allowing a more accurate determination of the decay rate (Lowe and Walker 1984). Exciting developments using accelerator mass spectrometry (AMS) since 1977 involve the direct counting of ^{14}C atoms, rather than the small quantities of beta-particles emitted during decay. The increased sensitivity enables the radiocarbon dating of samples 1000 times smaller than those required using conventional methods, and involves less time and fewer problems related to background radiation. AMS is applicable to a wide range of carbonaceous materials, and has been useful in the study of longer time scales using other isotopes, including ^{10}Be, ^{26}Al, ^{36}Cl, ^{41}Ca and ^{129}I (Linick *et al.* 1989).

Both natural and anthropogenic variations in ^{14}C levels modify the basic assumption that carbon ratios in the atmosphere, biosphere and hydrosphere remain constant. Materials formed between 1850 and 1954 should be treated with caution, if not disregarded, owing to the anomalous enrichment of ^{12}C resulting from the burning of fossil fuels, and the consequent dilution of the ^{14}C content. Radiocarbon dates for this period, peaking at the time of the Industrial Revolution, indicate an estimated age for samples which is older than the true age. Similarly, carbon-bearing materials formed since 1954 show artificially young dates, owing to the increase in ^{14}C in the atmosphere resulting from the detonation of nuclear weapons, peaking in 1963 with a 90 per cent increase in the $^{14}C:C$ ratio. Consequently, ^{14}C dates are presented as years before 1950.

Doubts have been expressed recently about the reliability of uncalibrated ^{14}C ages, encouraging the establishment of calibration curves for the entire radiocarbon time scale (Bard *et al.* 1990). Tree rings provide an annual record of variations in ^{14}C levels, allowing correction of the radiocarbon time scale and a means of calibrating radiocarbon ages to calendar years before 1950. Figure A.2 shows the radiocarbon calibration curve established for most of the Holocene. The long-term peak-to-trough change is attributed to variation in the strength of the earth's magnetic field which affects the cosmic ray flux, while the shorter-term fluctuations, with amplitudes of 100 to 200 years, are attributed to solar modulation of the cosmic ray flux (Barbetti 1991). Between 10 and 20 per cent of the age discrepancy is a result of changes in carbon reservoir sizes and the rates of exchange between them (Bard *et al.* 1990). A calibration curve for the late Pleistocene has been established from a comparison of conventional radiocarbon ages with other dating methods as shown in Figure A.3.

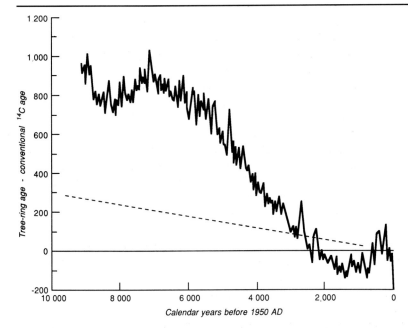

Fig. A.2 Radiocarbon calibration curve for most of the Holocene, based on data from dendrochronologically dated tree-ring samples by Stuiver and Pearson (1986); Pearson and Stuiver (1986); Pearson *et al.* (1986); and Stuiver and Reimer (1986). The solid line indicates the true age of the samples, as tree-ring age minus radiocarbon age equals zero. The departure of the dashed line from the solid line represents the 3 per cent error in the accepted ^{14}C half-life of 5568 ± 30 years. The remaining difference between the dashed line and the curve reflects past variations in the ^{14}C production rate and the possible ^{14}C transfer between the oceans and the atmosphere (Barbetti 1991)

Isotopic fractionation alters the carbon ratio, enriching plant tissues with ^{12}C through photosynthesis, and oceans with ^{14}C through preferential absorption. This is corrected by a normalization procedure using δ^{13}C (Gillespie 1982). Deep ocean waters are infrequently replenished with ^{14}C owing to a slow rate of mixing with surface water. Upwelling currents introduce lower ^{14}C:C ratios to the organisms with which they have contact (the reservoir effect). Groundwater in regions of carbonate rocks such as limestones and coal measures contains dissolved carbonates with no ^{14}C content, thus diluting the ^{14}C:C ratio and resulting in an older apparent age. Samples may be contaminated by younger carbon from rootlets, humic acids, bioturbation, or cigarette ash. These sources of variation and contamination may be minimized, and in some cases avoided, by careful selection of the sample and strict handling and laboratory procedures. Tables and graphs are available for some of the standard corrections (Stuiver 1986).

Radiocarbon dates are presented as ages in years before the present (BP) with a statistical error margin of one standard deviation and a laboratory code number. For example, the date of $19\,640 \pm 210$ years BP (SUA-819) was determined at Sydney University of Australia Radiocarbon Laboratory with a 68.27 per cent probability that the age of the sample lies between 19 430 and 19 850 years BP.

Thermoluminescence dating

Thermoluminescence dating is a technique applied directly to

Fig. A.3 Radiocarbon
calibration curve for the Late
Pleistocene, based on
comparison of dates derived
from conventional ^{14}C and
other methods such as U-Th
(Bard *et al.* 1990), varve data
(Stuiver 1971; Stuiver *et al.*
1986), thermoluminescence
data (Barbetti 1980; Aitken
1987), and tree-ring data from
southern Germany (pre-Boreal
pine) (Becker and Kromer
1986) and Tasmania (Stanley
River trees) (Barbetti 1991).
Tree-ring data are floating
sequences, not absolutely
dated but included assuming
that ^{14}C dates are 1 and 2 ka
too young respectively. They
show short-term wiggles
similar to those seen in the
Holocene. The German data
also suggest a plateau in the
calibration curve with nearly
constant radiocarbon ages
(9.6 ka BP) over several
centuries in the early
Holocene. The solid line
indicates the relationship
assuming that the ^{14}C age
equals true age. The departure
of the dashed line from the
solid line represents the 3 per
cent error in the accepted [14]C
half-life of 5568 ± 30 years
(Barbetti 1991)

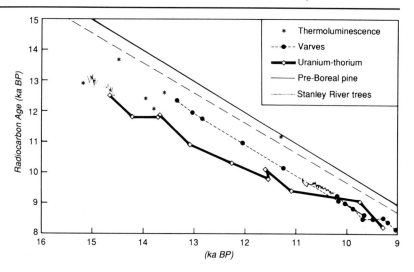

certain mineral grains, particularly quartz and feldspar, potentially providing numerical ages ranging from 100 a to 1 Ma (Singhvi
and Wagner 1986) with an error of 10–15 per cent (Forman 1989).

Radioactive isotopes of uranium, thorium and potassium in
soils and sediments continually bombard minerals with alpha,
beta and gamma radiation, damaging the crystal lattice and
displacing electrons which are subsequently trapped in other
defects of the lattice (Aitken 1985). Heating to greater than 500°C
or exposure to sunlight for more than 8 hours vibrates the crystal
lattice and evicts the trapped electrons, producing thermoluminescence (TL), a light additional to the red-hot incandescent
glow of heated materials. The radiation record is erased
(bleached) and the TL 'clock' is set to zero. The strength of the TL
signal is dependent upon the trapped electrons, which is in turn a
function of the time and intensity of the exposure to alpha, beta,
and gamma radiation. Thus, TL may be used to determine the
elapsed time since the sample was most recently set to zero by a
thermal event of greater than 500°C (e.g., pottery, ceramics, tephra
or sediment baked by lava flows); crystallization (e.g., biological
and chemical precipitates such as foraminifera, speleothems,
travertines and gypcrete); or sun bleaching prior to deposition and
burial (e.g., loess, dune sands, ocean sediments, shells, and
possibly fluvial sediments).

A simplified equation for the TL age (from Singhvi and Wagner
1986) is

$$\text{TL age (years)} = \frac{\text{TL acquired}}{\text{TL acquisition/year}}$$

$$= \frac{\text{natural TL} - \text{residual TL}}{(\text{TL/unit radiation dose}) \times (\text{radiation dose/year})}$$

The natural TL is measured by heating the sample and photo-electronically counting the photons emitted during thermo-luminescence. This is then adjusted for the residual TL present at zero-age, approximated from the TL intensity of modern surface samples from similar depositional environments or by simulation of the zeroing mechanism in the laboratory (Berger 1988), to determine the TL acquired. The TL/unit radiation dose is the induced TL obtained from exposing the sample to a known radiation dose in the laboratory. The radiation dose per year may be determined by either the chemical properties of the radioactive isotopes of the sample, or by direct measurement with dosimeters, devices placed in the deposit at the sample site for up to a year. This latter approach, although time consuming, has the significant advantage of recording site-specific influences (Berger 1988). For example, the accumulation of non-radiogenic precipitates such as gypsum and calcite within a 30 cm radius of the sample progressively dilutes the environmental dose rate, resulting in determinations from the sample or by dosimeters yielding an anomalously high dose rate (underestimate of the age) or an anomalously low dose rate (overestimate of the age), respectively (Chen *et al.* 1990). Use of both techniques will provide a check, identifying and allowing for dose rate variability through time and within heterogeneous stratigraphies.

TL dating is not a routine technique and must be applied to each sample uniquely for a number of reasons. First, each mineral, and minerals from different origins, display a variety of crystal defect types and concentrations and thus different TL responses. Concurrent with this is the problem that TL growth is non-linear, particularly at high dose rates (Readhead 1988). Second, as indicated by the age equation, zeroing may be incomplete, especially for waterlain sediments exposed to filtered sunlight. The level of residual TL is determined by the environment of deposition. Third, water in sediments absorbs part of the radiation dose, attenuating the TL acquired by the sample. Knowledge of seasonal and secular changes in moisture content is required for meaningful TL age estimation. Similarly, changes to the radiation dose result from the post-depositional processes of compaction, leaching and redeposition which alter the moisture content and the location of radioactive elements in the profile (Singhvi and Wagner 1986). Volcanic feldspars rapidly lose laboratory-induced TL as electrons escape without a TL signal. This is termed anomalous fading and is responsible for underestimates of the TL/unit radiation dose. Techniques have been developed to overcome this underestimation. Feldspars are the preferred mineral for TL dating because they are highly susceptible to radiation, present a strong TL signal, and are easily bleached (Forman 1989). Readhead (1988) observed anomalous fading of the TL signal in aeolian quartz sediments and avoided the problem by storing the sample for several weeks after irradiation before measuring the TL dose.

Sample collection must be conducted with extreme care, ensuring that the sample is not exposed for more than a few seconds before it is sealed in an opaque container. Equally important to an understanding of the TL technique is a knowledge of the pedological and geomorphological setting and processes, and the choice of a relevant sample site. The application of TL dating to aeolian sediments (loess and dunes) has encompassed the majority of TL research and success. These sediments typically exhibit the preferred characteristics of long light exposure, homogeneous accumulation (ideally over 50 cm) and relatively constant water content (Forman 1989). The results from TL dating of quartz grains in the lunette on the eastern border of Lake Mungo, south-eastern Australia (Readhead 1988), illustrate this success. The TL dating of waterlain sediments is more problematic. The estimation of residual TL is complicated by the attenuation of ultraviolet rays, especially in turbid waters, and variation in the speed and duration of sediment transport and settling. In addition, these environments typically have complex moisture histories related to events such as fluctuating sea level, channel abandonment, flood and drought. Despite this, a comparative study of TL and uranium-thorium dates by Nanson *et al.* (1991) shows great promise for TL dating of previously undatable waterlain sediments of tropical northern Queensland.

ESR dating

The electron spin resonance (ESR) technique for dating Quaternary events, similar to the TL technique, is based upon the measurement of time-dependent radiation damage to the crystals of minerals such as calcite (e.g., shells and coral), apatite (e.g., bones and teeth), sulphates and phosphates. The theory is similar to that of TL dating as previously discussed.

Electrons, in orbits surrounding the nucleus of an atom, rotate to produce a circulating current and magnetic field. Electrons of opposite spin direction usually form magnetically neutral pairs with no net magnetic moment. Natural radiation from the decay of uranium, thorium, and potassium isotopes in the material and its environment ionise, or damage, the crystals to produce an electron excess centre (trapped electron) and an electron deficient centre (trapped hole). Subsequent heating or recrystallization releases the trapped electrons which recombine with the electron deficient hole, emitting thermoluminescence.

As with thermoluminescence, the ESR signal intensity is an indicator of the concentration of trapped electrons that have accumulated since the material was formed, heated, or exposed to sunlight, and thus may be used to estimate the age of the sample. Each unpaired (trapped) electron possesses a net magnetic moment, with the different spin directions creating an energy termed the 'Zeeman Effect'. Exposure of the sample to microwave radiation of similar energy causes a transition in the direction of spin – electron spin resonance; a process which is recorded by the unpaired electrons through the absorption of microwave radiation, and is detected and measured in an ESR spectrometer (Partridge *et al.* 1984). The simplified age equation is similar to that for TL dating, i.e.,

$$\text{ESR age} = \frac{\text{electron spin concentration}}{(\text{electron spin/unit radiation}) \times (\text{radiation dose/year})} \tag{3}$$

Exposure of the sample to artificial radiation enables assessment of the concentration of spin produced per unit of radiation. The external radiation dose (ED) is measured by dosimeters placed at the site of sample collection and the internal dose rate is determined from the isotopic concentration in the sample. Allowance must be made for incomplete zeroing of the previous radiation record (see discussion of TL dating). Digital ESR techniques determine the age of the sample by directly measuring accumulated and accumulating defects, rendering the knowledge of annual dose unnecessary (Ikeya 1986).

As an ESR analysis does not require heating, it has the advantages over TL dating of being non-destructive, repeatable and applicable to a wide range of materials, such as teeth, bones and aragonitic shell, which change form when heated (Goede 1989). ESR, however, is less sensitive than TL and in most situations is not applicable to ages less than 2 ka BP (Goede 1989).

Difficulties in ESR analysis arise from differences in the nature of electron traps, conditions of radiation and the complexity of the ESR signal (Molodkov 1988). The ESR signal is illustrated as a spectrum, the shape of which is influenced by the processes of deposition, genesis and crystallization, as well as by impurities such as humic acids and clay minerals (Chen *et al.* 1988). The age determined for a sample may differ by up to three to ten times according to the choice of spectral line used in the analysis (Molodkov 1988). Figure A.4(A) shows a sample ESR spectrum and Figure A.4(B) an ESR spectrum complicated by varying degrees of contamination by detrital sediments.

The validity of ESR dating has been tested by analysing samples previously dated by ^{14}C (Chen *et al.* 1988) and uranium-thorium (Mangini 1986a, Smart *et al.* 1988) dating methods. The results have shown that ESR is not only a reliable chronological tool but also provides information on genesis and depositional environments. The age range of this method is a function of the stability of the crystal defects and thus is unique to the material properties

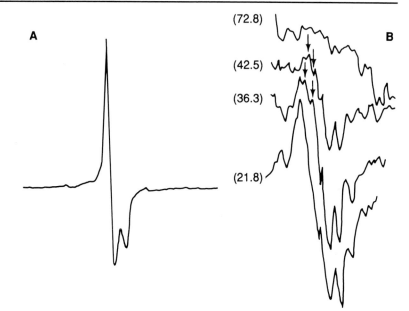

Fig. A.4
(A) Sample ESR spectrum from a horse molar (Ikeya 1986). (B) ESR spectra of calcrete core with varying detrital sediment content (percentage indicated). Arrows indicate interference from clay minerals (adapted after Chen *et al.* 1988)

and thermal history of each sample. Events as old as 0.1–10 Ma BP can be dated using ESR techniques, however the decomposition of organics younger than 1 ka BP may yield erroneous ESR signals.

Amino acid racemization

The amino acid racemization (AAR) dating technique may be applied to any fossiliferous material containing proteins and protein breakdown products. Examples include bones, teeth, egg shells, freshwater and marine shells, foraminifera, calcareous sediments and peats. The age range of AAR is dependent upon the diagenetic temperature history and the composition of the amino acids present in the sample, but may possibly extend to 2 Ma BP (Kimber 1987).

Twenty amino acids are commonly found in living matter and appear in the L-configuration. Proteins of living organisms consist primarily of peptides in which L-amino acids (left-handed) are bound; the ratio of D- (right-handed) to L-amino acids (D/L ratio) is approximately zero, providing the starting point for the AAR 'clock'. After death, proteins undergo complex diagenetic chemical reactions which include racemization and epimerization. These processes convert L-amino acids into the D-form as illustrated in Figure A.5.

D/L ratios of 1.0 and 1.30 represent the racemic (equilibrated) mixtures for racemized and epimerized amino acids respectively. This may take from several hundred years to millions of years,

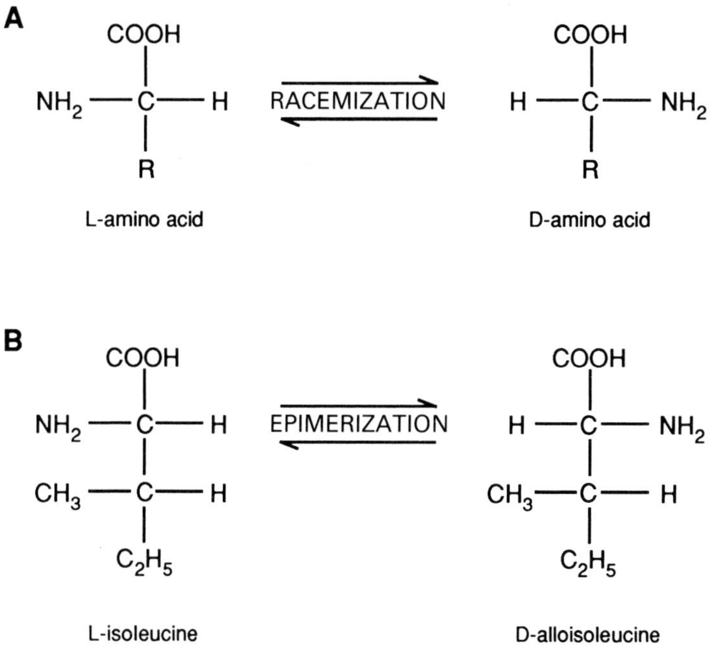

A

L-amino acid   D-amino acid

B

L-isoleucine   D-alloisoleucine

Fig. A.5
(A) Racemization of L- and D-amino acids with one carbon atom (e.g., leucine, aspartic acid).
(B) Epimerization of L- and D-amino acids with two carbon atoms (e.g., isoleucine) (adapted after Miller and Brigham-Grette 1989)

representing the limit of AAR as a chronological indicator (Masters 1986). The rate of racemization/epimerization is dependent upon the amino acid and its position in the protein chain, the species of the organism and various environmental factors including the temperature and moisture history, matrix material, pH and oxidation/reduction conditions (Rutter and Vlahos 1988). The most influential environmental factor is temperature. Samples from temperate and tropical environments attain equilibrium within 2 Ma, whereas samples from polar environments may not have attained equilibrium after 10 Ma (Murray-Wallace and Kimber 1990). The AAR technique cannot be routinely applied, nor can D/L ratios of different amino acids be directly compared.

The measurement of the D/L ratios is a relatively simple procedure using one of three methods, as outlined by Miller and Brigham-Grette (1989)

(i)   Ion-exchange liquid chromatography is a rapid, economical, sensitive and quantitative method suitable for small samples, but is limited to one amino acid per analysis.
(ii) Gas chromatography can potentially measure 18 amino acids per analysis, but is time-consuming and expensive.
(iii) Reversed-phase liquid chromatography has the advantages of rapid analysis and easy preparation, is inexpensive and capable of multiple amino acid analysis, but is not yet a routinely applied method.

The determination of relative ages, or the aminostratigraphy, of a region is the simplest, most reliable application of AAR, providing that the samples have experienced a similar thermal history, as only D/L ratios are required (Miller and Brigham-Grette 1989). To obtain numerical ages for this relative stratigraphy a knowledge of the rates of racemization/epimerization for the amino acids present in the sample and a reconstruction of the thermal history of the site is necessary. Reaction rates are complex, but may be calculated from AAR measurements of a nearby sample of suitable material dated by another technique (e.g., radiocarbon, uranium-series disequilibrium). The measurement of D/L ratios for samples of known age allows the calculation of reaction rates and associated temperature. This latter application is termed palaeothermometry and has been useful in reconstructing Quaternary palaeoclimates (Murray-Wallace *et al.* 1988).

The D/L ratio resulting from reactions of indigenous amino acids may be contaminated by the loss or gain of amino acids by leaching, diagenetic formation of amino acids, bacterial contamination, and/or contamination of the sample during collection and preparation (Murray-Wallace and Kimber 1990). Samples which are either well preserved or collected from depth remain unexposed to frequent temperature oscillation, therefore avoiding the enhancement of racemization rates (Wehmiller 1986).

Foraminifera and calcareous marine sediments experience complex racemization kinetics. Despite the variation in reaction rates according to species and different parts of shells, mollusc shells are particularly suitable for AAR dating as they are generally unaffected by loss of fossil proteins, pH or clay mineral catalysis (chemical changes promoted by association with clay minerals) (Murray-Wallace and Kimber 1990). Bones and teeth are preferred for AAR as they have a simple protein composition and, like wood, have non-species specific reaction rates, but care must be taken that samples have not been artificially heated.

Uranium-series disequilibrium

The family of uranium-series disequilibrium dating methods is based on the time-dependent growth or decay of uranium parent and daughter nuclides as they restore equilibrium in the uranium series. U-series dating is applicable to a wide variety of materials and can be used to date materials from the present to 1 Ma BP, depending upon the nuclides used in the analysis.

Uranium and thorium are incorporated into crustal rocks during the late crystallizing phase, and are therefore mainly associated with granites and pegmatites. Systems which contain uranium and have remained undisturbed for long periods of time

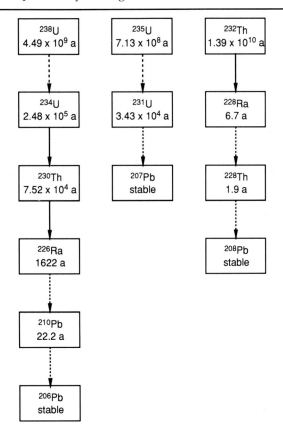

Fig. A.6 Simplified uranium and thorium decay series indicating nuclides relevant to dating. Dashed arrows indicate that intermediate nuclides have been omitted (adapted after Ku 1976)

relative to parent half-lives display secular radioactive equilibrium (Ivanovich and Harmon 1982). The decay rate of the daughter nuclide is equal to the production of the daughter by decay of the parent isotope. Disturbed systems which have experienced natural isotopic fractionation (separation) of parent and daughter nuclides do not show this equilibrium. By determining the degree to which equilibrium has been restored, the time since disturbance can be estimated (Bradley 1985; Ivanovich and Harmon 1982). Restoration of secular equilibrium sets the limit of U-series dating, typically at four times the half-life of the dating nuclide (Goldberg 1986). Nuclides of the uranium and thorium dating series which are relevant to dating are shown in Figure A.6.

Disturbance of the U-series is common through the preferential leaching of soluble and precipitation of insoluble members of the U-series during weathering, transportation and deposition (Ku 1976). Two forms of disturbance to the U-series can occur which, by definition, distinguish the two groups of U-series disequilibrium dating methods.

The first group of methods is based upon the decay of unsupported daughters of the U-series. For example, nearly all of the insoluble ^{230}Th and ^{231}Pa that forms in sea water from dissolved uranium undergoes rapid hydrolysis and is either adsorbed onto sinking particulates or incorporated into ferromanganese nodules (Ku 1976). Because the ^{230}Th and ^{231}Pa nuclides found in excess in bottom sediments are unsupported by uranium, their activity decays at a known rate. The activity of the deposit gives an indication of its age (Ivanovich and Harmon 1982). Another example of this U-series dating method is ^{210}Pb (lead-210) dating. Uranium present in the earth's crust decays via ^{222}Rn which, as an inert gas, diffuses into the atmosphere where it eventually decays to ^{210}Pb (Goldberg 1963). ^{210}Pb rapidly attaches itself to aerosol particles and settles to the earth's surface, either as fallout or in rain, where it decays at a known rate, unsupported by ^{226}Ra. The ^{210}Pb activity in sediments provides an indication of their age. Dating methods based on the decay of unsupported daughter nuclides are listed in Table A.2 with their dating ranges and applications.

Table A.2 U-series disequilibria dating methods based on the decay of unsupported daughter nuclides (modified after Ivanovich and Harmon 1982)

Method	Upper age limit	Applications
^{234}U/^{238}U	1.25 Ma	some corals and waters
^{230}Th	300 ka	deep-sea sediments, Mn nodules
^{231}Pa	150 ka	deep-sea sediments, Mn nodules
^{230}Th/^{232}Th	300 ka	deep-sea sediments, Mn nodules
^{231}Pa/^{230}Th	150 ka	deep-sea sediments, Mn nodules
^{228}Th/^{232}Th	10 a	nearshore marine deposits, lake deposits

The second group of U-series disequilibrium dating methods is based on the accumulation of decay products in systems where the parent isotope has been deposited without any daughters or with a known daughter deficiency. Ages are determined from the extent of equilibrium restoration by the accumulation of decay products (Ford and Schwarcz 1981). For example, soluble uranium isotopes in natural waters are incorporated biogenically into the skeletons of organisms (including shells and coral) and inorganically to form precipitates such as carbonates (Ivanovich and Harmon 1982). These formations are essentially free of thorium and protactinium. If the system is then closed, the age of the formation may be determined from the extent of growth of ^{230}Th and ^{231}Pa as they restore equilibrium with their uranium parents, ^{234}U and ^{235}U respectively (Harmon 1980; Ku and Liang

1984; and Bradley 1985). Table A.3 summarizes this group of methods with their age ranges and applications.

Table A.3 U-series disequilibria dating methods based on the accumulation of decay products of uranium (modified after Ivanovich and Harmon 1982)		
Method	*Upper age limit*	*Applications*
^{230}Th/^{234}U	350 ka	carbonates (mostly corals, molluscs, speleothems), volcanic rocks
^{231}Pa/^{235}U	150 ka	marine phosphates, bones, carbonates (mostly corals, molluscs, speleothems), volcanic rocks
^{231}Pa/^{230}Th	200 ka	marine phosphates, bones, carbonates (mostly corals, molluscs, speleothems), volcanic rocks
^{226}Ra/^{238}U	10 ka	main use as a check for closed systems, some use on carbonates and volcanic rocks

Ages may be determined from the concentration of the relevant isotopes (e.g., ^{230}Th method), requiring additional assumptions (outlined below), or from isotopic ratios (e.g., ^{230}Th/^{234}U method), where the isotope relevant to the age is normalized against an isotope considered stable, owing to its long half-life. Isotopic concentrations may be measured by counting the particle emissions of radioactive decay using alpha- or gamma-spectrometry. Direct atom counting using a mass spectrometer requires more costly equipment, but determines the U-isotope ratios of smaller samples with greater simplicity, precision and range than former methods (Schwarcz 1989). It is argued that U-Th techniques in particular have greater precision than ^{14}C for the 20–6 ka BP period (Bard *et al.* 1990).

Two primary assumptions apply to all U-series dating methods. (a) Since formation the material has been part of a chemically closed system involving no loss or gain of any isotope except by radioactive decay; and (b) at formation the isotope ratio used for the age determination was either zero or a known amount. When measuring the concentration of isotopes in deposits rather than isotopic ratios, two additional assumptions are required. Firstly, the supply of radioisotopes must have been constant through time, and secondly, the ratio of radioisotope to transportation mechanism (be it in solution or adsorbed onto particulate matter) has been constant (Mangini 1986b). The reliability of U-series results depends on the viability of these assumptions.

The removal of uranium from precipitates by groundwater and the preferential absorption of uranium from groundwater by bones are examples of violation of the closed-system assumption (a), yielding ages that are anomalously too old and too young respectively (Ford and Schwarcz 1981). Assumption (b) is violated by detrital contamination, that is, contamination by non-authigenic daughter isotopes (e.g.,[230]Th) at the time of deposition. In the case of carbonate deposits, the activity of [232]Th, derived only from detritus, gives an indication of the accompanying detrital [230]Th present at the time of deposition (Ford and Schwarcz 1981; Schwarcz 1989). Ratios of [230]Th/[232]Th greater than 20 indicate negligible detrital contamination. Correction schemes have been devised to date contaminated samples (e.g., Ku and Liang 1984; Schwarcz and Latham 1989). Contamination problems are reduced by the careful selection of samples which are dense, macrocrystalline, of a single formation event and without evidence of post-depositional alteration (Harmon 1980).

Potassium-argon dating

The potassium-argon dating technique (K-Ar, ^{40}K/^{40}Ar) is a radiometric method applicable to many materials containing potassium, a common constituent of volcanic, metamorphic and igneous rocks. Age determination is based on the accumulation of radiogenic argon, ^{40}Ar* (a decay product of ^{40}K) which has a half-life of 1.25×10^9 years, permitting the K-Ar dating of some of the oldest known rocks (Flisch 1986).

^{40}K experiences branching decay mostly by beta-particle, or electron, emission from the nucleus to form the daughter isotope ^{40}Ca, as illustrated in Figure A.7. The remainder decays to ^{40}Ar* by the process of electron capture, which involves the capture of an electron by the nucleus (see Figure A.7). ^{40}Ca is not useful as a geological dating tool as it is difficult to distinguish between ^{40}Ca produced by ^{40}K decay and ^{40}Ca present at the formation of the rock or mineral. In contrast, as there are no naturally occurring argon compounds, the argon present must either be a result of ^{40}K decay or atmospheric argon contamination, for which correction can be made (Dalrymple and Lanphere 1969). ^{40}Ar*, therefore, is easily recognized, and as an inert gas can be easily and accurately measured, even in small quantities.

At formation, minerals and rocks contain potassium in the ratio ^{40}K/K$_{TOTAL}$ = 1.19×10^{-4} (Dalrymple and Lanphere 1969), and contain no radiogenic argon, ^{40}Ar*. Through time ^{40}K decays at a known rate (half-life of 1.25×10^9 years) to ^{40}Ar*, regardless of the physical or chemical environment. The ^{40}K/^{40}Ar* ratio, corrected for atmospheric argon contamination, gives an indication of the time that has elapsed since the whole rock or mineral sample was formed or significantly altered to reset the K-Ar clock.

The first stage of argon measurement is to extract and purify the argon content by fusing the sample in a high vacuum system to produce an argon gas. The argon is then measured, commonly by the stable isotope dilution method. A known quantity of an

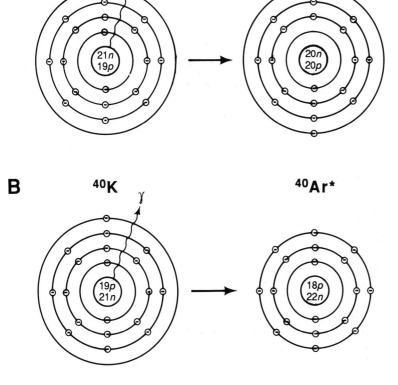

A ^{40}K ^{40}Ca

β^-

21n
19p

20n
20p

B ^{40}K ^{40}Ar*

γ

19p
21n

18p
22n

Fig. A.7 The branching decay of ^{40}K to the daughter isotopes, ^{40}Ca and ^{40}Ar*.
(A) Decay to ^{40}Ca occurs by beta-particle emission from the nucleus, resulting in the conversion of a neutron to a proton.
(B) Decay to ^{40}Ar* occurs by electron capture. An extranuclear electron, usually from the innermost shell, falls into the nucleus and a proton is converted to a neutron (Dalrymple and Lanphere 1969)

isotope of different composition to the sample (a tracer or spike) is mixed with the argon gas, and analysed using a mass spectrometer to determine the ratio of isotopic composition, spike:argon. The argon content is calculated by inference, from which the atmospheric argon is subtracted to leave a value for the amount of the ^{40}Ar* in the sample.

The potassium measurement method which best fits the criteria of speed, simplicity, precision and accuracy, is flame photometry. This is based on the principle that as electrons move from a high to a low energy state (as with the decay of ^{40}K by electron capture to ^{40}Ar*), the radiation emission has a wavelength characteristic of the element and a radiation intensity proportional to the quantity of the element that is present. Analysis of the sample using the flame photometer allows a comparison of the radiation of the sample with that of a known standard solution, to determine the total potassium content. Substitution into the ratio ^{40}K/K$_{TOTAL} = 1.19 \times 10^{-4}$ identifies the ^{40}K content. The decay rate of ^{40}K is so low that it would take 16 Ma to reduce the potassium ratio by 1 per cent (Dalrymple and Lanphere 1969).

There are five fundamental assumptions of K-Ar dating:

(i)   The decay rate of ^{40}K is the same for all materials regardless of the physical or chemical environment.

(ii) The ratio $^{40}K/K_{TOTAL}$ is constant for all materials, and varies minimally through time.

(iii) All argon in the sample is either radiogenic or atmospheric.

(iv) The sample has been part of a closed system, thus there has been no loss or gain of ^{40}K or $^{40}Ar^*$ except through radioactive decay.

(v)   The duration of mineral or rock formation must be short relative to the age (Dalrymple and Lanphere 1969).

Laboratory experiments have shown that the first two assumptions are reliable. Anomalously old K-Ar dates will result from extraneous argon in the rock or mineral sample, introduced by processes such as diffusion, inherited argon that survived a metamorphic event, or contamination by older grains (e.g., xenoliths) incorporated into the rock during formation or emplacement. Similarly, the K-Ar clock may be partially or completely reset by the processes of weathering and subsequent alteration, recrystallization of dissolved minerals, temperature elevations of several hundred degrees Celsius, metamorphism, melting, radiation or deformation. Both contamination and resetting of the K-Ar clock violate assumptions (iii) and (iv). The final assumption forms a limitation to sampling and definition of the age. As the K-Ar age generally indicates the elapsed time since temperatures fell sufficiently to trap argon in the crystal lattice of the mineral, lavas and tuffaceous sediments which cool rapidly yield ages which approximate the date of eruption, whereas slowly cooled intrusives will yield ages representative of the cooling period, which may span from hours to millions of years.

Erroneous ages may also result from conventional K-Ar dating of heterogeneous samples, as the measurements of potassium and argon are made on different parts of the sample. This problem is avoided by the $^{40}Ar/^{39}Ar$ age spectrum method which simultaneously measures the potassium and argon from the same trap in the crystal lattice (Bradley 1985). The sample is irradiated in a nuclear reactor, producing ^{39}Ar from the decay of ^{39}K by heating at progressively higher temperatures. Apparent ages, calculated directly from the $^{40}Ar/^{39}Ar$ ratio at each temperature increment, are plotted against cumulative argon release to form the age spectrum. This spectrum can be used to identify and date both thermal events which disturbed the sample and evidence for argon gain or loss (Hammerschmidt 1986). For example, Figure A.8 is representative of age spectra from rock or mineral samples from closed systems free of argon loss or gain (Lanphere

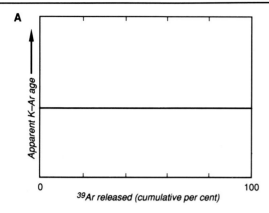

A

Apparent K-Ar age →

0   *39Ar released (cumulative per cent)*   100

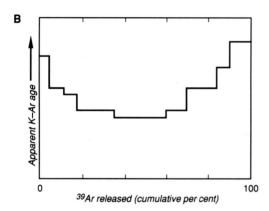

B

Apparent K-Ar age →

0   *39Ar released (cumulative per cent)*   100

Fig. A.8
(A) Constant $^{40}Ar/^{39}Ar$ ratio and age spectrum representative of rock or mineral samples from closed systems (modified after Lanphere and Dalrymple 1971).
(B) Saddle-shaped age spectrum diagnostic of rock or mineral samples which have experienced a thermal event since crystallization (modified after Lanphere and Dalrymple 1976)

and Dalrymple 1971), whereas saddle-shaped age spectra (Figure A.8) are diagnostic of excess ^{40}Ar resulting from a thermal event (Lanphere and Dalrymple 1976). The minimum apparent age approximates the maximum age of the thermal event and the apparent age determined at highest temperatures provides a minimum age for crystallization of the rock or mineral sample (Lanphere and Dalrymple 1971, 1976).

The K-Ar method is not limited to the dating of rock or mineral formation. For example, interbedded lava flows and tuff layers allow for the K-Ar dating of sedimentary sequences, the $^{40}Ar/^{39}Ar$ age spectrum provides information on the thermal evolution of sedimentary basins, and potassium-bearing minerals associated with slickensides, such as illites, may be used to date geological faults. K-Ar and $^{40}Ar/^{39}Ar$ methods are also useful for determining rates of geological processes and as an indicator of geological conditions. Co-operative research with geomorphologists and geologists will ensure suitable sample selection and evaluation of the K-Ar age.

Some minerals are inefficient or totally incapable of retaining argon, and will therefore indicate ages which are too recent.

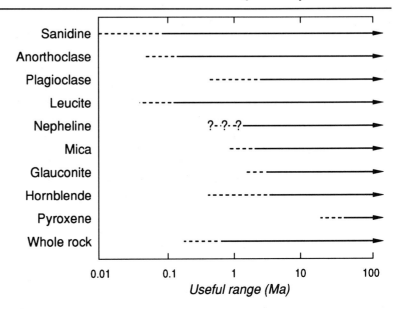

Fig. A.9 Common rock-forming materials considered suitable for K-Ar dating, with their approximate useful dating range (Dalrymple and Lanphere 1969)

Materials suitable for K-Ar dating must retain argon at 'normal' geological temperatures (i.e., less than 200°C), be resilient to weathering, alteration and dissolution, contain enough potassium to measure ^{40}K and ^{40}Ar*, and preferably be widespread in occurrence. Some of the datable materials and their useful age ranges are shown in Figure A.9.

Although it is difficult to date minerals younger than 100 ka BP owing to the overwhelming proportion of atmospheric argon relative to ^{40}Ar*, K-Ar ages as recent as a few thousand years are obtainable. There is practically no upper age limit for the K-Ar method because of the long half-life of ^{40}K.

Palaeomagnetism

The remanent magnetism of minerals in rocks or sediments records the nature of the earth's magnetic field at the time of cooling, deposition or alteration forming the basis of palaeomagnetism. This dating method allows the correlation of sedimentary sequences and the determination of numerical ages only when sequences can be calibrated with master curves which have been dated by another method.

Variations in the strength and direction of the earth's magnetic field occur irregularly, usually as a result of changes in the convection currents within the molten core of the earth (Barbetti and Hein 1989; Verosub 1988). As a result, the palaeomagnetic record is not unique around the globe. Variations may be classified into three main groups: polarity reversals, polarity excursions and secular variations, which provide three Quaternary stratigraphic and dating techniques. Polarity reversals, the changes between normal and reversed polarity, are global phenomena which may be used for international correlations and dating, spanning time scales ranging from 50 ka to 1 Ma (Easterbrook 1988; Løvlie 1989). At a regional scale, reversals spanning 10 to 50 ka are classified as polarity excursions, while secular variations refer to changes in the inclination, declination and intensity of the earth's magnetic field, and span time scales of between 2.5 and 3 ka (Løvlie 1989), each providing more precise results (Easterbrook 1988). The inclination is the vertical angle between the local horizontal and the line of magnetic force, and the declination is the horizontal angle between magnetic and true north.

Another palaeomagnetic tool is virtual geomagnetic poles (VGP). Using the simplifying assumption that the earth's magnetic field is defined by a dipole, similar to a bar magnet in the centre of the earth, the VGP is located at the intersection of the dipole and the surface of the earth (Verosub 1988).

VGPs provide a means of comparing locality specific secular variation data from different regions, as shown in Figure A.10, comparing Fish Lake (Oregon) with sites in the south-western United States.

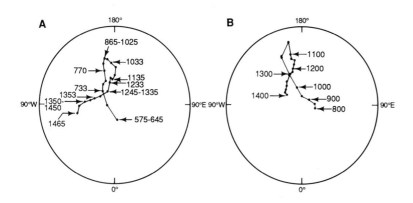

Fig. A.10
(A) VGP curves of the palaeomagnetic record from Fish Lake, Oregon, western North America, and (B) archaeomagnetic record from sites in the south-western United States; regions which are 900 km apart. Numerals in (A) and (B) represent calibrated radiocarbon and dendrochronological calendar dates respectively (Verosub 1988)

Most materials contain magnetic minerals such as magnetite which physically align themselves to the earth's magnetic field during heating or deposition. For heated materials, such as volcanics, hearths and kilns, heating to greater than 700°C mobilizes magnetic grains which are then free to align with the earth's magnetic field. Upon cooling the magnetic alignment is progressively locked as thermoremanent magnetism, indicating the nature of the earth's magnetic field at the time of cooling (Barbetti 1986).

For both windblown and waterlain sediments, minerals are free to align with the geomagnetic field as they settle out of suspension, recording depositional remanent magnetism. In saturated sediments with greater than 60 per cent sand content, small magnetic grains remain free to move within the water-filled pore spaces according to variations in the geomagnetic field, until compaction or dewatering is sufficient to restrict grain movement and fix the magnetic alignment as post-depositional remanent magnetism (Payne and Verosub 1982). Similarly, bioturbated deep-sea sediments will record the post-depositional remanent magnetism at the time when burrowing ceased (Løvlie 1989). In such cases it is important to recognize the event being dated.

The magnetism which is measured from the samples is the natural remanent magnetism (NRM), composed of the primary detrital remanent magnetism (DRM) related to the event being dated, and possibly more recent magnetic 'overprints' (Løvlie 1989). Spinner magnetometers are most commonly used to measure the NRM of samples. Cryogenic magnetometers operate up to four times faster and with greater sensitivity, but are more expensive and demand complex routine maintenance (Verosub 1986). To correct for magnetic overprints samples must be 'cleaned' by either alternating field or thermal demagnetization. Each of these procedures progressively randomize portions of the remanent magnetism, removing the magnetic overprints first and leaving behind the primary, more stable magnetism (Verosub 1986). Magnetic precipitates, which form in deposits, record a chemical remanent magnetism (Løvlie 1989), providing an example of the use of overprints in the dating of post-depositional events such as diagenesis.

Samples are obtained by drilling into the outcrop, rock sample or kiln wall, or collected in cylinders from unconsolidated sediments or cores. Samples must be accurately labelled with orientation and location details, stored in non-magnetic containers, and, if necessary, impregnated with wax to restrict distortion during sampling, transport or storage (Løvlie, 1989).

Erroneous remanent magnetism may result from the alignment of grains by flows, currents or glacial shearing rather than magnetism (Easterbrook 1988), the net shallowing of the inclination due to the rotation of elongate grains to the horizontal, rolling or compaction, or the oversteepening of inclinations from deposition on sloping surfaces (Verosub 1986). The best sediments for palaeomagnetism studies are silt- and clay-sized particles as they record a high coercivity remanent magnetism and are least susceptible to alignment by flows, glacial shearing or rotation. It has been demonstrated that poorly sorted deposits, such as till, which contain considerable amounts of silt and clay in their matrix show reliable net magnetism indicators despite the orientation of the larger grains (Easterbrook 1988).

A magnetopolarity time scale (MPTS) for the last 100 Ma has been constructed from lava flows dated by K-Ar and marine magnetic anomalies (Lowrie and Alvarez 1981), allowing global

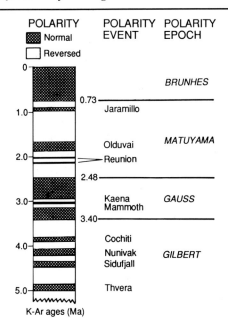

Fig. A.11 Magnetopolarity time scale for the last 5 Ma, dated by K-Ar (Mankinen and Dalrymple 1979)

chronostratigraphic correlation of polarity reversals. Figure A.11 shows the MPTS for the 5 Ma BP, dated by K-Ar. Independent time estimates (usually biostratigraphic) are required to identify particular polarity epochs or events. This technique has been successfully applied to deep-sea sediments and oxygen isotope records, however for the Quaternary, the Brunhes normal epoch yields poor resolutions of 700 ka (Løvlie 1989). The uncertainty of extent and duration, and the lack of synchroneity with neighbouring records has meant that master curves have not been established for polarity excursions, limiting the use of this technique to chronostratigraphic work. Secular variations have been particularly useful for high-resolution (less than 100 years) regional chronologies correlated with established Holocene master curves for most continents (Løvlie 1989), as shown in Figure A.12 for western North America. General master curves for North America, United Kingdom, Australia and Argentina are found in Creer and Tucholka (1983). Most applications have been to lake and near-shore marine sediments and hearths. In all applications, identifiable tephra layers or lava flows dated by K-Ar or fission track methods provide a method for correlating sequences from different regions.

As changes to the earth's magnetic field affect the flux of cosmic rays to the earth and thus the production of radioactive isotopes such as ^{14}C in the atmosphere, palaeomagnetism provides a means of correcting the radiocarbon time scale (Barbetti and Flude 1979).

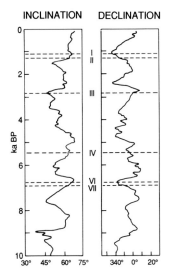

Fig. A.12 Proposed master curve of secular variation for western North America, based on agreement of secular variation records from three lakes including Fish Lake, Oregon (Verosub 1988)

Fission-track dating

Spontaneous fission of the ^{238}U atom explosively produces two fragments which pass through solids, disrupting the mineral lattice in the form of a single, narrow track of intense damage (Fleischer and Hart 1972; Naeser and Naeser 1988). The number of tracks is proportional to the uranium content which produced them, therefore the determination of a numerical age requires measurement of both uranium content and track density. Etchants are used to enhance the fission tracks for counting under an optical microscope. Calculation of ages younger than 100 ka BP involves long counting times and large analytical uncertainties owing to the low track density. The major advantage of fission-track dating over other techniques is that it is a grain discrete method. Ages are determined for each grain, therefore contamination is easily recognized. Fission-track dating has mainly been used to date tephra layers. Other applications include the dating of thermal events and archaeological materials, and the determination of ages and rates of landscape evolution and sea-floor spreading.

Other dating methods

Historical and archaeological evidence such as medieval manuscripts, remnants of past civilizations and artefacts may provide age indications for stratigraphic sequences (Roberts 1989).

Dendrochronology, the study of annual tree-rings for correlation and numerical age determination in calendar years, provides the most reliable linear time scale for calibration of ^{14}C ages in the Holocene (Ottaway 1983). A. E. Douglass, the father of dendrochronology in the early 1900s, also pioneered dendroclimatology and the use of cross-dating to extend the tree-ring record beyond the life of a single tree (Fritts 1976). Dendrochronology can be used to date events related to the death of the tree, including landslides, volcanic eruptions and fire (Fritts 1976) as well as climatic change within the life of the tree.

Similarly rythmites, alterations in the character of lake sediments over time scales of days to years, may form time sequences useful for dating (Kempe and Degens 1979) and reconstruction of past climates (Schove 1979). For example, varves, typically formed in proglacial lakes, consist of coarse-grained, light-coloured sand or silt, summer meltwater laminae and fine-grained, dark, organic laminae deposited in winter months (Sturm 1979). Counting of paired varves and cross-dating of varve and tree-ring sequences provides regional chronologies and climatic indicators.

Obsidian, a volcanic glass, chemically absorbs atmospheric

water on freshly exposed surfaces. The atmospheric water diffuses into the obsidian as a function of time, thermal history and chemical composition of the obsidian (Friedman and Long 1976; Ericson 1981). The thickness of the hydrated rind indicates the age of the sample, with time scales ranging from thousands to millions of years and resolutions of 10 per cent (calibrated) and 20 to 30 per cent (uncalibrated) (Goudie 1981). Stratigraphical correlation and relative age can also be determined from the degree of pedogenesis (Birkeland 1974) or weathering (e.g., Porter 1975). Numerical ages are obtained when calibrated using radiometric methods.

Worsley (1981) outlines the principles of lichenometry as a numerical and relative dating technique, based on the relationships between thallus size and age, with applications for glacier fluctuations, shorelines and river-flooding frequency.

Tephrochronology refers to the dating or correlating of sequences over large areas owing to the presence of distinctive tephra layers which provide time-equivalent marker horizons. These may be dated by fission track, K-Ar, ^{14}C, or a number of approximate dating methods (Naeser *et al.* 1981; Steen-McIntyre 1981; Westgate and Gorton 1981).

Bibliography

Abell, P. I. and Williams, M. A. J. 1989: Oxygen and carbon isotope ratios in gastropod shells as indicators of palaeoenvironments in the Afar region of Ethiopia. *Palaeogeography, Palaeoclimatology, Palaeoecology* 74, 265–278.

Adams, J. M., Faure, H., Faure-Denard, L., McGlade, J. M. and Woodward, F. I. 1990: Increases in terrestrial carbon storage from the Last Glacial Maximum to the present. *Nature* 348, 711–714.

Adams, R. M., Rosenzweig, C., Peart, R. M., Ritchie, J. T., McCarl, B. A., Glyer, J. D., Curry, R. B., Jones, J. W., Boote, K. J. and Allen, L. H. 1990: Global climate change and US agriculture. *Nature* 345, 219–224.

Adamson, D. A. and Fox, M. D. 1982: Change in Australian vegetation since European settlement. In Smith, J. M. B. (ed.), *A history of Australasian vegetation.* (Sydney: McGraw-Hill), 109–146.

Adamson, D. A. and Pickard, J. 1986: Cainozoic history of the Vestfold Hills. In Pickard, J. (ed.), *Antarctic Oasis: Terrestrial environments and history of the Vestfold Hills.* (Sydney: Academic Press), 63–97.

Adamson, D. A., Gasse, F., Street, F. A. and Williams, M. A. J. 1980: Late Quaternary history of the Nile. *Nature* 287, 50–55.

Adamson, D. A. and Williams M. A. J. 1987: Geological setting of Pliocene rifting in the Afar Depression of Ethiopia. *Journal of Human Evolution* 16, 597–610.

Adamson, D. A., Williams, M. A. J. and Baxter, J. T. 1985: Complex late Quaternary alluvial history in the Nile, Murray–Darling and Ganges basins: three river systems presently linked to the Southern Oscillation. In Gardiner, V. (ed.), *Proceedings of the First International Conference on Geomorphology.* Manchester, Sept. 1985. (Chichester: Wiley), 875–887.

Aharon, P. 1984: Implications of the coral reef record from New Guinea concerning the astronomical theory of ice ages. In Berger, A. L., Imbrie, J., Hays, J., Kukla, G. and Saltzman, B. (eds), *Milankovitch and climate: understanding the response to astronomical forcing.* (Dordrecht: Reidel), 379–389.

Aharon, P. and Chappell, J. 1986: Oxygen isotopes, sea level changes and the temperature history of a coral reef environment in New Guinea over the last 10^5 years. *Palaeogeography, Palaeoclimatology, Palaeoecology* 56, 337–379.

Aitken, M. J. 1985: *Thermoluminescence dating.* London: Academic Press.

Aitken, M. J. 1987: Archaeometrical dating: rappoteur review. In Aurenche, O., Evin, J. and Hours, F. (eds), *Chronologies in the Near East: relative chronologies and absolute chronology 16 000–4000 BP.* British Archaeological Reports International Series 379(i), 207–218.

Aitken, M. J. 1990: *Science-based dating in archaeology.* London: Longman.

Ambrose, S. H. and De Niro, M. J. 1986: Reconstruction of African human diet using bone collagen carbon and nitrogen isotope ratios. *Nature* 318, 321–324.

An, Z. and Ho, C. K. 1989: New magnetostratigraphic dates of Lantian *Homo erectus*. *Quaternary Research* 32, 213–221.

Andel, T. H. van 1985: *New Views on an old planet*. Cambridge: Cambridge University Press.

Anderson, B. G. 1981: Late Weichselian ice sheets in Eurasia and Greenland. In Denton, G. H. and Hughes, T. J. (eds), *The last great ice sheets*. (New York: Wiley), 1–65.

Anderson, N. R. and Malahoff, A. (eds) 1977: *The fate of fossil fuel CO_2 in the oceans*. New York: Plenum Press.

Andreas, E. L. and Ackley, S. F. 1982: On the differences in ablation seasons of Arctic and Antarctic sea ice. *Journal of Atmospheric Science* 39, 440–447.

Andrews, J. T. 1975: *Glacial systems*. Massachusetts: Duxbury Press.

Andrews, J. T. and Miller, G. H. 1985: Holocene sea level variations within Frobisher Bay. In Andrews, J. T. (ed.), *Quaternary environments: Eastern Canadian Arctic, Baffin Bay and Western Greenland*. (Boston: Allen and Unwin), 585–607.

Arensburg, B., Tillier, A. M., Vandermeersch, B., Duday, H., Hepartz, L. A. and Rak, Y. 1989: A Middle Palaeolithic human hyoid bone. *Nature* 338, 758–760.

Ash, J. E. and Wasson, R. J. 1983: Vegetation and sand mobility in the Australian desert dunefield. *Zeitschrift für Geomorphologie, Supplementband* 45, 7–25.

Assefa, G., Clark, J. D. and Williams, M. A. J. 1982: Late Cenozoic history and archaeology of the Upper Webi Shebele Basin, east central Ethiopia. *Sinet: Ethiopian Journal of Science* 5, 27–46.

Australian Academy of Science 1976: *Report of a committee on climatic change*. Canberra: Australian Academy of Science.

Axelrod, D. I. and Raven P. H. 1978: Late Cretaceous and Tertiary vegetation history of Africa. In Werger, M. A. J. (ed.), *Biogeography and ecology of southern Africa*. (The Hague: Junk), 77–130.

Bach, W. 1984: *Our threatened climate: ways of averting the CO_2 problem through rational energy use*. Boston: Reidel.

Bagnold, R. A. 1941: *The physics of blown sand and dunes*. London: Methuen.

Baird, R. F. 1989: Fossil bird assemblages from Australian caves: precise indicators of late Quaternary environments. *Palaeogeography, Palaeoclimatology, Palaeoecology* 69, 241–244.

Baird, R. F. 1991: The taphonomy of late Quaternary cave localities yielding vertebrate remains in Australia. In Vickers-Rich, P., Monaghan, J. M., Baird, R. F. and Rich, T. H. (eds), *Vertebrate palaeontology of Australasia*. (Melbourne: Pioneer Design Studio), 267–310.

Baker, V. R. 1983: Late Pleistocene fluvial systems. In Wright, H. E. and Porter, S. C. (eds), *Late Quaternary environments of the United States. Volume 1. The Late Pleistocene*. (London: Longman), 115–129.

Balout, L. 1955: *Préhistoire de L'Afrique du Nord*. Paris: Arts et Métiers Graphiques.

Barbetti, M. 1980: Geomagnetic strength over the last 50 000 years and changes in the atmospheric ^{14}C concentration: emerging trends. *Radiocarbon* 28, 192–199.

Barbetti, M. 1986: Traces of fire in the archaeological record, before one million years ago. *Journal of Human Evolution* 15, 771–781.

Barbetti, M. and Allen, H. 1972: Prehistoric man at Lake Mungo, Australia, by 32,000 years BP. *Nature* 240, 46–48.

Barbetti, M., Clark, J. D., Williams, F. M. and Williams M. A. J. 1980: Palaeomagnetism and the search for very ancient fireplaces in Africa. *Anthropologie* 18, 229–304.

Barbetti, M. and Flude, R. 1979: Geomagnetic variation during the late Pleistocene period and changes in the radiocarbon time scale. *Nature* 279, 202–205.

Barbetti, M. and Hein, D. 1989: Palaeomagnetism and high-resolution dating of ceramic kilns in Thailand: a progress report. *World Archaeology* 21(1), 51–70.

Bard, E., Hamelin, B. and Fairbanks, R. G. 1990: U-Th ages obtained by mass spectrometry in corals from Barbados: sea level during the past 130 000 years. *Nature* 346, 456–458.

Bard, E., Hamelin, B., Fairbanks, R. G. and Zindler, A. 1990: Calibration of the ^{14}C timescale over the past 30 000 years using mass spectrometric U-Th ages from Barbados corals. *Nature* 345, 405–410.

Barker, W. R. and Greenslade, P. J. M. (eds) 1982: *Evolution of the flora and fauna of arid Australia*. Adelaide: Peacock.

Barnola, J.-M., Pimienta, P., Raynaud, D. and Korotkevich, Y. S. 1991: CO_2-climate relationship as deduced from the Vostok ice core: a re-examination based on new measurements and on a re-evaluation of the air dating. *Tellus* 43B, 83–90.

Barnola, J. M., Raynaud, D., Korotkevich, Y. S. and Lorius, C. 1987: Vostok ice core provides 160 000 year record of atmospheric CO_2. *Nature* 329, 408–414.

Barraclough, G. (ed.) 1982: *The Times concise atlas of world history*. London: Angus and Robertson.

Bartholomew, J. C., Geelan, P. J. M., Lewis, H. A. C., Middleton, P. and Winkleman, B. (eds) 1980: *The Times atlas of the world. Comprehensive edition*. London: Times Books.

Baulin, V. V. and Danilova, N. S. 1984: Dynamics of late Quaternary permafrost in Siberia. In Velichko, A. A. (ed.), *Late Quaternary environments of the Soviet Union*. English language edition editors Wright, H. E. and Burnosky, C. W. (Essex: Longman), 69–77.

Bé, A. W. H. and Tolderlund, D. S. 1971: Distribution and ecology of living planktonic foraminifera in surface waters of the Atlantic and Indian Oceans. In Funnell, B. M. and Riedel, W. R. (eds), *The micropalaeontology of the oceans*. (Cambridge: Cambridge University Press), 105–149.

Beaumont, P. 1989: *Environmental management and development in drylands*. London: Routledge.

Becker, B. and Kromer, B. 1986: Extension of the Holocene dendrochronology by the Preboreal pine series, 8800 to 10 100 BP. *Radiocarbon* 28, 861–934.

Bednarik, R. G. 1989: On the Pleistocene settlement of South America. *Antiquity* 63, 101–111.

Behrensmeyer, A. K. 1982: The geological context of human evolution. *Annual Review of Earth and Planetary Science* 10, 39–60.

Behrensmeyer, A. K., Gordon, K. D. and Yanagi, G. T. 1986: Trampling as a cause of bone surface damage and pseudo-cutmarks. *Nature* 319, 768–771.

Behrensmeyer, A. K. and Kidwell, W. M. 1985: Taphonomy's contributions to paleobiology. *Paleobiology* 11, 105–119.

Bell, M. and Laine, E. P. 1985: Erosion of the Laurentide region of North America by glacial and glaciofluvial processes. *Quaternary Research* 23, 154–174.

Bennett, K. D. 1983a: Devensian late-glacial and Flandrian vegetational history at Hockham Mere, Norfolk, England. *New Phytologist* 95, 457–487.

Bennett, K. D. 1983b: Postglacial population expansion of forest trees in Norfolk, U.K. *Nature* 303, 164–167.

Benson, L. and Thompson, R. S. 1987: The physical record of lakes in the Great Basin. In Ruddiman, W. F. and Wright, H. E. (eds), *North America and adjacent oceans during the last deglaciation. The geology of North America*. Volume K-3. (Boulder, Colorado: Geological Society of America), 241–260.

Benson R. H. 1984: The Phanerozoic 'Crisis' as viewed from the Miocene. In Berggren, W. A. and Couvering, J. A. Van (eds), *Catastrophes and earth history*. (Princeton, N.J.: Princeton University Press), 437–446.

Bentley, H. W., Phillips, F. M., Davis, S. N., Habermehl, M. A., Airey, P. L., Calf, G. E., Elmore, D., Grove, H. E. and Torgerson, T. 1986: Chlorine 36 dating of very old groundwater 1. The Great Artesian Basin, Australia. *Water Resources Research* 20(13), 1991–2001.

Berger, A. (ed.) 1981: *Climatic variations and variability: facts and theories*. Dordrecht: Reidel.

Berger, A. 1989: Pleistocene climatic variability at astronomical frequencies. *Quaternary International* 2, 1–14.

Berger, W. H. 1988: Dating Quaternary events by luminescence. In Easterbrook, D. J. (ed.), Dating Quaternary sediments. *Geological Society of America Special Paper* 227, 13–50.

Berger, W. H. and Crowell, J. C. (Panel Co-chairmen) 1982: *Climate in earth history*. Washington D.C.: National Academy Press, Studies in Geophysics Series, National Research Council.

Berger, W. H., Killingley, J. S. and Vincent, E. 1985: Timing of deglaciation from an oxygen isotope curve for Atlantic deep-sea sediments. *Nature* 314, 156–158.

Berggren, W. A. and Couvering, J. A. Van (eds) 1984: *Catastrophes and earth history*. Princeton: Princeton University Press.

Berkey, C. P. and Morris, F. K. 1927: *Geology of Mongolia. Natural history of Central Asia, Volume II*. New York: The American Museum of Natural History.

Bigarella, J. J. and Ferreira, A. M. M. 1985: Amazonian geology and the Pleistocene and the Cenozoic environments and paleoclimates. In Prance, G. T. and Lovejoy, T. E. (eds), *Amazonia*. (Oxford: Pergamon), 49–71.

Billard, A. and Orombelli, G. 1986: Quaternary glaciations in the French and Italian piedmonts of the Alps. *Quaternary Science Reviews* 5, 407–419.

Binford, L. R. 1981: *Bones: ancient men and modern myths.* New York: Academic Press.

Binford, L. R. 1983: *In pursuit of the past: Decoding the archaeological record.* New York: Thames and Hutton.

Birchfield, G. E. 1987: Changes in the deep-ocean water $\delta^{18}0$ and temperature from the Last Glacial Maximum to the present. *Paleoceanography* 2, 431–442.

Birkeland, P. W. 1974: *Pedology, weathering, and geomorphological research.* New York: Oxford University Press.

Birks, H. J. B. 1981: The use of pollen analysis in the reconstruction of past climates: a review. In Wigley, T. M. L., Ingram, M. J. and Farmer, G. (eds), *Climate and history.* (Cambridge: Cambridge University Press), 111–138.

Birks, H. J. B. and Birks, H. H. 1981: *Quaternary palaeoecology.* London: Edward Arnold.

Blaikie, P. and Brookfield, H. 1987: *Land degradation and society.* London: Methuen.

Bleil, U. and Thiede, J. (eds) 1990: *Geological history of the Polar oceans: Arctic versus Antarctic.* Dordrecht: Kluwer Academic Publishers.

Bloemendal, J. and Menocal, P. de 1989: Evidence for a change in the periodicity of tropical climate cycles at 2.4 Myr from whole-core magnetic susceptibility measurements. *Nature* 342, 897–900.

Bloom, A. L. 1978: *Geomorphology: a systematic analysis of late Cenozoic landforms.* New Jersey: Prentice Hall.

Bloom, A. L. 1983: Sea level and coastal morphology of the United States through the Late Wisconsin glacial maximum. In Wright, H. E. and Porter, S. C. (eds), *Late Quaternary environments of the United States.* (Minneapolis: University of Minnesota Press), Volume 1, 215–229.

Boden, T. A., Kanciruk, P. and Farrell, M. P. 1990: *Trends '90..* Oak Ridge, Tennessee: Information Analysis Center, Oak Ridge National Laboratory.

Bonnefille, R. 1983: Evidence for a cooler and drier climate in the Ethiopian uplands towards 2.5 Myr ago. *Nature* 303, 487–491.

Bonnefille, R., Roeland, J. C. and Guiot, J. 1990: Temperature and rainfall estimates for the past 40 000 years in equatorial Africa. *Nature* 346, 347–349.

Boucher de Perthes, J. 1847: *Antiquités celtiques et antédiluviennes. Mémoire sur l'industrie primitive et les arts à leur origine.* Volume 1. Paris: Treuttel and Wertz.

Boucher de Perthes, J. 1857: *Antiquités celtiques et antédiluviennes. Mémoire sur l'industrie primitive et les arts à leur origine.* Volume 2. Paris: Treuttel and Wertz.

Boucher de Perthes, J. 1864: *Antiquités celtiques et antédiluviennes. Mémoire sur l'industrie primitive et les arts à leur origine.* Volume 3. Paris: Treuttel and Wertz.

Boulton, G. S., Smith, G. D., Jones, A. S. and Newsome, J. 1985: Glacial geology and glaciology of the last mid-latitude ice sheets. *Journal of the Geological Society of London* 142, 447–474.

Bowen D. Q., Rose, J., McCabe, A. M. and Sutherland, D. G. 1986:

Correlation of Quaternary glaciations in England, Ireland, Scotland and Wales. *Quaternary Science Reviews* 5, 299–340.

Bowen, R. 1991: *Isotopes and climates.* London: Elsevier.

Bowler, J. M. 1978a: Glacial age aeolian events at high and low altitudes: a Southern Hemisphere perspective. In Zinderen Bakker, E. M. Van (ed.), *Antarctic glacial history and world paleoenvironments.* (Rotterdam: Balkema), 149–172.

Bowler, J. M. 1978b: Quaternary climate and tectonics in the evolution of the riverine plain, southeastern Australia. In Davies, J. L. and Williams, M. A. J. (eds), *Landform evolution in Australasia.* Canberra: Australian National University Press, 70–112.

Bowler, J. M., Thorne, A. G. and Polach, H. A. 1972: Pleistocene man in Australia: age and significance of the Mungo skeleton. *Nature* 240, 48–50.

Bowler, J. M. and Wasson, R. J. 1984: Glacial age environments of inland Australia. In Vogel, J. C. (ed.), *Late Cainozoic environments of the Southern Hemisphere.* (Rotterdam: Balkema), 183–208.

Bowles, F. A. 1975: Paleoclimatic significance of quartz/illite variations in cores from the eastern equatorial North Atlantic. *Quaternary Research* 5, 225–235.

Boyle, E. A. 1988: Cadmium: chemical tracer of deepwater paleoceanography. *Paleoceanography* 3, 471–489.

Boyle, E. A. 1990: Quaternary deepwater paleoceanography. *Science* 249, 863–870.

Bradley, R. S. 1978: *Quaternary palaeoclimatology: methods of palaeoclimatic reconstruction.* Boston: Allen and Unwin.

Bradley, R. S. 1985: *Quaternary paleoclimatology.* London: Allen and Unwin.

Brain, C. K. 1981: The evolution of man in Africa: was it a consequence of Cainozoic cooling? *Alex L. du Toit Memorial Lectures No. 17, Geological Society of South Africa,* Annexure to Volume 84, 1–19.

Brain, C. K. and Sillen, A. 1988: Evidence from the Swartkrans cave for the earliest use of fire. *Nature* 336, 464–466.

Bray, J. R. 1977: Pleistocene volcanism and glacial initiation. *Science* 197, 251–254.

Bray, J. R. and Trump, D. 1970: *The Penguin dictionary of archaeology.* Harmondsworth, Australia: Penguin Books.

Broecker, W. S. 1987: Unpleasant surprises in the greenhouse? *Nature* 328, 123–126.

Broecker, W. S. and Denton, G. H. 1990: The role of ocean-atmosphere reorganizations in glacial cycles. *Quaternary Science Reviews* 9, 305–341.

Broecker, W. S., Kennett, J. P., Flower, B. P., Teller, J. T., Trumbone, S., Bonani, G. and Wolfi, W. 1989: Routing of meltwater from the Laurentide Ice Sheet during the Younger Dryas cold episode. *Nature* 341, 318–321.

Broecker, W. S. and Peng, T.-H. 1984: The climate-chemistry connection. In Hansen, J. E. and Takahashi, T. (eds), *Climate processes and climate sensitivity.* Geophysical Monograph 29. (Washington D.C.: American Geophysical Union), 327–336.

Broecker, W. S. and Takahashi, T. 1984: Is there a tie between atmospheric CO_2 content and ocean circulation? In Hansen, J. E. and

Takahashi, T. (eds), *Climate processes and climate sensitivity.* Geophysical Monograph 29. (Washington D.C.: American Geophysical Union), 314–326.

Broecker, W. S., Peteet, D. M. and Rind, D. 1985: Does the ocean-atmosphere system have more than one stable mode of operation? *Nature* 315, 21–25.

Brown, L. R. 1981: Eroding the base of civilisation. *Journal of Soil and Water Conservation* 36(5), 255–260.

Brown, L. R. 1984: The global loss of topsoil. *Journal of Soil and Water Conservation* 39(3), 162–165.

Brown, R. J. E. 1970: *Permafrost in Canada: Its influence on northern development.* Toronto: University of Toronto Press.

Bruenig, E. 1987: The forest ecosystem: tropical and boreal. *Ambio* 14(2–3), 68–79.

Bunn, H. T. 1981: Archaeological evidence for meat-eating by Plio-Pleistocene hominids from Koobi Fora and Olduvai Gorge. *Nature* 291, 574–577.

Bush, M. B. and Colinvaux, P. A. 1990: A pollen record of a complete glacial cycle from lowland Panama. *Journal of Vegetation Science* 1, 105–118.

Butler, B. E. 1956: Parna – an aeolian clay. *Australian Journal of Science* 18, 145–151.

Butzer, K. W. 1982: *Archaeology as human ecology.* Cambridge: Cambridge University Press.

Butzer, K. W. and Hansen, C. L. 1968: *Desert and river in Nubia: geomorphology and prehistoric environments at the Aswan Reservoir.* Madison: University of Wisconsin Press.

Butzer, K. W., Isaac, G. L., Richardson, J. L. and Washbourn-Kamau, C. 1972: Radiocarbon dating of East African lake levels. *Science* 175, 1069–1076.

Calaby, J. H. 1976: Some biogeographical factors relevant to the Pleistocene movement of man in Australasia: In Kirk, R. L. and Thorne, A. G. (eds), *The origin of the Australians.* (Canberra: Australian Institute of Aboriginal Studies), 23–28.

Carter, R. M. and Johnson, D. P. 1986: Sea-level controls on the post-glacial development of the Great Barrier Reef, Queensland. *Marine Geology* 71, 137–164.

Cess, R. D., Potter, G. L., Blanchet, J. P., Boer, G. J., Ghan, S. J., Kiehl, J. T., Le Treut, H., Li, Z.-X., Liang, X.-Z., Mitchell, J. F. B., Morcrette, J. J., Randall, D. A., Riches, M. R., Roeckner, E., Schlese, U., Slingo, A., Taylor, K. E., Washington, W. M., Wetherald, R. T. and Yagai, I. 1989: Interpretation of cloud-climate feedback as produced by 14 atmospheric general circulation models. *Science* 245, 513–516.

Chappell, J. 1983: Aspects of sea levels, tectonics, and isostasy since the Cretaceous. In Gardner, R. and Scoging, H. (eds), *Megageomorphology.* (Oxford: Clarendon), 56–72.

Chappell, J. 1987: Ocean volume change and the history of sea water. In Devoy, R. J. N. (ed.), *Sea surface studies: a global view.* (London: Croom-Helm), 33–56.

Chappell, J. and Polach, H. 1991: Post-glacial sea-level rise from a coral record at Huon Peninsula, Papua New Guinea. *Nature* 349, 147–149.

Chappell, J. and Shackleton, N. J. 1986: Oxygen isotopes and sea level. *Nature* 324, 137–140.

Chappellaz, J., Barnola, J. M., Raynaud, D., Korotkevich, Y. S. and Lorius, C. 1990: Ice-core record of atmospheric methane over the last 160 000 years. *Nature* 345, 127–131.

Charles, C. D. and Fairbanks, R. G. 1992: Evidence from Southern Ocean sediments for the effect of North Atlantic deep-water flux on climate. *Nature* 355, 416–419.

Chen, Y., Lu, J., Cao, T., He, R. and Zhang, R. 1988: A comparative examination of AD value estimated from ESR spectra of various carbonate sediments. *Quaternary Science Reviews* 7, 455–459.

Chen, X. Y., Prescott, J. R. and Hutton, J. T. 1990: Thermoluminescence dating on gypseous dunes of Lake Amadeus, central Australia. *Australian Journal of Earth Sciences* 37, 93–101.

Chivas, A. R. and De Deckker, P. (eds) 1991: Palaeoenvironments of salt lakes. *Palaeogeography, Palaeoclimatology, Palaeoecology* 84, 1–423.

Chivas, A. R., De Deckker, P. and Shelley, J. M. G. 1986: Magnesium and strontium in non-marine ostracod shells as indicators of palaeosalinity and palaeotemperature. *Hydrobiologia* 143, 135–142.

Chivas, A. R., Torgersen, T. and Bowler, J. M. (eds) 1986: Palaeoenvironments of salt lakes. *Palaeogeography, Palaeoclimatology, Palaeoecology* 54, 1–328.

Clark, G. 1977: *World prehistory in new perspective.* Cambridge: Cambridge University Press.

Clark, J. A. and Lingle, C. S. 1979: Predicted relative sea-level changes (18 000 years BP to present) caused by late-glacial retreat of the Antarctic ice sheet. *Quaternary Research* 11, 279–298.

Clark, J. D. 1971: A re-examination of the evidence for agricultural origins in the Nile Valley. *Proceedings of the Prehistoric Society* 37, 34–79.

Clark, J. D. 1975: Africa in prehistory: peripheral or paramount? *Man (N.S.)* 10, 175–198.

Clark, J.D. 1976a: African origins of man the toolmaker. In Isaac, G. L. and McCown, E. R. (eds), *Human origins: Louis Leakey and the East African evidence.* (Menlo Park, California: Benjamin), 1–53.

Clark, J. D. 1976b: Prehistoric populations and pressures favouring plant domestication in Africa. In Harlan, J. R., De Wet, J. M. J. and Stemler, A. B. L. (eds), *Origins of African plant domestication.* (The Hague: Mouton), 67–105.

Clark, J. D. 1980: Human populations and cultural adaptations in the Sahara and Nile during prehistoric times. In Williams, M. A. J. and Faure, H. (eds), *The Sahara and the Nile.* (Rotterdam: Balkema), 527–582.

Clark, J. D. 1984b: The domestication process in Northeast Africa: ecological change and adaptive strategies. In Krzyzaniak, L. and Kobusiewicz, M. (eds), *Origin and early development of food-producing culture in north-eastern Africa.* (Poznan: Polish Academy of Sciences), 25–41.

Clark, J. D. 1987: Transitions: *Homo erectus* and the Acheulian: the Ethiopian sites of Gadeb and the Middle Awash. *Journal of Human Evolution* 16, 809–826.

Clark, J. D. 1988: The Middle Stone Age of East Africa and the beginnings of regional identity. *Journal of World Prehistory* 2, 235–305.

Clark, J. D., Asfaw, B., Assefa, G., Harris, J. W. K., Hurashina, H., Walter, R. C., White, T. D. and Williams, M. A. J. 1984: Palaeoanthropological discoveries in the Middle Awash Valley, Ethiopia. *Nature* 307, 423–428.

Clark, J. D. and Harris, J. W. K. 1985: Fire and its roles in early hominid lifeways. *African Archaeological Review* 3, 3–27.

Clark, J. D. and Haynes, C. V. 1970: An elephant butchery site at Mwanganda's village, Karonga, Malawi and its relevance for Palaeolithic archaeology. *World Archaeology* 1(3), 390–411.

Clark, J. D. and Kurashina, H. 1979: Hominid occupation of the east-central highlands of Ethiopia in the Plio-Pleistocene. *Nature* 282, 33–39.

Clark, M. J. (ed.) 1988: *Advances in periglacial geomorphology.* New York: Wiley.

Clark, R. L. 1981: The prehistory of bushfires. In Stanbury, P. (ed.), *Bushfires: their effect on Australian life and landscape.* (Sydney: The Macleay Museum, University of Sydney), 61–74.

Clark, R. L. 1983: Pollen and charcoal evidence for the effects of Aboriginal burning on the vegetation of Australia. *Archaeology in Oceania* 18, 32–37.

Clifford, T. N. 1970: The structural framework of Africa. In Clifford, T. N. and Gass, I. G. (eds), *African magmatism and tectonics.* (Edinburgh: Oliver and Boyd), 1–26.

CLIMAP Project Members 1976: The surface of the ice-age Earth. *Science* 191 (4232), 1131–1137.

CLIMAP Project Members 1981: Seasonal reconstructions of the earth's surface at the Last Glacial Maximum. *Geological Society of America Map and Chart Series* MC-36, 1–18 and 9 maps.

Close, A. E. 1989: Lithic development in the Kubbaniyan (Upper Egypt). In Krzyzaniak, L. and Kobusiewicz, M. (eds), *Late prehistory of the Nile Basin and the Sahara.* (Poznan: Poznan Archaeological Museum), 117–125.

Cofer, W. R., Levine, J. S., Winstead, E. L. and Stocks, B. J. 1991: New estimates of nitrous oxide emissions from biomass burning. *Nature* 349, 689–691.

Cohen, M. N. 1977: *The Food crisis in prehistory. Overpopulation and the origins of Agriculture.* New Haven: Yale University Press.

COHMAP Members 1988: Climatic changes of the last 18 000 years: observations and model simulations. *Science* 241, 1043–1052.

Colinvaux, P. A. 1989: The past and future Amazon. *Scientific American* 260(5), 68–74.

Conroy, G. C., Jolly, C. J., Cramer, D. and Kalb, J. E. 1978: Newly discovered fossil hominid skull from the Afar depression, Ethiopia. *Nature* 276, 67–70.

Coope, G. R. 1970: Interpretations of Quaternary insect fossils. *Annual Review of Entomology* 15, 97–120.

Coope, G. R. 1975: Mid-Weichselian climate changes in Western Europe, reinterpreted from coleopteran assemblages. In Suggate, R. P. and Cresswell, M. M. (eds), *Quaternary studies.* (Wellington: Royal Society of New Zealand), 101–108.

Corliss, B. H. 1983: Quaternary circulation of the Antarctic Circumpolar Current. *Deep-Sea Research* 30: 47–61.

Corliss, B. H. and Fois, E. 1990: Morphotype analysis of deep-sea benthic foraminifera from the northwest Gulf of Mexico. *Palaios* 5, 589–605.

Cosgrove, R. 1989: Thirty thousand years of human colonisation in Tasmania: New Pleistocene dates. *Science* 243, 1706–1708.

Covey, C. 1984: The Earth's orbit and the Ice Ages. *Scientific American* 250, 42–50.

Cowie, J. W. and Bassett, M. G. 1989: 1989 global stratigraphic chart, with geochronometric and magnetostratigraphic calibration. Supplement to *Episodes* 12(2).

Crabb, P. 1980: Alluvium and agriculture in the semi-arid world. In Williams, M. A. J. and Adamson, D. A. (eds), *A land between two Niles*. (Rotterdam: Balkema), 1–11.

Crane, A. and Liss, P. 1985: Carbon dioxide, climate and the sea. *New Scientist* 108 (1483), 50–54.

Creer, K. M. and Tucholka, P. 1983: On the current state of lake sediment palaeomagnetic research. *Geophysical Journal of the Royal Astronomical Society* 74, 223–238.

Dalrymple, G. B. and Lanphere, M. A. 1969: *Potassium-Argon dating: principles, techniques and applications to geochronology*. San Francisco: Freeman.

Damuth, J. E. and Fairbridge, R. W. 1970: Equatorial Atlantic deep-sea arkosic sands and ice-age aridity in tropical South America. *Bulletin of the Geological Society of America* 81, 189–206.

Damuth, J. E. and Kumar, N. 1975: Amazon cone: morphology, sediments, age, and growth pattern. *Bulletin of the Geological Society of America* 86, 863–878.

Dansgaard, W., Clausen, H. B., Gundestrup, N., Johnsen, S. J. and Rygner, C. 1985: Dating and climate interpretation of two deep Greenland ice cores. In Langway, C. C., Oeschger, H. and Dansgaard, W. (eds), *Greenland ice core: geophysics, geochemistry, and the environment*. (Washington D.C.: American Geophysical Union), 71–76.

Dansgaard, W., White, J. W. C. and Johnsen, S. J. 1989: The abrupt termination of the Younger Dryas climate event. *Nature* 339, 532–534.

Darwin, C. 1871: *The Descent of Man*. New York: Random House.

Davies, J. L. 1969: *Landforms of cold climates*. Canberra: Australian National University Press.

Davis, M. B. 1967: Pollen accumulation rates at Rogers Lake, Connecticut, during late and postglacial time. *Review of Palaeobotany and Palaeoecology* 2, 219–230.

Davis, M. B. 1976: Pleistocene biogeography of temperate deciduous forests. *Geoscience and Man* 13, 13–26.

Davis, M. B., Woods, K. D., Webb, S. L. and Futyma, R. P. 1986: Dispersal versus climate: expansion of Fagus and Tsuga into the Upper Great Lakes region. *Vegetatio* 67, 93–103.

Davis, R. S. 1986: The Soan in Central Asia? Problems in Lower Palaeolithic cultural history. In Jacobsen, J. (ed.), *Studies in the archaeology of India and Pakistan*. (New Delhi: Oxford and IBH Publishing Company), 1–17.

De Deckker, P. 1988: Large Australian lakes during the last 20 million years: sites for petroleum source rocks or metal ore deposition or both? In Fleet, A. J., Kelts, K. R. and Talbot, M. R. (eds), *Lacustrine petroleum source rocks.* Geological Society of London, Special Publication 40, 45–58.

De Deckker, P., Kershaw, A. P. and Williams, M. A. J. 1988: Past environmental analogues. In Pearman, G. I. (ed.), *Greenhouse: planning for climate change.* (Melbourne: CSIRO), 473–488.

Del Genio, A. D., Lacis, A. A. and Ruedy, R. A. 1991: Simulations of the effect of a warmer climate on atmospheric humidity. *Nature* 351, 382–385.

Denton, G. H. and Hughes, T. J. 1981: The Arctic Ice Sheet: An outrageous hypothesis. In Denton, G. H. and Hughes, T. J. (eds), *The last great ice sheets.* (New York: Wiley Interscience), 440–467.

Denton, G. H. and Hughes, T. J. 1986: Global ice-sheet system interlocked by sea level. *Quaternary Research* 26, 3–26.

Denton, G. H. and Karlén, N. W. 1973: Holocene climatic variations – their pattern and possible cause. *Quaternary Research* 3, 155–205.

Deuser, W. G., Ross, E. H. and Waterman, L. S. 1976: Glacial and pluvial periods: their relationship revealed by Pleistocene sediments of the Red Sea and Gulf of Aden. *Science* 191, 1168–1170.

Devoy, R. J. N. (ed.) 1987a: *Sea surface studies: a global view.* London: Croom Helm.

Devoy, R. J. N. 1987b: Sea-level changes during the Holocene: the North Atlantic and Arctic Oceans. In Devoy R. J. N. (ed.), *Sea surface studies: a global view.* (London: Croom Helm), 294–347.

Diamond, J. M. 1989: Were Neanderthals the first humans to bury their dead? *Nature* 340, 344.

Diester-Haas, L., Heine, K., Rothe, P. and Schrader, H. 1988: Late Quaternary history of continental climate and the Benguela Current off South West Africa. *Palaeogeography, Palaeoclimatology, Palaeoecology* 65, 81–91.

Dillehay, T. D. and Collins, M. D. 1988: Early cultural evidence from Monte Verde in Chile. *Nature* 332, 150–152.

Dingle, R. V., Siesser, W. G. and Newton, A. R. 1983: *Mesozoic and Tertiary geology of southern Africa.* (Rotterdam: Balkema).

Dong Guangrong (ed.) 1991: *Quaternary environmental research on the deserts in China.* (Lanzhou: Institute of Desert Research, Academia Sinica), 116.

Douglas, R. G. and Savin, S. M. 1973: Oxygen and carbon isotope analysis of Cretaceous and Tertiary foraminifera from the central North Pacific. In Roth, P. H. and Herring, J. R. (eds), *Initial reports of the Deep Sea Drilling Project 17.* (Washington D.C.: U.S. Government Printing Office), 591–605.

Dregne, H. L. F. 1983: *Desertification of arid lands.* London: Harwood.

Drewry, D. J. 1986: *Glacial geologic processes.* London: Edward Arnold.

Dunwiddie, P. W. 1979: Dendrochronological studies of indigenous New Zealand trees. *New Zealand Journal of Botany* 17, 251–266.

Duplessy, J. C. 1982: Glacial to interglacial contrasts in the Northern Indian Ocean. *Nature* 295, 494–498.

Duplessy, J. C., Delibrias, G., Turon, J. L., Pujol, C. and Duprat, J. 1981: Deglacial warming of the northeastern Atlantic Ocean: correlation with the paleoclimatic evolution of the European continent. *Palaeogeography, Palaeoclimatology, Palaeoecology* 35, 121–144.

Easterbrook, D. J. 1988: Paleomagnetism of Quaternary deposits. *Geological Society of America Special Paper* 227, 111–122.

Emery, K. O. and Aubrey, D. G. 1991: *Sea levels, land levels and tide gauges.* Berlin: Springer-Verlag.

Emiliani, C. (ed.) 1981: The Oceanic Lithosphere. In *The Sea: Volume 7.* New York: John Wiley & Sons.

Erickson, J. 1990: *Ice Ages – past and future.* Blue Ridge Summit, Pa.: TAB Books.

Ericson, J. E. 1981: Exchange and production systems in Californian prehistory. *British Archaeological Reports International Series* 110.

Evans, J. 1860: On the occurrence of flint implements in undisturbed beds of gravel, sand, and clay. *Archaeologia* 38, 280–307.

Evenari, M., Shanan, L. and Tadmore, N. 1971: *The Negev: the challenge of a desert.* Cambridge, Massachusetts: Harvard University Press.

Faegri, K., Kaland, P. E. and Krzywinski, K. 1989: *Textbook of pollen analysis.* Fourth Edition. Chichester: Wiley.

Fagan, B. M. 1989: *People of the Earth.* Sixth Edition. Illinois: Scott, Foresman.

Fairbanks, R. G. 1989: A 17 000 year glacio-eustatic sea level record: influence of glacial melting rates on Younger Dryas event and deep-ocean circulation. *Nature* 342, 637–642.

Fairbridge, R. W. 1961: Eustatic changes in sea level. *Physics and Chemistry of the Earth* 4, 99–185.

Fairbridge, R. W. 1970: World palaeoclimatology of the Quaternary. *Revue de Géographie Physique et de Géologie Dynamique* 12(2), 97–104.

Faure, H. 1966: Évolution des grand lacs sahariens à l'Holocène. *Quaternaria* 8, 167–175.

Faure, H. 1990: Changes in the global continental reservoir of carbon. *Palaeogeography, Palaeoclimatology, Palaeoecology* 82, 47–52.

Faure, H. F., Faure-Denard, L. and Fairbridge, R. W. 1990: Possible effects of man on the carbon cycle in the past and in the future. In Paepe, R., Fairbridge, R. W. and Jelgersma, S. (eds), *Greenhouse effect, sea level and drought.* (Dordrecht: Kluwer Academic Publishers), 459–462.

Feibel, C. S., Brown, F. H. and McDougall, I. 1989: Stratigraphic context of fossil hominids from the Omo Group deposits: northern Turkana Basin, Kenya and Ethiopia. *American Journal of Physical Anthropology* 78, 595–622.

Fischer, A. G. 1984: The two phanerozoic supercycles. In Berggren, W. A. and Couvering, J. A. Van (eds), *Catastrophes and earth history.* (Princeton: Princeton University Press), 129–150.

Fleischer, R. L. and Hart, H. R. 1972: Fission tracking dating: techniques and problems. In Bishop, W. W. and Miller, J. A. (eds), *Calibration of hominoid evolution.* (Edinburgh: Scottish Academic Press), 135–170.

Flenley, J. R. 1978: *The equatorial rainforest: a geological history.* London: Butterworth.

Flint, R. F. 1957: *Glacial and Pleistocene geology.* New York: Wiley.

Flint, R. F. 1971: *Glacial and Quaternary geology.* New York: Wiley.

Flisch, M. 1986: K-Ar dating of Quaternary samples. In Hurford, A. J., Jäger, E. and TenCate, J. A. M. (eds), *Dating young sediments.* Proceedings of the Workshop, Beijing, People's Republic of China,

Sept. 1985. (Bangkok, Thailand: CCOP Technical Secretariat), 299–318.

Flood, J. 1989: *Archaeology of the dreamtime. The story of prehistoric Australia and its people.* Second Edition. Sydney: Collins.

Fontes, J.-C., Gasse, F., Callot, Y., Plaziat, J.-C., Carbonnel, P., Dupeuple, P. A. and Kaczmarska, I. 1985: Freshwater to marine-like environments from Holocene lakes in Northern Sahara. *Nature* 317, 608–610.

Ford, D. C. and Schwarcz, H. P. 1981: Uranium-series disequilibrium dating methods. In Goudie, A. (ed.), *Geomorphological techniques.* (London: Allen and Unwin), 284–287.

Forester, R. M. 1987: Late Quaternary paleoclimate records from lacustrine ostracodes. In Ruddiman, W. F. and Wright, H. E. (eds), *North America and adjacent oceans during the last deglaciation. The geology of North America.* Volume K-3. Boulder, Colorado: Geological Society of America.

Forman, S. L. 1989: Applications and limitations of thermoluminescence to date Quaternary sediments. *Quaternary International* 1, 47–59.

Foucault, A. and Stanley, D. J. 1989: Late Quaternary palaeoclimatic oscillations in East Africa recorded by heavy minerals in the Nile delta. *Nature* 339, 44–46.

Frakes, L. A. 1986: Mesozoic-Cenozoic climatic history and causes of glaciation. In Hsü, K. J. (ed.), *Mesozoic and Cenozoic oceans.* (Washington D.C.: American Geophysical Union/Boulder, Colorado: Geological Society of America), 33–48.

Frey, D. G. (ed.) 1969: Symposium on paleolimnology. *Internationale Vereinigung für theoretische und angewandte Limnologie, Mitteilungen* 17, 1–448.

Friedman, I. and Long, W. 1976: Hydration rate of obsidian. *Science* 191, 347–352.

Fritts, H. C. 1976: *Tree rings and climate.* London: Academic Press.

Fritz, P., Anderson, T. W. and Lewis, C. F. M. 1975: Late-Quaternary climatic trends and history of Lake Erie from stable isotope studies. *Science* 190, 267–269.

Fuji, N. and Horowitz, A. 1989: Brunhes epoch paleoclimates of Japan and Israel. *Palaeogeography, Palaeoclimatology, Palaeoecology* 72, 79–88.

Fulton, R. J. (ed.) 1989: *Quaternary geology of Canada and Greenland.* In Geological Survey of Canada No. 1/Geological Society of North America, Geology of North America. Volume K-1. (Ottowa: Canadian Government Publishing Centre).

Funnell, B. M. and Riedel, L. W. R. (eds) 1971: *The micropalaeontology of the oceans.* Cambridge: Cambridge University Press.

Fyfe, W. S. 1990: The International Geosphere/Biosphere Programme and global change: an anthropocentric or an ecocentric future? A personal view. *Episodes* 13(2), 100–102.

Gargett, R. H. 1989: Grave shortcomings. The evidence for Neanderthal burial. *Current Anthropology* 30, 157–190.

Gasse, F., Rognon, P. and Street, F. A. 1980: Quaternary history of the Afar and Ethiopian Rift Lakes. In Williams, M. A. J. and Faure, H. (eds), *The Sahara and the Nile.* (Rotterdam: Balkema), 361–400.

Gasse, F., Tehet, R., Durand, A., Gibert, E. and Fontes, J.-C. 1990: The arid-humid transition in the Sahara and the Sahel during the last deglaciation. *Nature* 346, 141–146.

Gates, W. L. 1976: Modelling the Ice-Age climate. *Science* 191, 1138–1144.

Gautier, A. 1988: The final demise of *Bos ibericus? Sahara* 1, 37–48.

Genthon, C., Barnola, J. M., Raynaud, D., Lorius, C., Jouzel, J., Barkov, N. I., Korotkevich, Y. S. and Kotlyakov, V. M. 1987: Vostok ice core: climatic response to CO_2 and orbital forcing changes over the last climatic cycle. *Nature* 329, 414–418.

Gilland, B. 1988: Population, economic growth, and energy demand 1985–2020. *Population and Development Review* 14(2), 233–244.

Gillespie, R. 1982: *Radiocarbon users' handbook.* Quaternary Research Unit, Occasional Paper No. 1. Sydney: Macquarie University.

Gillespie, R., Horton, D. R., Ladd, P., Macumber, P. G., Thorne, R. and Wright, R. V. S. 1978: Lancefield Swamp and the extinction of the Australian megafauna. *Science* 200, 1044–1048.

Godwin, H. and Tallantire, P. A. 1951: Studies in the post-glacial history of British vegetation. XII. Hockham Mere, Norfolk. *Journal of Ecology* 39, 285–307.

Goede, A. 1989: Electron spin resonance – a relative dating technique for Quaternary sediments near Warrnambool, Victoria. *Australian Geographical Studies* 27(1), 14–30.

Goldberg, E. D. 1963: Geochronology with lead-210. In *Radioactive dating.* IAEA STI-PUB-68, 121.

Goldberg, E. D. 1986: Uranium series disequilibrium techniques for dating Quaternary sediments. In Hurford, A. J., Jäger, E. and TenCate, J. A. M. (eds), *Dating young sediments.* Proceedings of the Workshop, Beijing, People's Republic of China, Sept. 1985. (Bangkok, Thailand: CCOP Technical Secretariat), 59–71.

Gorecki, P. P., Horton, D. R., Stern, N. and Wright, R. V. S. 1984: Co-existence of humans and megafauna in Australia: improved stratified evidence. *Archaeology in Oceania* 19, 117–119.

Gornitz, V., Lebedeff, S. and Hansen, J. 1982: Global sea level trend in the past century. *Science* 215, 1611–1614.

Gorschkov, S. G. (ed.) 1978: *World ocean atlas Volume 2: Atlantic and Indian Oceans.* Australia: Pergamon Press.

Goudie, A. (ed.) 1981: *Geomorphological techniques.* London: Allen and Unwin.

Goudie, A. 1983: *Environmental change.* Second Edition. Oxford: Clarendon Press.

Goudie, A., Allchin, B. and Hegde, K. T. M. 1973: The former extensions of the Great Indian Sand Desert. *Geographical Journal* 134, 243–257.

Gowlett, J. A. J. 1984a: *Ascent to civilization. The archaeology of early man.* London: Collins.

Gowlett, J. A. J. 1984b: Mental abilities of Early Man: a look at some hard evidence. In Foley, R. (ed.), *Homonid evolution and community ecology.* (London: Academic Press), 167–192.

Gowlett, J. A. J., Harris, J. W. K., Walton, D. A. and Wood, B. A. 1981: Early archaeological sites, further hominid remains and traces of fire from Chesowanja, Kenya. *Nature* 294, 125–129.

Gray, J. (ed.) 1988: Aspects of freshwater paleoecology and biogeography. *Palaeogeography, Palaeoclimatology, Palaeoecology* 62, 1–623.

Green, D. G. 1981: Time series and postglacial forest ecology. *Quaternary Research* 15, 265–277.

Green, D. G. and Dolman, G. S. 1988: Fine resolution pollen analysis. *Journal of Biogeography* 15, 685–701.

Gribbin, J. (ed.) 1986: *The breathing planet*. Oxford: Blackwell and New Scientist.

Gross, M. G. 1982: *Oceanography: a view of the Earth*. New Jersey: Prentice Hall.

Grove, A. T. 1980: Geomorphic evolution of the Sahara and the Nile. In Williams, M. A. J. and Faure, H. (eds), *The Sahara and the Nile*. (Rotterdam: Balkema), 7–16.

Grove, A. T. and Warren, A. 1968: Quaternary landforms and climate on the south side of the Sahara. *Geographical Journal* 134, 194–208.

Grove, J. M. 1988: *The Little Ice Age*. London: Methuen.

Guidon, N. and Delibrias, G. 1985: Inventaire des sites sud-américains antérieurs à 12 000 ans. *L'Anthropologie* 89, 385–407.

Guidon, N. and Delibrias, G. 1986: Carbon-14 dates point to man in the Americas 32 000 years ago. *Nature* 321, 769–771.

Habermehl, M. A. 1985: Groundwater in Australia. In Downing, R. A. and Jones, G. P. (eds), *Hydrogeology in the service of man*. Memoirs of the 18th Congress of the International Association of Hydrogeologists, Cambridge, 1985. (Cambridge: International Association of Hydrological Sciences), IAHS Publication No. 154, 31–52.

Hall, C. M., Walter, R. C., Westgate, J. A. and York, D. 1984: Geochronology, stratigraphy and geochemistry of Cindery Tuff in Pliocene hominid-bearing sediments of the Middle Awash, Ethiopia. *Nature* 308, 26–31.

Hallam, S. J. 1977: The relevance of Old World archaeology to the first entry of man into New Worlds: colonization seen from the Antipodes. *Quaternary Research* 8, 128–148.

Hamilton, T. D. and Thorson, R. M. 1983: The Cordilleran Ice Sheets in Alaska. In Wright, H. E. and Porter, S. C. (eds), *Late Quaternary environments of the United States. Volume 1. The Late Pleistocene*. (Essex: Longman), 38–52.

Hammerschmidt, K. 1986: ^{40}Ar-^{39}Ar dating of young samples. In Hurford, A. J., Jäger, E. and TenCate, J. A. M. (eds), *Dating young sediments*. Proceedings of the Workshop, Beijing, People's Republic of China, Sept. 1985. (Bangkok, Thailand: CCOP Technical Secretariat), 339–357.

Hansen, J. E. and Lacis, A. A. 1990: Sun and dust versus greenhouse gases: an assessment of their relative roles in global climatic change. *Nature* 346, 713–719.

Haq, B. U., Hardenbol, J. and Vail, P. R. 1987: Chronology of the fluctuating sea levels since the Triassic. *Science* 235, 1156–1167.

Harmon, R. S. 1980: Uranium-series geochronology: a review of its application to absolute age dating of archaeological deposits. In Burleigh, R. (ed.), *Progress in scientific dating methods*. British Museum Occasional Paper No. 21, 53–64.

Harris, J. M., Brown, F. H., Leakey, M. G., Walker, A. C. and Leakey, R. E. 1988: Pliocene and Pleistocene hominid-bearing sites from west of Lake Turkana, Kenya. *Science* 239, 27–33.

Harris, J. W. K. 1980: Early Man. In Sherratt, A. (ed.), *The Cambridge encyclopedia of archaeology*. (New York: Cambridge University Press), 62–70.

Harris, J. W. K. 1983: Cultural beginnings: Plio-Pleistocene archaeological occurrences from the Afar, Ethiopia. *African Archaeological Review* 1, 2–31.

Harris, S. A. 1985: Distribution and zonation of permafrost along the eastern ranges of the Cordillera of North America. *Biuletyn Peryglacjalny* 30, 107–118.

Harris, S. A. 1988: The Alpine Periglacial Zone. In Clark, M. J. (ed.), *Advances in periglacial geomorphology*. (Chichester: Wiley), 369–413.

Harrison, S. P., Metcalfe, S. E., Street-Perrott, F. A., Pittock, A. B., Roberts, C. N. and Salinger, M. J. 1984: A climatic model of the last glacial/interglacial transition based on palaeotemperature and palaeohydrological evidence. In Vogel, J. (ed.), *Late Cainozoic environments of the Southern Hemisphere*. (Rotterdam: Balkema), 21–34.

Harwell, M. A. 1984: *Nuclear winter: the human and environmental consequences of nuclear war*. Berlin: Springer-Verlag.

Hassan, F. A. 1980: Prehistoric settlement along the Main Nile. In Williams, M. A. J. and Faure, H. (eds), *The Sahara and the Nile*. (Rotterdam: Balkema), 421–450.

Hasselmann, K. 1991: Ocean circulation and climate change. *Tellus* 43AB, 82–103.

Hastenrath, S. and Kutzbach, J. E. 1983: Palaeoclimatic estimates from water and energy budgets of East African lakes. *Quaternary Research* 19, 141–153.

Hay, W. W. 1988: Paleoceanography: a review for the GSA Centennial. *Bulletin of the Geological Society of America* 100: 1934–1956.

Haynes, C. V. 1991: Geoarchaeological and palaeohydrological evidence for a Clovis-age drought in North America and its bearing on extinction. *Quaternary Research* 35, 438–450.

Hays, J. D. 1978: A review of the Late Quaternary climatic history of Antarctic Seas. In Zinderen Bakker, E. M. Van (ed.), *Antarctic glacial history and world palaeoenvironments*. (Rotterdam: Balkema), 57–71.

Hays, J. D., Imbrie, J. and Shackleton, N. J. 1976: Variations in the earth's orbit: pacemaker of the Ice Ages. *Science* 194, 1121–1132.

Hays, J. D., Lozano, J. A., Shackleton, N. J. and Irving, G. 1976: Reconstruction of the Atlantic and Western Indian Ocean sectors of the 18 000 BP Antarctic Ocean. *Geological Society of America Memoir* 145, 337–372.

Hayworth, E. Y. and Lund, J. W. (eds) 1984: *Lake sediments and environmental history*. Exeter: Leicester University Press.

Hedgpeth, J. W. 1969: Introduction to Antarctic zoogeography. *Antarctic zoogeography, Antarctic map folio series*, Folio 11: 1. New York: American Geographical Society.

Heller, F. and Liu, T.-S. 1982: Magnetostratigraphical dating of loess deposits in China. *Nature* 300, 431–433.

Higham, C. 1989: *The archaeology of mainland southeast Asia: from 10 000 BC to the fall of Angkor*. Cambridge: Cambridge University Press.

Hill, A. and Ward, S. 1988: Origin of the Hominidae: the record of African large hominoid evolution between 14 My and 4 My. *Yearbook for Physical Anthropology* 31, 49–83.

Hillaire-Marcel, C. and Occhietti, S. 1980: Chronology, paleogeography, and paleo-climatic significance of the late and post-glacial events in eastern Canada. *Zeitschrift für Geomorphologie N.F.* 24, 373–392.

Holland, H. D. and Trendall, A. F. (eds) 1984: *Patterns of change in earth evolution.* Berlin: Springer-Verlag.

Hollin, J. T. and Schilling, D. H. 1981: Late Wisconsin–Weichselian mountain glaciers and small ice caps. In Denton, G. H. and Hughes, T. J. (eds), *The last great ice sheets.* (New York: Wiley), 179–206.

Hooghiemstra, H. 1988: The orbital-tuned marine oxygen isotope record applied to the Middle and Late Pleistocene pollen record of Funza (Colombian Andes). *Palaeogeography, Palaeoclimatology, Palaeoecology* 66, 9–17.

Hooghiemstra, H. 1989: Quaternary and Upper-Pliocene glaciations and forest development in the tropical Andes: evidence from a long high-resolution pollen record from the sedimentary basin of Bogota, Colombia. *Palaeogeography, Palaeoclimatology, Palaeoecology* 72, 11–26.

Hooghiemstra, H., Bechler, A. and Beug, H.-J. 1987: Isopollen maps for 18 000 years BP of the Atlantic offshore of northwest Africa: evidence for paleowind circulation. *Paleoceanography* 2, 561–582.

Horton, D. R. 1980: A review of the extinction question: man, climate and megafauna. *Archaeology and Physical Anthropology in Oceania* 15, 86–97.

Horton, D. R. 1982: The burning question: Aborigines, fire and Australian ecosystems. *Mankind* 13, 237–251.

Horton, D. R. 1984: Red kangaroos: last of the megafauna. In Martin, P. S. and Klein, R. G. (eds), *Quaternary extinctions.* (Tucson: University of Arizona Press), 639–680.

Houghton, J. T., Jenkins, G. J. and Ephraums, J. J. (eds) 1990: *Climate change: The Intergovernmental Panel on Climate Change.* Cambridge: Cambridge University Press.

Houghton, R. A. and Woodwell, G. M. 1989: Global climatic change. *Scientific American* 260(4): 18–26.

Hsü, K. J. 1983: *The Mediterranean was a desert. A voyage of the Glomar Challenger.* Princeton: Princeton University Press.

Hsü, K. J., Montaderts, L., Bernoulli, D., Cita, M. B., Erikson, A., Garrison, R. E., Kidd, R. B., Mélières, F., Müller, C. and Wright, R. 1977: History of the Mediterranean salinity crisis. *Nature* 267, 399–403.

Huggett, R. J. 1991: *Climate, earth processes and earth history.* Berlin: Springer-Verlag.

Hughes, T. J., Denton, G. H., Andersen, B. G., Schilling, D. H., Fastook, J. L. and Lingle, C. S. 1981: The last great ice sheets: a global view. In Denston, G. H. and Hughes, T. J. (eds), *The last great ice sheets.* (New York: Wiley), 263–317.

Huntley, B. 1990: European post-glacial forests: compositional changes in response to climatic change. *Journal of Vegetation Science* 1, 507–518.

Huntley, B. and Webb, T. 1988: *Vegetation history.* Dordrecht: Kluwer Academic Publishers.

Husen, D. Van 1989: The last interglacial-glacial cycle in the eastern Alps. *Quaternary International* 3/4, 115–121.

Ikeya, M. 1986: Electron spin resonance. In Zimmerman, M. R. and Angel, J. L. (eds), *Dating and determination of biological materials.* (London: Croom and Helm), 59–125.

Ikeya, M. 1987: Electron spin resonance (ESR) dating in archeometry: a review with Australian prospects. In Ambrose, W. R. and Mummery, J. M. J. (eds), *Archaeometry: further Australasian studies.* (Canberra: Australian National University), 156–166.

Imbrie, J., Hays, J. D., Martinson, D. G., McIntyre, A., Mix, A. C., Morley, J. J., Pisias, N. G., Prell, W. L. and Shackleton, N. J. 1984: The orbital theory of Pleistocene climate: support from a revised chronology of the marine ^{18}O record. In Berger, A. L., Imbrie, J., Hays, J., Kukla, G. and Saltzman, B. (eds), *Milankovitch and climate: understanding the response to astronomical forcing.* (Dordrecht: Reidel), 269–305.

Imbrie, J. and Imbrie, K. P. 1979: *Ice ages: solving the mystery.* London: Macmillan.

Intergovernmental Panel on Climate Change 1990: *Scientific assessment of climate change.* The Policymakers' Summary of the Report of Working Group 1. WMO/UNEP.

International Geosphere-Biosphere Programme 1989: *Effects of atmospheric and climate changes on terrestrial ecosystems.* IGBP Global Change Report 5.

International Geosphere-Biosphere Programme 1990: A study of global change. The initial core projects. Report No. 12, Sweden.

Irion, G. 1984: Sedimentation and sediments of Amazonian rivers and evolution of the Amazonian landscape since Pliocene times. In Sioli, H. (ed.), *The Amazon.* (Dordrecht, Junk), 201–214.

Israel Land Development Authority 1990: Redeeming the Negev Desert in Israel. In *United Nations Environment Programme, Exchange of Environment Experience Series,* Book 3. (Nairobi, Kenya: Infoterra Programme Activity Centre), 1–16.

Issawi, B. 1983: Ancient rivers of the eastern Egyptian desert. *Episodes* 1983(2), 3–6.

Ivanovich, M. and Harmon, R. S. (eds) 1982: *Uranium-series disequilibrium: applications to environmental problems.* Oxford: Clarendon Press.

Jansen, E. 1989: The use of stable oxygen and carbon isotope stratigraphy as a dating tool. *Quaternary International* 1, 151–166.

Jansen, E. and Veum, T. 1990: Evidence of two-step deglaciation and its impact on North Atlantic deep-water circulation. *Nature* 343, 612–616.

Jansson, C. R. 1966: Recent pollen spectra from the deciduous and coniferous-deciduous forests of northeastern Minnesota: a study in pollen dispersal. *Ecology* 47, 804–825.

Jansson, C. R. 1973: Local and regional pollen deposition. In Birks, H. J. B. and West, R. G. (eds), *Quaternary plant ecology.* (Oxford: Blackwell), 31–42.

Jelgersma, S. 1961: Holocene sea level changes in the Netherlands. *Mededelingen Geologische Stichting* Series C-VI, No. 7.

Johanson, D. C. 1989: The current status of Australopithecus. In Giacobini, G. (ed), *Hominidae: Proceedings of the 2nd International Congress of Human Paleontology.* (Milan: Jaca), 77–96.

Johanson, D. C., Masao, F. T., Eck, G. G., White, T. D., Walter, R. C.,

Kimbel, W. H., Asfaw, B., Manega, P., Ndessokia, P. and Suwa, G. 1987: New partial skeleton of *Homo habilis* from Olduvai Gorge, Tanzania. *Nature* 372, 205–209.

Johnson, R. G. 1991: Major northern hemisphere deglaciation caused by a moisture deficit 140 ka. *Geology* 19, 686–689.

Jones, P. 1980: Experimental butchery with modern stone tools and its relevance for Palaeolithic archaeology. *World Archaeology* 12(2), 153–165.

Jones, P. D. and Wigley, T. M. L. 1990: Global warming trends. *Scientific American* 263(2), 66–73.

Jones, P. D., Wigley, T. M. L. and Wright, P. B. 1986: Global temperature variations between 1861 and 1984. *Nature* 322, 430–434.

Jones, R. 1968: The geographical background to the arrival of man in Australia and Tasmania. *Archaeology and Physical Anthropology in Oceania* 3, 186–215.

Jones, R. 1969: Fire-stick farming. *Australian Natural History* 16, 224–228.

Jones, R. 1979: The fifth continent: problems concerning the human colonization of Australia. *Annual Review of Anthropology* 8, 445–466.

Jouzel, J., Barkov, N. I., Barnola, J. M., Genthon, C., Korotkevich, Y. S., Kotlyakov, V. M., Legrand, M., Lorius, C., Petit, J. P., Petrov, V. N., Raisbeck, G., Raynaud, D., Ritz, C. and Yiou, F. 1989: Global change over the last climatic cycle from the Vostok ice core record (Antarctica). *Quaternary International* 2, 15–24.

Jouzel, J., Lorius, J. C., Petit, J. R., Genthon, C., Barkov, N. I., Kotlyakov, V. M. and Petrov, V. N. 1987: Vostok ice core: a continuous isotope temperature record over the last climatic cycle (160 000 years). *Nature* 329, 403–408.

Jutson, J. T. 1934: The physiography (geomorphology) of Western Australia. *Western Australia Geological Survey Bulletin* 95, 1–366 (Revised Second Edition).

Kadomura, H. (ed.) 1989: *Savannization processes in tropical Africa 1.* Department of Geography, Tokyo Metropolitan University and Zambia Geographical Association Occasional Study 17.

Keany, J. 1976: Diachronous deposition of ice-rafted debris in sub-Antarctic deep-sea sediments. *Geological Society of America*, Bulletin 87, 873–882.

Keeley, L. 1980: *Experimental determination of stone tool use.* Chicago: University of Chicago Press.

Keeley, L. and Toth, N. 1981: Microwear polishes on early stone tools from Koobi Fora, Kenya. *Nature* 293, 464–465.

Kellogg, T. B. 1987: Glacial-interglacial changes in global deepwater circulation. *Paleoceanography* 2, 259–271.

Kempe, S. and Degens, E. T. 1979: Varves in the Black Sea and in Lake Van (Turkey). In Schlüchter, C. (ed.), *Moraines and varves: origin, genesis and classification.* (Rotterdam: Balkema), 309–318.

Kennedy-Sutherland, E. 1983: The effects of fire exclusion on growth in mature Ponderosa Pine in Northern Arizona. Unpublished. Arizona: University of Arizona.

Kennett, J. P. 1982: *Marine geology.* New Jersey, Prentice Hall.

Kerr, R. A. 1986: Mapping orbital effects on climate. *Science* 234, 283–284.

Kershaw, A. P. 1973: *Late Quaternary vegetation of the Atherton Table-land, north-east Queensland, Australia.* Unpublished Ph.D. thesis. Canberra: Australian National University.

Kershaw, A. P. 1978: Record of last interglacial-cycle from north-eastern Queensland. *Nature* 272, 159–162.

Kershaw, A. P. 1984: Late Cenozoic plant extinctions in Australia. In Martin, P. S. and Klein, R. G. (eds), *Quaternary extinctions: a prehistoric revolution.* (Tucson: University of Arizona Press), 601–709.

Kershaw, A. P. 1986: The last two glacial-interglacial cycles from northeastern Queensland: implications for climatic change and Aboriginal burning. *Nature* 322, 47–49.

Kershaw, A. P. and Hyland, B. P. M. 1975: Pollen transport and periodicity in a rainforest situation. *Review of Palaeobotany and Palynology* 19, 129–138.

Kimber, R. W. L. 1987: The use of amino acid racemization in establishing time frameworks and correlations for geological and archaeological materials. In Ambrose, W. R. and Mummery, J. M. J. (eds), *Archaeometry: further Australasian studies.* (Canberra: Australian National University), 150–155.

Kirk, R. C. and Thorne, A. G. (eds) 1976: *The Origin of the Australians.* Canberra: Australian Institute of Aboriginal Studies.

Klammer, G. 1984: The relief of the extra-Andean Amazon basin. In Sioli, H. (ed.), *The Amazon.* (Dordrecht: Junk), 47–83.

Klein, R. 1975: First entry of man into the New World. *Quaternary Research* 5, 391–394.

Klein, R. 1986: Carnivore size and Quaternary climatic change in southern Africa. *Quaternary Research* 26, 153–170.

Knox, J. C. 1984a: Responses of river systems to Holocene climates. In Wright, H. E. (ed.), *Late Quaternary environments of the United States, Volume 2. The Holocene.* (London: Longman), 26–41.

Knox, J. C. 1984b: Fluvial responses to small scale climate changes. In Costa, J. E. and Fleisher, P. J. (eds), *Developments and applications of geomorphology.* (Berlin: Springer-Verlag), 318–342.

Kohl, H. 1986: Pleistocene glaciations in Austria. *Quaternary Science Reviews* 5, 421–427.

Krebs, J. R. and Coe, H. J. 1985: Sahel famine: an ecological perspective. *Nature* 317, 13–14.

Ku, T.-L. 1976: The uranium-series methods of age determination. *Annual Review of Earth and Planetary Science* 4, 347–379.

Ku, T.-L. and Liang, Z.-C. 1984: The dating of impure carbonates with decay series isotopes. *Nuclear Instruments and Methods in Physics Research* 223, 563–571.

Kukla, G. 1989: Long continental records of climate – an introduction. *Palaeogeography, Palaeoclimatology, Palaeoecology* 72, 1–9.

Kukla, G. and Zhisheng A. 1989: Loess stratigraphy in central China. *Palaeogeography, Palaeoclimatology, Palaeoecology* 72, 203–225.

Kutzbach, J. E. and Guetter, P. J. 1986: The influence of changing orbital parameters and surface boundary conditions on climate simulations for the past 18 000 years. *Journal of the Atmospheric Sciences* 43, 1726–1759.

Kutzbach, J. E. and Street-Perrott, F. A. 1985: Milankovich forcing of fluctuations in the level of tropical lakes from 18 to 0 kyr BP. *Nature* 317, 130–134.

Kutzbach, J. E. and Wright, H. E. 1985: Simulation of the climate of 18 000 years BP. Results for the North American/North Atlantic/European sector and comparison with the geologic record of North America. *Quaternary Science Reviews* 4(3), 147–187.

Labeyrie, L. D., Duplessy, J. C. and Blanc, P. L. 1987: Variations in mode of formation and temperature of oceanic deep waters over the past 125 000 years. *Nature* 327, 477–482.

Labeyrie, L. D., Pichon, J. J., Labracherie, M., Ippolito, P., Duprat, J. and Duplessy, J. C. 1986: Melting history of Antarctica during the past 60 000 years. *Nature* 322, 701–706.

Lambert, M. R. K. 1984: Amphibians and reptiles. In Cloudsley-Thompson, J. L. (ed.), *Sahara Desert.* (Oxford: Pergamon), 205–227.

Lamplugh, G. W. 1902: Calcrete. *Geological Magazine* 9, 575.

Lanphere, M. A. and Dalrymple, G. B. 1971: A test of the ^{40}Ar/^{39}Ar age spectrum technique on some terrestrial materials. *Earth and Planetary Science Letters* 12, 359–372.

Lanphere, M. A. and Dalrymple, G. B. 1976: Identification of excess ^{40}Ar by the ^{40}Ar/^{39}Ar age spectrum technique. *Earth and Planetary Science Letters* 32, 141–148.

Laporte, L. F. and Zihlman, A. L. 1983: Plates, climate and hominoid evolution. *South African Journal of Science* 79, 96–110.

Le Houérou, H. N. 1989: *The grazing land ecosystems of the African Sahel.* Berlin: Springer-Verlag.

Le Roy Ladurie, E. 1972: *Times of feast, times of famine. A history of climate since the year 1000.* London: Allen and Unwin.

Lea, D. W. and Boyle, E. A. 1989: Barium content of benthic foraminifera controlled by bottom-water composition. *Nature* 338, 751–753.

Lea, D. W., Shen, G. T. and Boyle, E. A. 1989: Coralline barium records temporal variability in equatorial Pacific upwelling. *Nature* 340, 373–376.

Leakey, M. D. and Hay, R. L. 1979: Pliocene footprints in the Laetoli Beds at Laetoli, Northern Tanzania. *Nature* 278, 317–323.

Lee, L. K. 1984: Land use and soil loss: a 1982 update. *Journal of Soil and Water Conservation* 39(4), 226–228.

Legge, A. J. and Rowley-Conway, P. A. 1987: Gazelle killing in Stone Age Syria. *Scientific American* 257, 76–83.

Leggett, J. (ed.) 1990: *Global warming: the Greenpeace report.* Oxford: Oxford University Press.

Legrand, M. R., Delmas, R. J. and Charlston, R. J. 1988: Climate forcing implications from Vostok ice-core sulphate data. *Nature* 334, 418–420.

Lehman, S. J., Jones, G. A., Keigwin, L. D., Andersen, E. S., Butenko, G. and Ostmo, S.-R. 1991: Initiation of Fennoscandian ice-sheet retreat during the last deglaciation. *Nature* 349, 513–516.

Leventer, A., Williams, D. F. and Kennett, J. P. 1982: Dynamics of the Laurentide ice sheet during the last glaciation: evidence from the Gulf of Mexico. *Earth and Planetary Science Letters* 59, 11–17.

Lezine, A.-M. 1991: West African paleoclimates during the last climatic cycle inferred from an Atlantic deep-sea pollen record. *Quaternary Research* 35, 456–463.

Lhote, H. 1959: *The Search for the Tassili Frescoes.* London: Hutchinson and Co.

Libby, W. F., Anderson, E. C. and Arnold, J. R. 1949: Age determination by radiocarbon content: world-wide assay of natural radiocarbon. *Science* 109, 227–228.

Linick, T. W., Damon, P. E., Donahue, D. J. and Jull, A. J. T. 1989: Accelerator mass spectrometry: The new revolution in radiocarbon dating. *Quaternary International* 1, 1–6.

Lister, G. S. 1988: Stable isotopes from lacustrine ostracoda as tracers for continental palaeoenvironments. In De Deckker, P., Colin, J. P. and Peypouquet, J. P. (eds), *Ostracoda in the earth sciences.* (Amsterdam: Elsevier), 201–218.

Liu, T.-S. (ed.) 1987: *Aspects of loess research.* Beijing: China Ocean Press.

Liu, T.-S., Ding, Z., Chen, M. and An, Z. 1989: The global surface energy system and the geological role of wind stress. *Quaternary International* 2, 43–54.

Liu, T.-S., Fei, G. X., Sheng, A. Z. and Xiang, F. Y. 1981: The dust fall on Beijing, China, on April 18, 1980. *Geological Society of America. Special Paper* 186, 149–158.

Liu, Z. 1983: Peking Man's cave yields new finds. *Geographical Magazine* 55, 297–300.

Livingstone, D. A. 1980: Environmental changes in the Nile headwaters. In Williams, M. A. J. and Faure, H. (eds), *The Sahara and the Nile.* (Rotterdam: Balkema), 339–359.

Löffler, E. 1972: Pleistocene glaciation in Papua and New Guinea. *Zeitschrift für Geomorphologie N.F. Supplementband* 13, 32–58.

Löffler, H. (ed.) 1987: Paleolimnology IV. *Developments in Hydrobiology* 37, 1–431.

Lorius, C., Jouzel, J., Ritz, C., Merlivat, L., Barkov, N. I., Korotkevich, Y. S. and Kotlyakov, V. M. 1985: A 150 000-year climatic record from Antarctic ice. *Nature* 316, 591–596.

Lovlie, R. 1989: Paleomagnetic stratigraphy: a correlation method. *Quaternary International* 1, 129–149.

Lowe, J. J. and Walker, M. J. C. 1984: *Reconstructing Quaternary environments.* London: Longman.

Lowrie, W. and Alvarez, W. 1981: One hundred million years of geomagnetic polarity history. *Geology* 9, 392–397.

Lundqvist, J. 1986: Late Weichselian glaciation and deglaciation in Scandinavia. *Quaternary Science Reviews* 5, 269–292.

Lyell, C. 1873: *Geological evidence of the antiquity of man, with an outline of glacial and post-Tertiary geology and remarks on the origin of species, with special reference to man's first appearance on earth.* Fourth Edition. London: John Murray.

Maarleveld, G. C. 1976: Periglacial phenomena and the mean annual temperature during the last glacial time in the Netherlands. *Biuletyn Peryglacjalny* 26, 57–78.

Mabbutt, J. A. 1977: *Desert landforms.* Canberra: Australian National University Press.

Mabbutt, J. A. 1985: Desertification of the world's rangelands. *Desertification control bulletin.* UNEP, Nairobi, 12, 1–11.

Macquarie 1984: *The Macquarie illustrated world atlas.* McMahons Point, Sydney: Macquarie Library.

McAndrews, J. H. and Boyko-Diakonow, M. 1989: Pollen analysis of varved sediments at Crawford Lake, Ontario: evidence of Indian and European farming. In Fulton, R. J. (eds), *Quaternary geology of Canada and Greenland.* Geological Survey of Canada, No. 1/Geological Society of America, The Geology of North America, Volume K-1. (Ottawa: Canadian Government Publishing Centre), 528–530.

McEwen-Mason, J. R. C. 1991: The late Cainozoic magnetostratigraphy and preliminary palynology of Lake George, New South Wales. In Williams, M. A. J., De Deckker, P. and Kershaw, A. P. (eds), *The Cainozoic in Australia: a re-appraisal of the evidence.* Geological Society of Australia, Special Publication No. 18, 195–205.

McIntyre, A. 1989: Surface water response of the equatorial Atlantic Ocean to orbital forcing. *Paleoceanography* 4, 19–55.

McIntyre, A. and Kipp, N. G., With Bé, A. W. H., Crowley, T., Kellogg, T., Gardner, J. V., Prell, W. and Ruddiman, W. F. 1976: Glacial North Atlantic 18 000 years ago: a CLIMAP reconstruction. *Geological Society of America Memoir* 145, 43–75.

McIntyre, M. L. and Hope, J. H. 1978: Procoptodon fossils from the Willandra Lakes, western New South Wales. *The Artefact* 3(3), 117–132.

McClure, H. A. 1976: Radiocarbon chronology of the late Quaternary lakes in the Arabian desert. *Nature* 263, 755–756.

McTainsh, G. 1980: Harmattan dust deposition in northern Nigeria. *Nature* 286, 587–588.

McTainsh, G. 1985: Dust processes in Australia and West Africa: a comparison. *Search* 16, 104–106.

Mainguet, M., Canon, L. and Chemin, M. C. 1980: Le Sahara: géomorphologie et paléogeomorphologie éoliennes. In Williams, M. A. J. and Faure, H. (eds), *The Sahara and the Nile.* (Rotterdam: Balkema), 17–35.

Mainguet, M. and Cossus, L. 1980: Sand circulation in the Sahara: geomorphological relations between the Sahara desert and its margins. In Sarnthein, M., Seibold, E. and Rognon, P. (eds), *Sahara and surrounding areas: sediments and climatic changes.* Proceedings of an international symposium, Akademie der Wissenschaften und der Literatur Mainz, 1–4 April 1979. Palaeoecology of Africa and surrounding islands, 12, 69–78.

Maley, J. 1980: Les changements climatiques de la fin du Tertiaire en Afrique: leur conséquence sur l'apparition du Sahara et de sa végétation. In Williams, M. A. J. and Faure, H. (eds), *The Sahara and the Nile.* (Rotterdam: Balkema), 63–86.

Manabe, S. and Broccoli, A. J. 1985: The influence of continental ice sheets on the climate of an ice age. *Journal of Geophysical Research* 90, 2167–2190.

Mangerud, J. 1989: Correlation of the Eemian and Weichselian with deep sea oxygen isotope stratigraphy. *Quaternary International* 3/4, 1–4.

Mangini, A. 1986a: Dating of marine carbonates using electron spin resonance (ESR) and ^{230}Th-^{234}U methods. In Hurford, A. J., Jäger, E. and TenCate, J. A. M. (eds), *Dating young sediments.* Proceedings of the Workshop, Beijing, People's Republic of China, Sept. 1985. (Bangkok, Thailand: CCOP Technical Secretariat), 73–84.

Mangini, A. 1986b: Application of the ^{230}Th, ^{231}Pa and ^{10}Be radioisotopes in sedimentary geology. In Hurford, A. J., Jäger, E. and TenCate, J. A. M. (eds), *Dating young sediments*. Proceedings of the Workshop, Beijing, People's Republic of China, Sept. 1985. (Bangkok, Thailand: CCOP Technical Secretariat), 59–71.

Mankinen, E. A. and Dalrymple, G. B. 1979: Revised geomagnetic polarity time scale for the interval 0–5 My BP. *Journal of Geophysical Research* 84 (B2), 615–626.

Martin, P. S. 1967: Prehistoric overkill. In Martin, P. S. and Wright, H. E. (eds), *Pleistocene extinctions: the search for a cause.* (New Haven: Yale University Press), 75–120.

Martin, P. S. 1973: The discovery of America. *Science* 179, 969–974.

Martin, P. S. 1984: Prehistoric overkill: the global model. In Martin, P. S. and Klein, R. G. (eds), *Quaternary extinctions: a prehistoric revolution.* (Tucson: University of Arizona Press), 354–403.

Martin, P. S. and Klein, R. (eds) 1984: *Quaternary extinctions: a prehistoric revolution.* Tucson: University of Arizona Press.

Martin, P. S. and Wright, H. E. (eds) 1967: *Pleistocene extinctions: the search for a cause.* New Haven: Yale University Press.

Martinelli, L. A., Devol, A. H., Victoria, R. L. and Ritchey, J. E. 1991: Stable carbon isotope variation in C_3 and C_4 plants along the Amazon River. *Nature* 353, 57–59.

Martinson, D. G. 1990: Evolution of the Southern Ocean winter mixed layer and sea ice: open ocean deepwater formation and ventilation. *Journal of Geophysical Research* 95 (C7), 641–654.

Masters, P. M. 1986: Amino acid racemization dating. In Zimmerman, M. R. and Angel, J. L. (eds), *Dating and age determination of biological materials.* (London: Croom and Helm), 39–58.

May, R. M. 1978: Human reproduction reconsidered. *Nature* 272, 491–495.

Meier, M. F. 1984: Contribution of small glaciers to global sea level. *Science* 226, 1418–1421.

Melilo, J. M., Callaghan, T. V., Woodward, F. I., Salati, E. and Sinha, S. K. 1990: Effects on ecosystems. In Houghton, J. T., Jenkins, G. J. and Ephraums, J. J. (eds), *Climate change: the IPCC assessment.* (Cambridge: Cambridge University Press for the Intergovernmental Panel on Climatic Change), 283–310.

Mellars, P. 1989: Major issues in the emergence of modern humans. *Current Anthropology* 30, 349–385.

Mercer, J. H. 1978: West Antarctic ice sheet and CO_2 greenhouse effect: a threat of disaster. *Nature* 271, 321–325.

Mercier, N., Valladas, H., Joron, J.-L., Reyss, J.-L., Lévêque, F. and Vandermeersch, B. 1991: Thermoluminescence dating of the late Neanderthal remains from Saint-Césaire. *Nature* 351, 737–739.

Meriläinen, J., Huttunen, P. and Battarbee, R. W. (eds) 1983: Paleolimnology. *Developments in Hydrobiology* 15, 1–318.

Merrilees, D. 1968: Man the destroyer: later Quaternary changes in the Australian marsupial fauna. *Journal of the Royal Society of Western Australia* 51, 1–24.

Merwe, N. J. van der 1982: Carbon isotopes and archaeology. *South African Journal of Science* 78, 14–16.

Messerli, B., Winiger, M. and Rognon, P. 1986: The Saharan and East African uplands during the Quaternary. In Williams, M. A. J. and Faure, H. (eds), *The Sahara and the Nile*. Rotterdam: Balkema, 87–132.

Mikolajewicz, U., Santer, B. D. and Maier-Reimer, E. 1990: Ocean response to greenhouse warming. *Nature* 345, 589–593.

Miller, G. H. and Brigham-Grette, J. 1989: Amino acid geochronology: Resolution and precision in carbonate fossils. *Quaternary International* 1, 111–128.

Mix, A. C. 1987: The oxygen-isotope record of glaciation. In Ruddiman, W. F. and Wright, H. E. (eds), *North America and adjacent oceans during the last deglaciation*. (Boulder, Colorado: Geological Society of America) The Geology of North America Volume K-3, 111–136.

Mix, A. C. and Ruddiman, W. F. 1985: Structure and timing of the last deglaciation: oxygen-isotope evidence. *Quaternary Science Reviews* 4, 59–108.

Molnar, P. and England, P. 1990: Late Cenozoic uplift of mountain ranges and global climate change: chicken or egg? *Nature* 346, 29–34.

Molodkov, A. 1988: ESR dating of Quaternary shells: recent advances. *Quaternary Science Reviews* 7, 477–484.

Monod, T. 1963: The Late Tertiary and Pleistocene in the Sahara. In Howell, F. C. and Bourlière, F. (eds), *African ecology and human evolution*. (Chicago: Aldine), 117–229.

Moore, P. D., Webb, J. A. and Collinson, M. E. 1991: *Pollen analysis*. Second Edition. Oxford: Blackwell.

Morales, C. (ed.) 1979: *Saharan dust – mobilization, transport, deposition*. New York: Wiley.

Mörner, N.-A. 1978: Low sea levels, droughts and mammalian extinctions. *Nature* 271, 738–739.

Mörner, N.-A. 1980: The Fennoscandian uplift: geological data and their geodynamical implication. In Mörner, N.-A. (ed.), *Earth rheology, isostasy and eustasy*. (Chichester: Wiley), 251–284.

Mörner, N.-A. 1987: Pre-Quaternary long-term changes in sea level. In Devoy, R. J. N. (ed.), *Sea surface studies: a global view*. (London: Croom Helm), 233–241.

Mörner, N.-A. 1989: Global changes: the lithosphere: internal processes and earth's dynamicity in view of Quaternary observational data. *Quaternary International* 2, 55–61.

Mulvaney, D. J. 1975: *The prehistory of Australia*. Victoria: Dominion Press.

Munson, P. J. 1976: Archaeological data on the origins of cultivation in the south-western Sahara and its implications for West Africa. In Harlan, J. R., de Wet, J. M. J. and Stemler, A. B. L. (eds), *Origins of African plant domestication*. (The Hague: Mouton), 187–210.

Murray-Wallace, C. V. and Kimber, R. W. L. 1990: Amino acid racemization dating. *Chemistry in Australia* 57, 68–70.

Murray-Wallace, C. V., Kimber, R. W. L. and Belperio, A. P. 1988: Holocene palaeotemperature studies using amino acid racemization reactions. *Australian Journal of Earth Sciences* 35, 575–577.

Naeser, C. W. and Naeser, N. D. 1988: Fission-track dating of Quaternary events. *Geological Society of America Special Paper* 227, 1–11.

Naeser, C. W., Briggs, N. D., Obradovich, J. D. and Izett, G. A. 1981: Geochronology of Quaternary tephra deposits. In Self, S. and Sparks, R. S. J. (eds), *Tephra studies.* (Dordrecht: Reidel), 13–47.

Nanson, G. S., Price, D. M., Short, S. A., Young, R. W. and Jones, B. J. 1991: Comparative uranium-thorium and thermoluminescence dating of weathered Quaternary alluvium in the tropics of northern Australia. *Quaternary Research* 35, 346–366.

National Academy of Sciences 1975: *Understanding climatic change: a program for action.* Washington D.C.: National Academy of Sciences.

National Academy of Sciences 1986: *Soil conservation: assessing the National Resources Inventory.* Volume 1. Washington D.C.: National Academy Press.

National Research Council 1982: *Studies in geophysics. Climate in earth history.* Washington National Academy Press.

Neftel, A., Moor, E., Oeschger, H. and Stauffer, B. 1985: Evidence from polar ice cores for the increase in atmospheric CO_2 in the past two centuries. *Nature* 315, 45–57.

Newby, J. E. 1984: Large mammals. In Cloudsley-Thompson, J. L. (ed.), *Sahara Desert.* (Oxford: Pergamon), 277–290.

Newman, W. S. and Fairbridge, R. W. 1986: The management of sea level rise. *Nature* 320, 319–321.

Nicholson, S. 1989: African drought: characteristics, causal theories and global teleconnections. In Berger, A., Dickinson, R. E. and Kidson, J. W. (eds), *Understanding climatic change.* (Washington D.C.: American Geophysical Union), 79–100.

Nicholson, S. and Flohn, H. 1980: African environmental and climatic changes and the general atmospheric circulation in Late Pleistocene and Holocene. *Climatic Change* 2(4), 313–348.

Nilsson, T. 1983: *The Pleistocene – geology and life in the Quaternary ice age.* Dordrecht: Reidel.

Odum, E. P. 1989: *Ecology and our endangered life-support systems.* Sunderland, Massachusetts: Sinauer.

Oeschger, H. and Langway, C. C. (eds) 1989: *The environment record in glaciers and ice sheets.* New York: Wiley.

Open University Course Team 1989a: *Ocean chemistry and deep-sea sediments.* Oxford: Pergamon Press.

Open University Course Team 1989b: *Ocean circulation.* Oxford: Pergamon Press.

Open University Course Team 1989c: *The ocean basins: their structure and evolution.* Oxford: Pergamon Press.

Ottaway, B. S. 1983: *Archaeology, dendrochronology and the radiocarbon calibration curve.* University of Edinburgh, Department of Archaeology, Occasional Paper No. 9.

Owen, H. G. 1983: *Atlas of continental displacement: 200 million years to present.* Melbourne: Cambridge University Press.

Owen, R. 1870: On the fossil mammals of Australia – Part III. *Diprotodon australis. Philosophical Transactions of the Royal Society* 162, 241–258.

Parkin, D. W. 1974: Trade-winds during the glacial cycles. *Proceedings of the Royal Society of London* A337, 73–100.

Parkin, D. W. and Shackleton, N. J. 1973: Trade-winds and temperature correlations down a deep-sea core off the Saharan coast. *Nature* 245, 455–457.

Parmenter, C. and Folger, D. W. 1974: Eolian biogenic detritus in deep sea sediments: a possible index of equatorial Ice Age aridity. *Science* 185, 695–698.

Partridge, T. C., Netterberg, F., Vogel, J. C. and Sellschop, J. P. F. 1984: Absolute dating methods for the southern African Cainozoic. *South African Journal of Science* 80, 394–400.

Pastouret, L., Charnley, H., Delibrias, G., Duplessy, J. C. and Thiede, J. 1978: Late Quaternary climatic changes in western tropical Africa deduced from deep-sea sedimentation off the Niger Delta. *Oceanologica Acta* 1, 217–232.

Paterson, W. S. B. and Hammer, C. U. 1987: Ice core and other glaciological data. In Ruddiman, W. F. and Wright, H. E. (eds), *North American and adjacent oceans during the last deglaciation*. (Boulder, Colorado: Geological Society of America), The Geology of North America Volume K-3, 91–109.

Patterson, W. A. III, Edwards, K. J. and Maguire, D. J. 1987: Microscopic charcoal as a fossil indicator of fire. *Quaternary Science Reviews* 6, 3–23.

Payne, M. A. and Verosub, K. L. 1982: The acquisition of post-depositional detrital remanent magnetization in a variety of natural sediments. *Geophysical Journal Royal Astronomical Society* 68, 625–642.

Pearce, R. H. and Barbetti, M. 1981: A 38 000 year-old site at Upper Swan, W.A. *Archaeology in Oceania* 16, 173–178.

Pearman, G. I. (ed.) 1988: *Greenhouse: planning for climate change.* Australia: CSIRO Publications.

Pearson, G. W., Pilcher, J. R., Baillie, M. G. L., Corbett, D. M. and Qua, F. 1986: High-precision ^{14}C measurement of the Irish oaks to show the natural ^{14}C variations from AD 1840–5210 BC. *Radiocarbon* 28, 911–934.

Pearson, G. W. and Stuiver, M. 1986: High-precision calibration of the radiocarbon time-scale, 500–2500 BC. *Radiocarbon* 28, 839–862.

Peltier, W. R. 1982: Dynamics of the ice age earth. *Advances in Geophysics* 24, 1–144.

Peltier, W. R. 1987: Glacial isostasy, mantle viscosity and Pleistocene climatic change. In Ruddiman, W. F. and Wright, H. E. (eds), *North America and adjacent oceans during the last deglaciation*. (Boulder, Colorado: Geological Society of America), The Geology of North America Volume K-3, 155–182.

Peltier, W. R. 1988: Lithospheric thickness, Antarctic deglaciation history and ocean basin discretization effects in a global model of postglacial sea level change: a summary of some sources of nonuniqueness. *Quaternary Research* 29, 93–112.

Penck, A. and Brückner, E. 1909: *Die Alpen im Eiszeitalter.* Leipzig: Tauchnitz.

Pennington, W. 1986: Lags in adjustment of vegetation to climate caused by the pace of soil development: evidence from Britain. *Vegetatio* 67, 105–118.

Petit, J. R., Mounier, L., Jouzel, J., Korotkevich, Y. S., Kotlyakov, V. I. and Lorius, C. 1990: Palaeoclimatological and chronological implications of the Vostok core dust record. *Nature* 343, 56–58.

Petit-Marie, N. 1990a: Will greenhouse green the Sahara? *Episodes* 13(2), 103–107.

Petit-Marie, N. 1990b: Natural aridification or man-made desertification? A question for the future. In Paepe, R., Fairbridge, R. W. and Jelgersma, S. (eds), *Greenhouse effect, sea level and drought.* (Dordrecht: Kluwer Academic Publishers), 281–285.

Péwé, T. L. 1981: Desert dust: an overview. *Geological Society of America Special Paper* 186, 1–10.

Péwé, T. L. 1983a: Alpine permafrost in the contiguous United States: A review. *Arctic and Alpine Research* 15(2), 145–156.

Péwé, T. L. 1983b: The periglacial environment in North America during Wisconsin time. *Late Quaternary environments of the United States. Volume 1,* (Porter, S. C. (ed.)) *The Late Pleistocene.* (Essex: Longman), 157–189.

Philander, S. G. H. 1983: El Niño Southern Oscillation phenomena. *Nature* 302, 295–301.

Pilbeam, D. and Gould, S. J. 1974: Size and scaling in human evolution. *Science* 186, 892–901.

Pittock, A. B. 1987: *Beyond darkness: nuclear winter in Australia and New Zealand.* Melbourne: Sun.

Pittock, A. B. and Nix, H. A. 1986: The effect of changing climate on Australian biomass production – a preliminary study. *Climate Change* 8, 243–255.

Pokras, E. M. and Mix, A. C. 1985: Eolian evidence for spatial variability of late Quaternary climates in tropical Africa. *Quaternary Research* 24, 137–149.

Polach, H. 1987: Radiocarbon dating techniques: a status report. In Ambrose, W. R. and Mummery, J. M. J. (eds), *Archaeometry: further Australasian studies.* (Canberra: Australian National University), 213–224.

Polach, H. and Golson, J. 1968: The collection and submission of radiocarbon samples. In Mulvaney, D. J. (ed.), *Australian archaeology: a guide to field techniques.* (Canberra: Australian Institute of Aboriginal Studies), 211–239.

Porter, S. C. 1975: Weathering rinds as a relative-age criterion: application to subdivision of glacial deposits in the Cascade Range. *Geology* 3, 101–104.

Potts, R. and Shipman, P. 1981: Cutmarks made by stone on bones from Olduvai Gorge, Tanzania. *Nature* 291, 577–580.

Prell, W. L. 1984: Monsoonal climate of the Arabian Sea during the Late Quaternary: a response to changing solar radiation. In Berger, A. L. and Labeyrie, L. (eds), *Milankovitch and climate.* (Dordrecht: Riedel), 349–366.

Prell, W. L., Hutson, W. H., Williams, D. F., Be, A. W. H., Geitzenauer, K. and Molfino, B. 1980: Surface circulation of the Indian Ocean during the last glacial maxima, approximately 18 000 yr BP. *Quaternary Research* 14, 309–336.

Prentice, K. C. and Fung, I. Y. 1990: The sensitivity of terrestrial carbon storage to climate change. *Nature* 346, 48–51.

Prentice, M. L. and Denton, G. H. 1988: The deep-sea oxygen isotope record, the global ice sheet system and hominid evolution. In Grine, F. E. (ed.), *Evolutionary history of the "Robust" Australopithecines.* (New York: Aldine de Gruyter), 383–403.

Press, F. and Siever, R. 1986: *Earth.* San Francisco: Freeman.

Prest, V. K. 1969: *Retreat of Wisconsin and recent ice in North America: Speculative ice marginal positions during recessions of the last ice sheet complex.* Map 1257A. Geological Survey of Canada.

Prestwich, J. 1860: On the occurrence of flint-implements, associated with remains of animals of extinct species in beds of a late geological period, in France at Amiens, and in England at Hoxne. *Philosophical Transactions of the Royal Society* 150, 277–317.

Pye, K. 1984: Loess. *Progress in physical geography* 8, 176–217.

Pye, K. 1987: *Aeolian dust and dust deposits.* London: Academic Press.

Quade, J., Cerling, T. E. and Bowman, J. R. 1989: Development of Asian monsoon revealed by marked ecological shift during the latest Miocene in northern Pakistan. *Nature* 342, 163–166.

Quilty, P. G. 1984: Phanerozoic climates and environments of Australia. In Veevers, J. J. (ed.), *Phanerozoic earth history of Australia.* (Oxford: Clarendon Press), 48–55.

Rasmusson, E. M. 1987: Global climate change and variability: effects on drought and desertification in Africa. In Glantz, M. H. (ed.), *Drought and hunger in Africa: denying famine a future.* (Cambridge: Cambridge University Press), 3–22.

Raymo, M. E., Ruddiman, W. F. and Froelich, P. N. 1988: Influence of Late Cenozoic mountain building on ocean geochemical cycles. *Geology* 16, 649–653.

Readhead, M. L. 1988: Thermoluminescence dating of quartz in aeolian sediments from southeastern Australia. *Quaternary Science Reviews* 7, 257–264.

Reeh, N. 1989: Dynamic and climatic history of the Greenland ice sheet. In Fulton, R. J. (ed.), *Quaternary geology of Canada and Greenland.* Geological Survey of Canada, Geology of Canada. No. 1/Geology Society of America, Geology of North America K1. (Ottawa: Canadian Government Publishing Centre), 793–822.

Retzer, J. L. 1974: Alpine Soils. In Ives, J. D. and Barry, R. G. (eds), *Arctic and alpine environments.* (London: Methuen), 771–802.

Revelle, R. R. (Chairman) 1990: *Sea-level change.* Washington: National Academy Press, Studies in Geophysics Series.

Rice, R. A. and Vandermeer, J. 1990: Climate and the geography of agriculture. In Carroll, C. R., Vandermeer, J. H. and Rosset, P. (eds), *Agroecology.* (New York: McGraw-Hill), 21–63.

Rind, D., Chiou, E.-W., Chu, W., Larsen, J., Oltmans, S., Lerner, J., McCormick, M. P. and McMaster, L. 1991: Positive water vapour feedback in climate models confirmed by satellite data. *Nature* 349, 500–503.

Rind, D. and Peteet, D. 1985: Terrestrial conditions at the last glacial maximum and CLIMAP sea surface temperature estimates: are they consistent? *Quaternary Research* 24, 1–22.

Roberts, L. 1988: Is there life after climate change? *Science* 242, 1010–1012.

Roberts, N. 1989: *The Holocene: an environmental history.* Oxford: Blackwell.

Roberts, R. G., Jones, R. and Smith, M. A. 1990: Thermoluminescence dating of a 50 000-year-old human occupation site in northern Australia. *Nature* 345, 153–156.

Roche, H. 1980: *Premiers Outils Taillés d'Afrique.* Paris: Société d'Ethnographie.

Rognon, P. 1967: *Le massif de l'Atakor et ses bordures (Sahara central): Etude géomorphologique.* Paris: CNRS and CRZA.

Rognon, P. 1989: *Biographie d'un désert.* Paris: Plon.

Rognon, P. and Williams, M. A. J. 1977: Late Quaternary climatic change in Australia and North Africa: a preliminary interpretation. *Palaeogeography, Palaeoclimatology, Palaeoecology* 21, 285–327.

Roset, J.-P. 1984: The prehistoric rock paintings of the Sahara. *Endeavour* 8, 75–84.

Rossignol-Strick, M. 1983: African monsoons, an immediate climate response to orbital insolation. *Nature* 304, 46–49.

Rossignol-Strick, M., Nesteroff, W., Olive, P. and Vergnaud-Grazzini, C. 1982: After the deluge: Mediterranean stagnation and sapropel formation. *Nature* 295, 105–110.

Roth, E. and Poty, B. (eds) 1989: *Nuclear methods of dating.* Dordrecht: Kluwer Academic Publishers.

Ruddiman, W. F. (co-ordinator and compiler) 1984: The Last Interglacial Ocean. CLIMAP Project Members. *Quaternary Research* 21, 123–224.

Ruddiman, W. F. 1987: Northern oceans. In Ruddiman, W. F. and Wright, H. E. (eds), *North America and adjacent oceans during the last deglaciation.* (Boulder, Colorado: Geological Society of America), The Geology of North America Volume K-3, 137–154.

Ruddiman, W. F. and Duplessy, J.-C. 1985: Conference on the last deglaciation: timing and mechanism. *Quaternary Research* 23, 1–17.

Ruddiman, W. F. and McIntyre, A. 1981: The mode and mechanism of the last deglaciation: oceanic evidence. *Quaternary Research* 16, 125–134.

Ruddiman, W. F., McIntyre, A., Niebler-Hunt, V. and Durazzi, J. T. 1980: Oceanic evidence for the mechanism of rapid northern hemisphere glaciation. *Quaternary Research* 13, 33–64.

Ruddiman, W. F. and Raymo, M. E. 1988: Northern hemisphere climate regimes during the past 3 Ma: possible tectonic connections. *Philosophical Transactions of the Royal Society of London* B318, 411–430.

Ruddiman, W. F., Raymo, M. E. and McIntyre, A. 1986. Matuyama 41,000-year cycles: North Atlantic Ocean and northern hemisphere ice sheets. *Earth and Planetary Science Letters* 80, 117–129.

Ruddiman, W. F. and Wright, H. E. (eds) 1987: *North America and adjacent oceans during the last deglaciation.* (Boulder, Colorado: Geological Society of America), The Geology of North America Volume K-3.

Rust, B. R. and Nanson, G. C. 1986: Contemporary and palaeochannel patterns and the Late Quaternary stratigraphy of Cooper Creek, southwest Queensland, Australia. *Earth Surface Processes and Landforms* 11, 581–590.

Rutter, N. W. and Vlahos, C. K. 1988: Amino acid racemization kinetics in wood – applications to geochronology and geothermometry. *Geological Society of America Special Paper* 227, 51–67.

Ryan, W. B. F. 1973: Geodynamic implications of the Messinian crisis of salinity. In Drooger, C. W. (ed.), *Messinian events in the Mediterranean*. (Amsterdam: North Holland Publishing), 26–38.

Sagan, C., Toon, O. B. and Pollack, J. B. 1979: Anthropogenic albedo changes and the Earth's climate. *Science* 206, 1363–1368.

Sanson, G. D., Riley, S. J. and Williams, M. A. J. 1980: A late Quaternary *Procoptodon* fossil from Lake George, New South Wales. *Search* 11, 39–40.

Sarich, V. M. and Wilson, A. C. 1967: Immunological time scale for hominid evolution. *Science* 158, 1200.

Sarnthein, M. 1978: Sand deserts during glacial maximum and climatic optimum. *Nature* 272, 43–46.

Sarnthein, M., Tetzlaff, G., Koopmann, B., Wolter, K. and Pflaumann, U. 1981: Glacial and interglacial wind regimes over the eastern subtropical Atlantic and North-West Africa. *Nature* 293, 193–196.

Sarnthein, M., Thiede, J., Pflaumann, U., Erlenkeuser, H., Fütterer, D., Koopmann, B., Lange, H. and Seibold, E. 1982: Atmospheric and oceanic circulation patterns off North West Africa during the past 25 million years. In von Rad, U., Hinz, K., Sarnthein, M. and Seibold, E. (eds), *Geology of the northwest African continental margin*. (Berlin: Springer-Verlag), 545–604.

Savin, S. M. and Yeh, H.-W. 1981: Stable isotopes in ocean sediments. In Emiliani, C. (ed.), *The sea, Volume 7: the oceanic lithosphere*. (New York: Wiley), 1521–1554.

Scarre, C., Bray, W., Cook, J., Daniel, G. E. and Sabloff, J. A. (eds) 1988: *Past worlds: The Times atlas of archaeology*. London: Times Books.

Schild, R. and Wendorf, F. 1989: The Late Pleistocene Nile in Wadi Kubbaniya. In Wendorf, F., Schild, R. with Close, A. E. (eds), *The prehistory of Wadi Kubbaniya. Volume 2: Stratigraphy, paleoeconomy and environment*. (Dallas: Southern Methodist University Press), 15–100.

Schneider, S. H. 1989: *Global warming: Are we entering the greenhouse century?* San Francisco: Sierra Club.

Schnitker, D. 1980: Global paleoceanography and its deepwater linkage to the Antarctic glaciation. *Earth-Science Reviews* 16, 1–20.

Schove, D. J. 1979: Varve-chronologies and their teleconnections, 14 000–750 BC. In Schlücter, Ch. (ed.), *Moraines and varves: origin, genesis and classification*. (Rotterdam: Balkema), 319–326.

Schumm, S. A. 1977: *The fluvial system*. New York: Wiley.

Schumm, S. A. and Parker, R. S. 1973: Implications of complex response of drainage systems for Quaternary alluvial stratigraphy. *Nature (Physical Science)* 243, 99–100.

Schwarcz, H. P. 1989: Uranium series dating of Quaternary deposits. *Quaternary International* 1, 7–17.

Schwarcz, H. P. and Latham, A. G. 1989: Dirty calcites: 1. Uranium-series dating of contaminated calcites using leachates alone. *Chemical Geology* 80, 35–43.

Schwarcz, H. P. and Skoflek, I. 1982: New dates for the Tata, Hungary archaeological site. *Nature* 295, 590–591.

Schwengruber F. H. 1988: *Tree rings: basics and applications of dendro-chronology*. Dordrecht: Reidel.

Scientific Assessment of Climatic Change, WMO/UNEP International Panel in Climate Change. Geneva 1990.

Servant, M. 1973: *Séquences continentales et variations climatiques: évolution du bassin du Tchad au Cénozoique supérieur*. D.Sc. Thesis, University of Paris.

Servant, M. and Servant-Vildary, S. 1980: L'environnement Quaternaire du bassin du Tchad. In Williams, M. A. J. and Faure, H. (eds), *The Sahara and the Nile*. (Rotterdam: Balkema), 133–162.

Shackleton, N. J. 1977: The oxygen isotope stratigraphic record of the late Pleistocene. *Philosophical Transactions of the Royal Society* 280, 169–179.

Shackleton, N. J. 1987: Oxygen isotopes, ice volume and sea level. *Quaternary Science Reviews* 6, 183–190.

Shackleton, N. J., Backman, J., Zimmerman, H., Kent, D. V., Hall, M. A., Robert, D. G., Schnitker, D., Baldauf, J. G., Desprairies, A., Homrighausen, R., Huddlestun, P., Keene, J. B., Kaltenback, A. J., Krumsiek, K. A. O., Morton, A. C., Murray, J. W. and Westberg-Smith, J. 1984: Oxygen isotope calibration of the onset of ice-rafting and history of glaciation in the North Atlantic region. *Nature* 307, 620–623.

Shackleton, N. J. and Kennett, J. P. 1975: Paleotemperature history of the Cenozoic and the initiation of Antarctic glaciation: oxygen and carbon analyses of DSDP sites 277, 279, 281. *Initial reports of the Deep Sea Drilling Project* 29, (Washington D.C.: U.S. Government Printing Office), 743–755.

Shackleton, N. J. and Opdyke, N. D. 1973: Oxygen isotope and paleomagnetic stratigraphy of equatorial Pacific core V28-238: oxygen isotope temperatures and ice volumes on a 10^5 year and a 10^6 scale. *Quaternary Research* 3, 39–55.

Shackleton, N. J. and Opdyke, N. D. 1977: Oxygen isotope and palaeomagnetic evidence for early Northern Hemisphere glaciation. *Nature* 270, 216–219.

Sharp, R. P. 1988: *Living ice – understanding glacier and glaciation*. Cambridge: Cambridge University Press.

Simmons, I. G. 1989: *Changing the face of the earth: culture, environment, history*. Oxford: Blackwell.

Sinclair, A. R. E. and Fryxell, J. M. 1985: The Sahel of Africa: ecology of a disaster. *Canadian Journal of Zoology* 63, 987–994.

Singer, S. F. 1989: Nuclear winter or nuclear summer? In Singer, S. F. (ed.), *Global climate change: human and natural influences*. (New York: Paragon House), 327–336.

Singh, G. 1988: History of arid land vegetation and climate: a global perspective. *Biological Review* 63, 159–195.

Singh, G. and Geissler, E. A. 1985: Late Cainozoic history of fire, vegetation, lake levels and climate at Lake George, New South Wales, Australia. *Philosophical Transactions of the Royal Society London* B311, 379–447.

Singh, G., Joshi, R. D., Chopra, S. K. and Singh, A. B. 1974: Late Quaternary history of vegetation and climate of the Rajasthan Desert, India. *Philosophical Transactions of the Royal Society of London* 267B, 467–501.

Singhvi, A. K. and Wagner, G. A. 1986: Thermoluminescence dating and its application to young sedimentary deposits. In Hurford, A. J., Jäger, E. and TenCate, J. A. M. (eds), *Dating young sediments*. Proceedings of the Workshop, Beijing, People's Republic of China, Sept. 1985. (Bangkok: Thailand, CCOP Technical Secretariat), 159–198.

Sioli, H. 1984: The Amazon and its main effluents: hydrography, morphology of the river courses, and river types. In Sioli, H. (ed.), *The Amazon: limnology and landscape ecology of a mighty tropical river and its basin*. (Dordrecht: Junk), 127–165.

Sissons, J. B. 1983: Shorelines and isostasy in Scotland. In Smith, D. E. and Dawson, A. G. (eds), *Shorelines and isostasy*. Institute of British Geographers Special Publication 16. (London: Academic Press), 209–225.

Smart, P. L., Smith, B. W., Chandra, H., Andrews, J. N. and Symons, M. R. C. 1988: An intercomparison of ESR and uranium series ages for Quaternary speleotherm calcites. *Quaternary Science Reviews* 7, 411–416.

Smil, V. 1990: Planetary warming: realities and responses. *Population and Development Review* 16(1), 1–29.

Smith, A. B. 1980: Domesticated cattle in the Sahara and their introduction into West Africa. In Williams, M. A. J. and Faure, H. (eds), *The Sahara and the Nile*. (Rotterdam: Balkema), 489–501.

Smith, A. B. 1984: The origins of food production in northeast Africa. *Palaeoecology of Africa* 16, 317–324.

Smith, J. B. and Tirpak, D. A. (eds) 1990: *The potential effects of global climate change on the United States*. New York: Hemisphere.

Smith, J. D. and Hamilton, T. F. 1985: Modelling of ^{20}Pb behaviour in the catchment and sediment of Lake Tali Karng, Victoria, and estimation of recent sedimentation rates. *Australian Journal Marine Freshwater Research* 36, 15–22.

Smith, J. P., Appleby, P. G., Battarbee, R. W., Dearing, J. A., Haworth, E. Y., Oldfield, F. and O'Sullivan, P. E. 1991: Environmental history and palaeolimnology. *Developments in Hydrobiology* 67, 1–382.

Snead, R. E. 1980: *World Atlas of geomorphic features*. New York: Kreiger.

Soule, J., Carré, D. and Jackson, W. 1990: Ecological impact of modern agriculture. In Carroll, C. R., Vandermeer, J. H. and Rosset, P. (eds), *Agroecology*. (New York: McGraw-Hill), 165–188.

Sparks, B. W. and West, R. G. 1972: *The Ice Age in Britain*. London: Methuen.

Spencer, R. J., Baldecker, M. J., Eugster, H. P., Forester, R. M., Goldhaber, M. B., Jones, B. F., Kelts, K., McKenzie, J., Madsen, D. B., Rettig, S. L., Rubin, M. and Bowser, C. J. 1984: Great Salt Lake, and precursors. Utah: the last 30 000 years. *Contributions to Mineralogy and Petrology* 86, 321–334.

Stahl, A. B. 1984: Hominid dietary selection before fire. *Current Anthropology* 25, 151–168.

Steen-McIntyre, V. 1981: Approximate dating of tephra. In Self, S. and Sparks, R. S. J. (eds), *Tephra studies*. (Dordrecht: Reidel), 49–64.

Steinen, R. P., Harrison, R. S. and Matthews, R. K. 1973: Eustatic low stand of sea level between 125 000 and 105 000 BP: evidence from the subsurface of Barbados, West Indies. *Geological Society of America Bulletin* 84, 63–70.

Stemler, A. B. L. 1980: Origins of plant domestication in the Sahara and Nile Valley. In Williams, M. A. J. and Faure, H. (eds), *The Sahara and the Nile*. (Rotterdam: Balkema), 503–526.

Stiles, D. 1984: Desertification in prehistory: the Sahara. *Sahara* 1, 85–92.

Strahler, A. N. and Strahler, A. H. 1987: *Modern Physical Geography*. Third Edition. Brisbane: Wiley.

Strakhov, N. M. 1967: *Principles of lithogenesis*, Volume 1. Edinburgh: Oliver and Boyd.

Straus, L. G. 1985: Stone Age prehistory of northern Spain. *Science* 230, 501–507.

Street, F. A. and Grove, A. T. 1979: Global maps of lake-level fluctuations since 30 000 year BP. *Quaternary Research* 10, 83–118.

Street-Perrott, F. A., Marchand, D. S., Roberts, N. and Harrison, S. P. 1989: *Global lake-level variations from 18 000 to 0 years ago: a paleoclimatic analysis*. US Department of Energy Technical Report (TR046), 1–213.

Street-Perrott, F. A., Roberts, N. and Metcalfe, S. 1985: Geomorphic implications of late Quaternary hydrological and climatic changes in the Northern Hemisphere tropics. In Douglas, I. and Spencer, T. (eds), *Environmental change and tropical geomorphology*. London: Allen and Unwin, 165–183.

Stringer, C. B. and Grün, R. 1991: Time for the last Neanderthals. *Nature* 351, 701–702.

Stringer, C. B., Grün, R., Schwarcz, H. P. and Goldberg, P. 1989: ESR dates for the hominid burial site of Es Skhul in Israel. *Nature* 338, 756–758.

Stuiver, M. 1971: Evidence for the variation of atmospheric ^{14}C content in the late Quaternary. In Turekian, K. K. (ed.), *The late Cenozoic glacial ages*. (New Haven: Yale University Press), 57–70.

Stuiver, M. 1986: ^{14}C dating, timescale calibration and application to geological problems. In Hurford, A. J., Jäger, E. and TenCate, J. A. M. (eds), *Dating young sediments*. Proceedings of the Workshop, Beijing, People's Republic of China, Sept. 1985. (Bangkok, Thailand: CCOP Technical Secretariat), 97–109.

Stuiver, M. and Braziunas, T. F. 1989: Atmospheric ^{14}C and century-scale solar oscillations. *Nature* 338, 405–408.

Stuiver, M., Kromer, B., Becker, B. and Fergusen, C. W. 1986: Radiocarbon age calibration back to 13 300 years BP and the ^{14}C age matching of the German Oak and US Bristlecone Pine chronologies. *Radiocarbon* 28, 969–979.

Stuiver, M. and Pearson, G. W. 1986: High-precision calibration of the radio-carbon time-scale, AD 1950–500 BC. *Radiocarbon* 28, 805–838.

Stuiver, M. and Reimer, P. J. 1986: A computer program for radiocarbon age calibration. *Radiocarbon* 28, 1022–1030.

Sturm, M. 1979: Introductory remarks. In Schlücter, Ch. (ed), *Moraines and varves: origin, genesis and classification*. (Rotterdam: Balkema), 279–280.

Suc, J.-P. 1984: Origin and evolution of the Mediterranean vegetation and climate in Europe. *Nature* 307, 429–432.

Sugden, D. and John, B. 1976: *Glaciers and landscape: a geomorphological approach*. London: Edward Arnold.

Sutton, J. E. G. 1977: The African Aqualithic. *Antiquity* 51, 25–34.

Swineford, A. and Frye, J. C. 1945: A mechanical analysis of wind-blown dust compared with analyses of loess. *American Journal of Science* 243, 249–255.

Takahashi, T. 1975: Carbonate chemistry of seawater and the calcium carbonate compensation depth in the oceans. *Journal of Foraminiferal Research, Special Publication* 13, 11–26.

Talbot, M. R. 1980: Environmental responses to climatic change in the West African Sahel over the past 20 000 years. In Williams, M. A. J. and Faure, H. (eds), *The Sahara and the Nile*. (Rotterdam: Balkema), 37–62.

Tanke, M. and Gulik, J. van 1989: *The global climate*. Amsterdam: Mirage Publishing.

Tauber, H. 1967: Investigations of the mode of pollen transfer in forested areas. *Review of Palaeobotany and Palynology* 3, 277–286.

Tchernia, P. 1980: *Descriptive regional oceanography*. Oxford: Pergamon Press.

Tchernov, E. 1987: The Age of the 'Ubeidiya Formation, an Early Pleistocene hominid site in the Jordan Valley, Israel. *Israel Journal of Earth Sciences* 36, 3–30.

Thom, B. G. and Roy, P. S. 1985: Relative sea levels and coastal sedimentation in southeast Australia in the Holocene. *Journal of Sedimentary Petrology* 55(2), 257–264.

Thomas, D. S. G. 1989: Aeolian sand deposits. In Thomas, D. S. G. (ed.), *Arid zone geomorphology*. (London: Belhaven Press), 232–261.

Thomas, D. S. G. 1989: Reconstructing ancient arid environments. In Thomas, D. S. G. (ed.), *Arid zone geomorphology*. (London: Belhaven), 311–334.

Thomas, D. S. G. (ed.) 1989: *Arid zone geomorphology*. (London: Belhaven), 372.

Thompson, R. S. 1988: Western North America. In Huntley, B. and Webb, T. (eds), *Vegetation history*. (Dordrecht: Kluwer Academic Publishers), 415–458.

Thompson, R. S. and Krautz, R. R. 1983: Pollen analysis. *Anthropological Papers of the American Museum of Natural History* 59, 136–151.

Thompson, L. G., Mosley-Thompson, E., Dansgaard, W. and Grootes, P. M. 1986: The Little Ice Age as recorded in the stratigraphy of the tropical Quelccaya Ice Cap. *Science* 234, 361–364.

Thwaites, F. T. 1963: *Outline of glacial geology*. University of Wisconsin.

Tivy, J. and O'Hare, G. 1981: *Human impact on the ecosystem*. Edinburgh: Oliver & Boyd.

Tobias, P. V. 1979: Men, minds and hands: cultural awakenings over two million years of humanity. *South African Archaeological Bulletin* 34, 92–95.

Tricart, J. and Caillieux, A. 1972: *Introduction to climatic geomorphology*. London: Longman.

Trinkhaus, E. 1983: *The Shanidar Neanderthals*. New York: Academic Press.

Trinkhaus, E. 1986: The Neanderthals and modern human origins. *Annual Review of Anthropology* 15, 193–218.

Trinkhaus, E. and Howells, W. W. 1979: The Neanderthals. *Scientific American* 241, 118–133.

Turner, J. and Peglar, S. M. 1988: Temporally-precise studies of vegetation history. In Huntley, B. and Webb, T. (eds), *Vegetation history.* (Dordrecht: Kluwer Academic Publishers), 753–777.

Tuttle, R. H. 1988: What's new in African paleoanthropology? *Annual Review of Anthropology* 17, 391–426.

U.S. Navy Hydrographic Office 1961: *Oceanographic atlas of the polar seas. Part 1: Antarctic.* Washington D.C.: U.S. Navy hydrographic office publication number 705.

Van Campo, E., Duplessy, J. C., Prell, W. L., Barratt, N. and Sabatier, R. 1990: Comparison of terrestrial and marine temperature estimates for the past 135 kyr off southeast Africa: a test for GCM simulations of palaeoclimate. *Nature* 348, 209–212.

Van der Merwe, N. J. 1982: Carbon isotopes and archaeology. *South African Journal of Science* 78, 14–16.

Verosub, K. L. 1986: Principles and applications of palaeomagnetism in the dating of young sediments. In Hurford, A. J., Jäger, E. and TenCate, J. A. M. (eds), *Dating young sediments.* Proceedings of the Workshop, Beijing, People's Republic of China, Sept. 1985. (Bangkok, Thailand: CCOP Technical Secretariat), 247–267.

Verosub, K. L. 1988: Geomagnetic secular variation and the dating of Quaternary sediments. *Geological Society of America Special Paper* 227, 123–138.

Vincent, E. and Berger, W. H. 1981: Planktonic Foraminifera and their use in paleoceanography. In Emiliani, C. (ed.), *The Sea. Volume 7: The oceanic lithosphere.* (New York: Wiley), 1025–1119.

Vogel, J. C. 1978: Isotopic assessment of the dietary habits of ungulates. *South African Journal of Science* 74, 298–301.

Vogel, J. C., Fuls, A. and Ellis, R. P. 1978. The geographical distribution of Kranz grasses in South Africa. *South African Journal of Science* 74, 209–215.

Volk, T. 1989: Rise of angiosperms as a factor in long-term climatic cooling. *Geology* 17, 107–110.

Vrba, E. S. 1988: Late Pliocene climatic events and hominid evolution. In Grine, F. E. (ed.), *Evolutionary history of the "Robust" Australopithecines.* (New York: Aldine de Gruyter), 405–426.

Walcott, R. I. 1972: Past sea levels, eustasy and deformation of the earth. *Quaternary Research* 2, 1–14.

Walker, D. and Chen, Y. 1987: Palynological light on tropical rainforest dynamics. *Quaternary Science Reviews* 6, 77–92.

Walker, D. and Guppy, J. C. 1978: *Biology and Quaternary environments.* Canberra: Australian Academy of Science.

Walker, A., Leakey, R. E., Harris, J. M. and Brown, F. H. 1986: 2.5 Myr *Australopithecus boisei* from west of Lake Turkana, Kenya. *Nature* 322, 517–522.

Walther, J. 1900: *Das Gesetz der Wüstenbildung in Gegenwart und Vorzeit.* Berlin.

Warrick, R. and Oerlemans, J. 1990: Sea level rise. In Houghton, J. T., Jenkins, G. J. and Ephraums, J. J. (eds), *Climate change: The IPCC assessment.* (Cambridge: Cambridge University Press for the Intergovernmental Panel on Climatic Change), 257–281.

Washburn, A. L. 1973: *Periglacial processes and environments.* London: Edward Arnold.

Washburn, A. L. 1979: *Geocryology – a survey of periglacial processes and environments.* London: Edward Arnold.

Washburn, A. L. 1980: Permafrost features as evidence of climatic change. *Earth Science Reviews* 15, 327–402.

Wasson, R. J. 1990: Palaeoenvironmental research and global warming. In Bishop P. (ed.), *Lessons for human survival: nature's record from the Quaternary.* Geological Society of Australia Symposium Proceedings 1, 83–92.

Wasson, R. J. and Hyde, R. 1983: Factors determining desert dune type. *Nature* 304, 337–339.

Wasson, R. J. and Nanninga, P. M. 1986: Estimating wind transport of sand on vegetated surfaces. *Earth Surface Processes and Landforms* 11, 505–514.

Wasson, R. J., Rajaguru, S. N., Misra, V. N., Agrawal, D. P., Ohir, R. P., Singhvi, A. K. and Rao, K. K. 1983: Geomorphology, Late Quaternary stratigraphy and palaeoclimatology of the Thar Desert. *Zeitschrift für Geomorphologie, Supplementband* 45, 117–152.

Weaver, K. F. 1985: Stones, bones and early man. The search for our ancestors. *National Geographic Magazine* 168, 560–623.

Webb, T. 1980: The reconstruction of climatic sequences from botanical data. *Journal of Interdisciplinary History* 10, 749–772.

Webb, T. 1986: Is vegetation in equilibrium with climate? How to interpret late-Quaternary pollen data. *Vegetatio* 67, 75–91.

Wehmiller, J. F. 1986: Amino acid racemization geochronology. In Hurford, A. J., Jäger, E. and TenCate, J. A. M. (eds), *Dating young sediments.* Proceedings of the Workshop, Beijing, People's Republic of China, Sept. 1985. (Bangkok, Thailand: CCOP Technical Secretariat), 139–158.

Wendorf, F. and Schild, R. 1980: *Prehistory of the Eastern Sahara.* New York: Academic Press.

Wendorf, F. and Schild, R. 1989: Summary synthesis. In Wendorf, F. and Schild, R. with Close, A. E. (eds), *The prehistory of Wadi Kubbaniya Volume 3: Late Paleolithic archaeology.* (Dallas: Southern Methodist University Press), 768–824.

West, R. G. 1977: *Pleistocene geology and biology with special reference to the British Isles.* London: Longman Group.

Westgate, J. A. and Gorton, M. P. 1981: Correlation techniques in tephra studies. In Self, S. and Sparks, R. S. J. (eds), *Tephra studies.* (Dordrecht: Reidel), 73–94.

Westoby, J. 1989: *Introduction to world forestry: people and their trees.* Oxford: Blackwell.

Wheeler, D. A. 1985: An analysis of the aeolian dustfall on eastern Britain. November 1984. *Proceedings of the Yorkshire Geological Society* 45(4), 307–310.

Whetton, P., Adamson, D. A. and Williams, M. A. J. 1990: Rainfall and river flow variability in Africa, Australia and East Asia linked to El Niño–Southern Oscillation events. In Bishop P. (ed.), *Lessons for human survival: nature's record from the Quaternary.* Geological Society of Australia Symposium Proceedings 1, 71–82.

White, J. P. and O'Connell, J. F. 1978: Australian prehistory: new aspects of antiquity. *Science* 203, 21–28.

White, J. P. and O'Connell, J. F. 1982: *A prehistory of Australia, New Guinea and Sahul.* Sydney: Academic Press.

White, T. D. 1986: Cutmarks on the Bodo cranium: a case of prehistoric defleshing. *American Journal of Physical Anthropology* 69, 503–509.

White, T. D. and Harris, J. M. 1977: Suid evolution and correlation of African hominid localities. *Science* 198, 13–21.

White, T. D. and Suwa, G. 1987: Hominid footprints at Laetoli: facts and interpretations. *American Journal of Physical Anthropology* 72, 485–514.

White, T. D., Johanson, D. C. and Kimbel, W. H. 1981: *Australopithecus afarensis:* its phyletic position reconsidered. *South African Journal of Science* 77, 445–470.

Whitemore, T. C. and Prance, G. T. (eds) 1987: *Biogeography and Quaternary history in tropical America.* Oxford: Clarendon Press.

Widstrand, C. G. 1975: The rationale of nomad economy. *Ambio* 4, 146–153.

Wigley, T. M. L. and Raper, S. C. B. 1987: Thermal expansion of sea water associated with global warming. *Nature* 330, 127–131.

Williams, D. F., Moore, W. S. and Fillon, R. H. 1981: Role of glacial Arctic Ocean ice sheets in Pleistocene oxygen isotope and sea level records. *Earth and Planetary Science Letters* 56, 157–166.

Williams, G. P. 1984: Palaeohydrologic equations for rivers. In Coasta, J. E. and Fleisher, P. J. (eds), *Developments and applications of geomorphology.* Berlin: Springer-Verlag, 343–367.

Williams, M. A. J. 1975: Late Quaternary tropical aridity synchronous in both hemispheres? *Nature* 253, 617–618.

Williams, M. A. J. 1982: Quaternary environments in North Africa. In Williams, M. A. J. and Adamson, D. A. (eds), *A land between Two Niles. Quaternary geology and biology of the Central Sudan.* (Rotterdam: Balkema), 13–22.

Williams, M. A. J. 1984a: Geology. In Cloudsley-Thompson, J. L. (ed.), *Sahara Desert.* (Oxford: Pergamon), 31–39.

Williams, M. A. J. 1984b: Late Quaternary environments in the Sahara. In Clark, J. D. and Brandt, S. A. (eds), *The causes and consequences of food production in Africa.* (Berkeley: University of California Press), 74–83.

Williams, M. A. J. 1984c: Cenozoic evolution in arid Australia. In Cogger, H. G. and Cameron, E. E. (eds), *Arid Australia.* (Sydney: Australia Museum), 59–78.

Williams, M. A. J. 1985a: Pleistocene aridity in Africa, Australia and Asia. In Douglas, I. and Spencer, T. (eds), *Environmental change and tropical geomorphology.* (London: Allen and Unwin), 219–233.

Williams, M. A. J. 1985b: On becoming human: geographical background to cultural evolution. 11th Griffith Taylor Memorial Lecture. *Australian Geographer* 16, 175–184.

Williams, M. A. J. 1986: The creeping desert: what can be done? *Current Affairs Bulletin* 63, 24–31.

Williams, M. A. J. 1990: Contemporary issues in physical geography:

challenges and possibilities. In Dyer, J. (ed.), *Interaction Journal of the Geography Teachers Association of Victoria* 18(1).

Williams, M. A. J. and Clarke, M. F. 1984: Late Quaternary environments in north central India. *Nature* 308, 633–635.

Williams, M. A. J. and Royce, K. 1982: Quaternary geology of the Middle Son Valley, north central India: implications for prehistoric archaeology. *Palaeogeography, Palaeoclimatology, Palaeoecology* 38, 139–162.

Williams, M. A. J., Abell, P. I. and Sparks, B. W. 1987: Quaternary landforms, sediments, depositional environments and gastropod isotope ratios at Adrar Bous, Tenere Desert of Niger, south-central Sahara. In Frostick, L. and Reid, I. (eds), *Desert sediments: ancient and modern*. Geological Society Special Publication No. 35, 105–125.

Williams, M. A. J., De Deckker, P. and Kershaw, A. P. (eds) 1991: *The Cainozoic in Australia: a re-appraisal of the evidence*. Geological Society of Australia Special Publication No. 18.

Williams, M. A. J., Adamson, D. A., Williams, F. M., Morton W. H. and Parry, D. E. 1980: Jebel Marra volcano: a link between the Nile Valley, the Sahara and Central Africa. In Williams, M. A. J. and Faure, H. (eds), *The Sahara and the Nile*. (Rotterdam: Balkema), 305–337.

Williams, R. B. G. 1969: Permafrost and temperature conditions in England during the last glacial period. In Péwé, T. L. (ed.), *The periglacial environment: past and present*. (Montreal: McGill-Queen's University Press), 399–410.

Wind, H. G. (ed.) 1987: *Impact of sea level rise on society*. Rotterdam: Balkema.

Wolfe, J. A. 1978: A paleobotanical interpretation of Tertiary climates in the Northern Hemisphere. *American Scientist* 66, 694–703.

Woodwell, G. M., Hobbie, J. E., Houghton, R. A., Melillo, J. M., Moore, B., Peterson, B. J. and Shaver, G. R. 1983: Global deforestation: contribution to atmospheric carbon dioxide. *Science* 222, 1081–1086.

Worsley, P. 1981: Lichenometry. In Goudie, A. (ed.), *Geomorphological techniques*. (London: Allen and Unwin), 302–305.

Worsley, T. R., Nance, D. and Moody, J. B. 1984: Global tectonics and eustasy for the past 2 billion years. *Marine Geology* 58, 373–400.

Wright, H. E. (ed.) 1983: Late Quaternary environments of the United States. Volume 1: (Porter, S. C. (ed.)) *The Late Pleistocene* (ed.). Minneapolis: University of Minnesota Press.

Wu, R. and Lin, S. H. 1983: Peking Man. *Scientific American* 248, 78–86.

Yiou, F., Raisbeck, G. M., Bourles, D., Lorius, C. and Barkov, N. I. 1985: [10]Be in ice at Vostok Antarctica during the last climatic cycle. *Nature* 316, 616–617.

Zagwijn, W. H. 1957: Vegetation, climate and time-correlations in the Early Pleistocene of Europe. *Geologie en Mijinbouw* 19, 233–244.

Zarate, M. A. and Fasano, J. L. 1989: The Plio-Pleistocene record of the central eastern Pampas, Buenos Aires province, Argentina: the Chapadmalal case study. *Palaeogeography, Palaeoclimatology, Palaeoecology* 72, 27–52.

Zinderen Bakker, E. M. van (ed.) 1978: *Antarctic glacial history and world palaeoenvironments*. Rotterdam: Balkema.

Zinderen Bakker, E. M. van and Mercer, J. H. 1986: Major late Cainozoic climatic events and palaeoenvironmental changes in Africa viewed in a world wide context. *Palaeogeography, Palaeoclimatology, Palaeoecology* 56, 217–235.

Index